# Springer Theses

Recognizing Outstanding Ph.D. Research

## Aims and Scope

The series "Springer Theses" brings together a selection of the very best Ph.D. theses from around the world and across the physical sciences. Nominated and endorsed by two recognized specialists, each published volume has been selected for its scientific excellence and the high impact of its contents for the pertinent field of research. For greater accessibility to non-specialists, the published versions include an extended introduction, as well as a foreword by the student's supervisor explaining the special relevance of the work for the field. As a whole, the series will provide a valuable resource both for newcomers to the research fields described, and for other scientists seeking detailed background information on special questions. Finally, it provides an accredited documentation of the valuable contributions made by today's younger generation of scientists.

## Theses are accepted into the series by invited nomination only and must fulfill all of the following criteria

- They must be written in good English.
- The topic should fall within the confines of Chemistry, Physics, Earth Sciences, Engineering and related interdisciplinary fields such as Materials, Nanoscience, Chemical Engineering, Complex Systems and Biophysics.
- The work reported in the thesis must represent a significant scientific advance.
- If the thesis includes previously published material, permission to reproduce this must be gained from the respective copyright holder.
- They must have been examined and passed during the 12 months prior to nomination.
- Each thesis should include a foreword by the supervisor outlining the significance of its content.
- The theses should have a clearly defined structure including an introduction accessible to scientists not expert in that particular field.

More information about this series at http://www.springer.com/series/8790

Benjamin Wallisch

# Cosmological Probes
# of Light Relics

Doctoral Thesis accepted by
the University of Cambridge, Cambridge, UK

 Springer

*Author*
Dr. Benjamin Wallisch
School of Natural Sciences
Institute for Advanced Study
Princeton, NJ, USA

Department of Physics
University of California
San Diego La Jolla, CA, USA

*Supervisor*
Prof. Daniel Baumann
Institute for Theoretical Physics
University of Amsterdam
Amsterdam, The Netherlands

ISSN 2190-5053                    ISSN 2190-5061    (electronic)
Springer Theses
ISBN 978-3-030-31100-1            ISBN 978-3-030-31098-1    (eBook)
https://doi.org/10.1007/978-3-030-31098-1

This Springer imprint is published by the registered company Springer Nature Switzerland AG
The registered company address is: Gewerbestrasse 11, 6330 Cham, Switzerland

*For my parents*

# Supervisor's Foreword

The oldest light in the universe was released 13.8 billion years ago, when the first stable atoms coalesced out of the primordial plasma. Since then, this ancient light has propagated freely through the vastness of space, carrying with it a treasure trove of information about the physics of the early universe. By decoding the pattern of fluctuations in the cosmic microwave background (CMB), cosmologists have obtained amazing insights into the origin, structure and evolution of our universe. We have learned that the universe today is filled mostly with dark matter and dark energy, and that ordinary atoms are only a small fraction of the cosmic energy budget. Moreover, we now understand in great detail how all of the structures we see around us (galaxies, stars, planets) grew out of the tiny inhomogeneities observed in the CMB. Finally, we have strong evidence that the very early universe went through a brief period of very rapid expansion called inflation, during which quantum fluctuations were amplified and the primordial density fluctuations were created.

Although the standard cosmological model is consistent with all current observations, it is theoretically unsatisfying in a number of ways. We do not understand the nature of dark matter and dark energy, and we do not have a fundamental theory explaining the origin of the inflationary expansion. Similarly, although the Standard Model of particle physics explains all available data, we have good reasons to believe that it is incomplete. Various fine-tuning problems suggest new physics that is either too heavy or too weakly coupled to be produced by collider experiments.

Benjamin Wallisch's remarkable Ph.D. thesis is an important step in using cosmology as a probe of physics beyond the Standard Model (BSM). The focus of the thesis was on very light particles that can have escaped detection in terrestrial experiments because they are so weakly interacting with ordinary matter. These particles can nevertheless be produced in large numbers in the early universe, making cosmology a unique way to look for them. Benjamin showed that free-streaming relativistic particles (such as neutrinos and light axions) leave a unique signature in the CMB and the clustering of galaxies. He then forecasted how well this effect can be measured by current and future observations. A specific prediction of Benjamin's work is a subtle imprint of the cosmic neutrino background on the statistical distribution of galaxies in the sky. Together with members of the BOSS collaboration, Benjamin then analysed data from the Sloan Digital Sky Survey (SDSS) and provided the first detection of this effect. This new measurement constitutes an interesting confirmation of the standard cosmological model which links the production of neutrinos one second after the big bang to the clustering of galaxies billions of years later.

Benjamin further showed that future cosmological observations will be highly sensitive to this type of BSM physics. He demonstrated that any light particle that has ever been in thermal equilibrium with the Standard Model will be detectable with these observations. This new experimental target has since been adopted by the CMB-S4 collaboration as one of the main science objectives of the next generation of CMB experiments. Even the absence of a detection of new thermal relics would be informative, since it would put strong constraints on their couplings to the Standard Model. Benjamin investigated the special case of axions and showed that the constraints achievable from cosmology will surpass existing bounds from laboratory experiments and astrophysical observations by orders of magnitude.

Benjamin Wallisch has produced a Ph.D. thesis of the highest quality. His work is a rare combination of theoretical insights, numerical computations and data analysis. It is a testimony to Benjamin's unique talent that he was able to combine these distinct aspects of the research effortlessly. It has been a great privilege to have been Benjamin's Ph.D. supervisor, and I wish him all the best in his future career.

Amsterdam, The Netherlands                                                           Prof. Daniel Baumann
June 2019

# Abstract

One of the primary targets of current and especially future cosmological observations are light thermal relics of the hot big bang. Within the Standard Model of particle physics, an important thermal relic are cosmic neutrinos, while many interesting extensions of the Standard Model predict new light particles which are even more weakly coupled to ordinary matter, and therefore hard to detect in terrestrial experiments. On the other hand, these elusive particles may be produced efficiently in the early universe and their gravitational influence could be detectable in cosmological observables. In this thesis, we describe how measurements of the cosmic microwave background (CMB) and the large-scale structure (LSS) of the universe can shed new light on the properties of neutrinos and on the possible existence of other light relics.

These cosmological observations are remarkably sensitive to the amount of radiation in the early universe, partly because free-streaming species such as neutrinos imprint a small phase shift in the baryon acoustic oscillations (BAO), which we study in detail in the CMB and LSS power spectra. Building on this analytic understanding, we provide further evidence for the cosmic neutrino background by independently confirming its free-streaming nature in different, currently available datasets. In particular, we propose and establish a new analysis of the BAO spectrum beyond its use as a standard ruler, resulting in the first measurement of this imprint of neutrinos in the clustering of galaxies.

Future cosmological surveys, such as the next generation of CMB experiments (CMB-S4), have the potential to measure the energy density of relativistic species at the sub-percent level and will, therefore, be capable of probing physics beyond the Standard Model. We demonstrate how this improvement in sensitivity can indeed be achieved and present an observational target which would allow the detection of any extra light particle that has ever been in thermal equilibrium. Interestingly, even the absence of a detection would result in new insights by providing constraints on the couplings to the Standard Model. As an example, we show that existing bounds on additional scalar particles, such as axions, may be surpassed by orders of magnitude.

**Parts of this thesis have been published in the following journal articles:**

[1]  D. Baumann, D. Green, J. Meyers and B. Wallisch, *Phases of New Physics in the CMB*, JCAP **01** (2016) 007. arXiv:1508.06342 [astro-ph.CO].

[2]  D. Baumann, D. Green and B. Wallisch, *New Target for Cosmic Axion Searches*, Phys. Rev. Lett. **117** (2016) 171301. arXiv:1604.08614 [astro-ph.CO].

[3]  D. Baumann, D. Green and B. Wallisch, *Searching for Light Relics with Large-Scale Structure*, JCAP **08** (2018) 029. arXiv:1712.08067 [astro-ph.CO].

[4]  D. Baumann, F. Beutler, R. Flauger, D. Green, A. Slosar, M. Vargas-Magaña, B. Wallisch and C. Yèche, *First Constraint on the Neutrino-Induced Phase Shift in the Spectrum of Baryon Acoustic Oscillations*, Nat. Phys. **15** (2019) 465. arXiv:1803.10741 [astro-ph.CO].

# Acknowledgements

First and foremost, I wish to express my sincere gratitude to my supervisor Daniel Baumann for his phenomenal guidance, tireless effort and unwavering support. His enthusiasm and passion for physics continue to amaze me and are truly contagious. He is a remarkable mentor with endless patience and dedication. I cannot thank him enough for all his contributions at many different levels to this work and beyond.

Second, I am most grateful to Daniel Green who was my other main collaborator and has essentially become a second supervisor in a number of ways. I could always count on his help and support, and the fact that our many discussions were enlightening and insightful. His imagination, excitement and approach to physics are an inspiration. It has been a true privilege and great pleasure to work with and learn from both Daniels throughout my Ph.D.

Thanks are also due to my other collaborators, Florian Beutler, James Fergusson, Raphael Flauger, Helge Gruetjen, Joel Meyers, Paul Shellard, Anže Slosar, Mariana Vargas-Magaña and Christophe Yèche, for their valuable input, amazing ideas and enriching discussions. Working with this great group of cosmologists has helped me to understand many different aspects of our field. In addition, I am grateful to my Ph.D. examiners Anthony Challinor and Jens Chluba.

Furthermore, I thank my departmental colleagues in both Cambridge and Amsterdam, especially Valentin Assassi, Matteo Biagetti, Horng Sheng Chia, Garrett Goon, Hayden Lee, Guilherme Pimentel and John Stout, for numerous conversations and providing such a friendly and stimulating environment. Special thanks go to my Cambridge officemates Chandrima Ganguly and Will Cook.

I am thankful to my friends and fellow students in Heidelberg, Cambridge, Amsterdam and elsewhere for spending countless fun and enriching moments together, for their understanding when I have neglected them, and for sharing numerous adventures.

Finally, I want to thank my brother. More than anyone else, I, however, owe my deepest gratitude to my parents for their continuous and unconditional support in so many ways. I would not be where I am now without them.

I acknowledge support by a Cambridge European Scholarship of the Cambridge Trust, by the Department of Applied Mathematics and Theoretical Physics, by a Research Studentship Award of the Cambridge Philosophical Society, from a Starting Grant of the European Research Council (ERC STG Grant 279617), by an STFC Studentship, by Trinity Hall and by a Visiting Ph.D. Fellowship of the Delta-ITP consortium, a program of the Netherlands Organisation for Scientific Research (NWO) that is funded by the Dutch Ministry of Education, Culture and Science (OCW). I am also grateful to the CERN theory group, the Institute for Theoretical Physics at the University of Heidelberg and, in particular, the Institute of Physics at the University of Amsterdam for their hospitality.

This work uses observations obtained by the Planck satellite (http://www.esa.int/Planck), an ESA science mission with instruments and contributions directly funded by ESA Member States, NASA and Canada. This research is also partly based on observations obtained by the Sloan Digital Sky Survey III (SDSS-III, http://www.sdss3.org/). Funding for SDSS-III has been provided by the Alfred P. Sloan Foundation, the Participating Institutions, the National Science Foundation and the U.S. Department of Energy Office of Science. Parts of this work were undertaken on the COSMOS Shared Memory System at DAMTP (University of Cambridge), operated on behalf of the STFC DiRAC HPC Facility. This equipment is funded by BIS National E-Infrastructure Capital Grant ST/J005673/1 and STFC Grants ST/H008586/1, ST/K00333X/1. Some analyses also used resources of the HPC cluster Atócatl at IA-UNAM, Mexico, and of the National Energy Research Scientific Computing Center, which is supported by the Office of Science of the U.S. Department of Energy under Contract No. DE-AC02-05CH11231.

The results presented in this thesis made use of CAMB [1], CLASS [2], CosmoMC/GetDist [3], FORM [4], IPython [5], MontePython [6], and the Python packages Astropy [7], emcee [8], Matplotlib [9], nbodykit [10] and NumPy/SciPy [11].

# References

1. A. Lewis, A. Challinor, A. Lasenby, Efficient computation of CMB anisotropies in closed FRW models. Astrophys. J. **538**, 473 (2000). arXiv:astro-ph/9911177 [astro-ph]
2. D. Blas, J. Lesgourgues, T. Tram, The Cosmic Linear Anisotropy Solving System (CLASS) II: Approximation schemes. JCAP **07**, 034 (2011). arXiv:1104.2933 [astro-ph.CO]
3. A. Lewis, S. Bridle, Cosmological parameters from CMB and other data: A Monte Carlo approach. Phys. Rev. D **66**, 103511 (2002). arXiv:astroph/0205436 [astro-ph]
4. J. Kuipers, T. Ueda, J. Vermaseren, J. Vollinga, FORM version 4.0. Comput. Phys. Commun. **184**, 1453 (2013). arXiv:1203.6543 [cs.SC]
5. F. Pérez, B. Granger, IPython: a system for interactive scientific computing. Comput. Sci. Eng. **9**, 21 (2007).
6. B. Audren, J. Lesgourgues, K. Benabed, S. Prunet, Conservative constraints on early cosmology: an illustration of the MontePython cosmological parameter inference code. JCAP **02**, 001 (2013). arXiv:1210.7183 [astro-ph.CO]

7. T. Robitaille et al. (Astropy Collaboration), Astropy: a community Python package for astronomy. Astron. Astrophys. **558**, A33 (2013). arXiv:1307.6212 [astro-ph.IM]
8. D. Foreman-Mackey, D. Hogg, D. Lang, J. Goodman, emcee: the MCMC hammer. Publ. Astron. Soc. Pac. **125**, 306 (2013). arXiv:1202.3665 [astro-ph.IM]
9. J. Hunter, Matplotlib: a 2D graphics environment. Comput. Sci. Eng. **9**, 90 (2007).
10. N. Hand, Y. Feng, F. Beutler, Y. Li, C. Modi, U. Seljak, Z. Slepian, nbodykit: An open-source, massively parallel toolkit for large-scale structure. Astron. J. **156**, 160 (2018). arXiv:1712.05834 [astro-ph.IM]
11. S. van der Walt, S. Colbert, G. Varoquaux, The NumPy array: a structure for efficient numerical computation. Comput. Sci. Eng. **13**, 22 (2011). arXiv:1102.1523 [cs.MS]

# Contents

# Chapter 1
# Introduction

Cosmology is a sensitive probe of particle physics, both within the Standard Model and beyond it. In fact, cosmological observations have now become precise enough to start complementing laboratory and collider experiments. For example, by measuring the radiation density of the universe, future observations may provide further insights into the properties of neutrinos. Moreover, if these measurements reach sub-percent level, they have the potential to discover particles that are more weakly coupled than neutrinos, which are predicted in many interesting models of physics beyond the Standard Model. In this thesis, we are searching for these elusive particles by identifying and extracting their robust signatures in cosmological observables.

Experiments at particle accelerators have established the Standard Model (SM) of particle physics as the description of the elementary building blocks of matter and their non-gravitational interactions. At the same time, observations of the anisotropies in the cosmic microwave background (CMB), the fossil radiation from the beginning of the universe, have led to the standard model of cosmology, which captures the entire evolution of the universe from the hot big bang until today. Despite the great successes of the standard models of both cosmology and particle physics, many questions remain unanswered. Strikingly, solutions to some of the cosmological puzzles may influence those in particle physics and vice versa. For instance, some ingredients of the cosmological model, such as dark matter and inflation, ask for new microscopic descriptions. Concurrently, many extensions of the Standard Model give rise to new particles that can be efficiently produced at the high temperatures in the early universe and may therefore be detectable in cosmological observables.

About 373 000 years after the big bang, photons decoupled from the rest of the primordial plasma and the cosmic microwave background was released carrying a treasure-trove of information. Most of our knowledge about the early universe comes from observations of these relic photons. As a matter of fact, cosmologists generally study the history and contents of the universe by detecting relics from the past or extracting their imprints. To probe the time before the epoch of recombination, we therefore rely either on theoretical extrapolations or the existence of further relics to get a snapshot of our cosmos. This has been very successfully employed by measur-

© Springer Nature Switzerland AG 2019
B. Wallisch, *Cosmological Probes of Light Relics*, Springer Theses,
https://doi.org/10.1007/978-3-030-31098-1_1

ing the relic abundances of the light elements, which were synthesized about three minutes after the beginning of the universe during big bang nucleosynthesis (BBN).

Roughly one second after the big bang, a thermal background of relic neutrinos was released when the rate of neutrino interactions dropped below the expansion rate of the universe and neutrinos were no longer in thermal equilibrium with the rest of the Standard Model. Measuring this cosmic neutrino background (CνB) would establish a window back to this time, when the universe was at nearly nuclear densities. Since these neutrinos were a dominant component of the energy density in the early universe, they played an important role in the evolution of cosmological perturbations. Extracting the imprints of neutrinos in observations may therefore provide new insights into the least understood sector of the Standard Model. In fact, as we will show in this thesis, one of the most remarkable results of the Planck satellite is the detection of cosmic neutrinos and a confirmation of their free-streaming nature. At present, we can therefore use cosmological measurements to explore the Standard Model and investigate the history of the universe back to a time when it was one second old.

Probing even earlier times requires detecting new particles that are more weakly coupled than neutrinos. Since neutrinos are the most feebly interacting SM particles, these new species necessarily lie beyond the Standard Model (BSM). There is indeed a lot of circumstantial evidence from both theoretical considerations and experimental measurements that the Standard Model is incomplete. In addition to new massive particles, an interesting consequence of many proposals for BSM physics are extra light species [1], such as axions [2–4], axion-like particles (ALPs) [5], dark photons [6, 7] and light sterile neutrinos [8]. The search for these particles is one of the main objectives of particle physics, but detecting them could be difficult in terrestrial experiments because their couplings might be too small or their masses too large. Interestingly, the temperature in the early universe was high enough to make the production of weakly-coupled and/or massive particles efficient. Their gravitational influence could then be detected if the energy density carried by these particles was significant. This will be the case for light relics which were in thermal equilibrium with the Standard Model at early times and subsequently decoupled from the SM degrees of freedom. This sensitivity to extremely weakly interacting particles is a unique advantage of cosmological probes of BSM physics.

Another advantage of cosmological observations is that they can provide broad constraints on phenomenological descriptions, whereas particle physics searches can be blind to unknown or incompletely specified forms of new physics. This means that terrestrial experiments may give strong constraints on specific scenarios, while cosmological measurements are less sensitive to the details of the models and can compress large classes of BSM physics into broad categories. This approach has led to important discoveries in the past: by comparing observations against simple phenomenological parametrizations, the existence of dark matter ($\Omega_m$) and dark energy ($\Omega_\Lambda$) was established, the baryon asymmetry ($\eta$) was identified, and evidence for cosmological inflation ($n_s$) was presented. We will take a similar path by theoretically describing light thermal relics within an effective field theory (EFT) framework [9] and experimentally searching for their contribution to the radiation density in the early universe ($N_{\text{eff}}$).

Rather remarkably, future cosmological observations, in particular the next generation of CMB experiments, such as the CMB Stage-4 (CMB-S4) mission [10], have the potential to determine the value of $N_{\text{eff}}$ at the sub-percent level. We find it intriguing that this level of sensitivity exactly corresponds to the thermal abundance of light relics in minimal scenarios. An important aspect of this thesis is the realization that the improvement of current constraints by one order of magnitude would allow us to either detect any particles which have ever been in thermal equilibrium with the Standard Model, or put strong constraints on their SM couplings. This sensitivity would therefore not only provide precision tests of the Standard Model, but also a new window into the very early universe and BSM physics.

Since reaching this threshold may have far-reaching implications, we contribute analytical and numerical insights into the main cosmological observables to facilitate the optimization of upcoming experiments. The cosmic microwave background anisotropies, which are the remnants of sound waves in the primordial plasma, are particularly interesting in this respect. The presence of neutrinos and other light relics is imprinted in these fluctuations in two notable ways: Their contribution to the background energy density leads to a characteristic damping of the CMB power spectra on small scales [11] and their free-streaming nature causes a coherent shift in the locations of the acoustic peaks [12]. This subtle shift in the temporal phase of the primordial sound waves has recently been extracted from Planck data [13]. We will present an alternative detection of this effect and establish a precise link to the underlying particle properties, which will in turn allow us to probe SM extensions in a complementary way. Furthermore, it will become clear that future CMB observations will be extremely sensitive to both the damping and the phase shift of the anisotropy spectrum mainly through measurements of the small-scale anisotropies and polarization.

The same physics that is imprinted in the CMB also contributed to the initial conditions for the clustering of matter and may therefore be observable in the large-scale structure (LSS) of the universe as well. The sound waves in the primordial plasma manifest themselves as baryon acoustic oscillations (BAO), which we can observe in the distribution of galaxies. This implies that we should be able to extract the same neutrino-induced phase shift in the BAO spectrum. As a matter of fact, we will present the first such measurement based on a newly developed BAO analysis. Moreover, near-future LSS surveys are projected to map more than ten times as many objects in a much larger cosmic volume than currently operating telescopes. With such remarkable improvements in sensitivity on the horizon, it is timely to re-assess how the wealth of incoming CMB and LSS data could sharpen our understanding of the early universe and, particularly, how they will inform our view of extensions of the standard models of particle physics and cosmology.

**Outline of the Thesis**

The rest of this thesis is organized as follows. We first provide a review of the important aspects of both cosmology and particle physics which underlie the research presented in this work. This also serves as background material for the later chapters. In Chap. 2, we focus on the cosmological standard model. We discuss the thermal history of the universe and introduce the two main observational windows employed in this thesis: the anisotropies of the cosmic microwave background and the large-scale structure of the universe with its distinct BAO signal. In Chap. 3, we concentrate on the Standard Model of particle physics and some of its well-motivated extensions. We collect a few current hints for BSM physics and present an effective field theory of light species as an efficient tool to study the additional particles that arise beyond the Standard Model. Furthermore, we unveil the main cosmological parameter studied in this thesis, the effective number of relativistic species $N_{\mathrm{eff}}$, and analyse the primary signatures of light relics in cosmological observables.

The remaining chapters consist of the main research results. In Chap. 4, which is based on [14], we derive new constraints on light thermal relics from precise measurements of the radiation density in the early universe and explicitly demonstrate the sensitivity of future cosmological observations to the SM couplings of light scalar particles, such as axions. The constraints achievable from cosmology have the potential to surpass existing bounds from laboratory experiments and astrophysical searches by orders of magnitude. In Chap. 5, which is based on [15], we examine the phase shift in the acoustic peaks of the CMB as a robust probe of both free-streaming Standard Model neutrinos and new physics. We find that the physical origin of this signature is limited to either free-streaming relativistic particles or isocurvature fluctuations. In addition, we provide observational constraints from Planck data which establish the free-streaming nature of cosmic neutrinos. In Chap. 6, which is based on [16], we explore to what degree these CMB observations can be enhanced by upcoming large-scale structure surveys. We carefully isolate the information encoded in the shape of the galaxy power spectrum and in the spectrum of baryon acoustic oscillations. In particular, we propose a new analysis of the BAO signal and show that the neutrino-induced phase shift can be detected at high significance in future experiments. In Chap. 7, which is based on [17], we implement this analysis and report on the first measurement of this coherent shift in the peak locations of the BAO spectrum at more than 95 % confidence in galaxy clustering data collected by the Baryon Oscillation Spectroscopic Survey (BOSS). Besides being a new measurement of the cosmic neutrino background and its free-streaming nature, it is also the first application of the BAO signal to early universe physics. In Chap. 8, we conclude with a brief summary of our results and an outlook.

A series of appendices contains technical details underlying the results presented in the main part of this thesis. In Appendix A, we provide further details on the calculations underlying the findings of Chap. 4. In Appendix B, we comment on extensions of the analytic treatment of the phase shift in Chap. 5. In Appendix C, we collect details of the CMB and LSS forecasts conducted in Chap. 6. Appendix D finally includes our methods for extracting the broadband power spectrum of matter fluctuations and the neutrino-induced phase shift as employed in Chaps. 6 and 7.

## Notation and Conventions

For ease of reference, we provide a collection of the employed notation and conventions. Table 1.1 contains a list of abbreviations that will commonly be used. Throughout this thesis, we work in natural units in which the speed of light $c$, the reduced Planck constant $\hbar$ and the Boltzmann constant $k_B$ are set to unity, $c = \hbar = k_B = 1$, and the reduced Planck mass is given by $M_{\mathrm{pl}}^2 = (8\pi G)^{-1}$, with Newton's constant $G$. Our metric signature is $(-+++)$ and we use Greek letters for four-dimensional spacetime indices, $\mu, \nu, \ldots = 0, 1, 2, 3$. Spatial vectors are written in boldface, $\boldsymbol{x}$, or in index notation, $x^i$, with Latin letters, $i, j, \ldots = 1, 2, 3$. Repeated indices are summed over except in cases where the same index appears unpaired on the other side of the equation as well. The magnitude of a vector is defined as $x \equiv |\boldsymbol{x}|$ and unit vectors are hatted, $\hat{\boldsymbol{x}} \equiv \boldsymbol{x}/x$. Our Fourier convention is

**Table 1.1** Common acronyms used throughout this thesis

| Acronym | Expression |
|---------|------------|
| BAO | Baryon acoustic oscillations |
| BBN | Big bang nucleosynthesis |
| BOSS | Baryon Oscillation Spectroscopic Survey |
| BSM | Beyond the Standard Model (of particle physics) |
| CDM | Cold dark matter |
| CMB | Cosmic microwave background |
| CMB-S4 | CMB Stage-4 (a future CMB experiment) |
| C$\nu$B | Cosmic neutrino background |
| CVL | Cosmic variance limit |
| DES | Dark Energy Survey |
| DESI | Dark Energy Spectroscopic Instrument |
| EFT | Effective field theory |
| EWSB | Electroweak symmetry breaking |
| $\Lambda$CDM | Standard cosmological model |
| LSS | Large-scale structure (of the universe) |
| LSST | Large Synoptic Survey Telescope |
| pNGB | Pseudo-Nambu-Goldstone boson |
| QCD | Quantum chromodynamics |
| SM | Standard Model (of particle physics) |

$$\tilde{f}(\pmb{k}) = \int \mathrm{d}^3 x \, f(\pmb{x}) \, \mathrm{e}^{-\mathrm{i}k \cdot x} \,, \qquad f(\pmb{x}) = \int \frac{\mathrm{d}^3 k}{(2\pi)^3} \, \tilde{f}(\pmb{k}) \, \mathrm{e}^{\mathrm{i}k \cdot x} \,, \qquad (1.1)$$

where we commonly drop the tilde, $\tilde{f}(\pmb{k}) \to f(\pmb{k})$, for ease of notation. Overdots and primes denote derivatives with respect to conformal time $\tau$ and physical time $t$, respectively. We use $\tau_0$ for the present time, $\tau_{\mathrm{rec}}$ for the time of recombination and photon decoupling, $\tau_{\mathrm{eq}}$ for matter-radiation equality, and $\tau_{\mathrm{in}}$ for the time at which we set the initial conditions. We will use a subscript '$\alpha$' to denote quantities evaluated at the time $\tau_\alpha$, such as the radiation energy density today, $\rho_{r,0} = \rho_r(\tau = \tau_0)$. As for the radiation content, individual components of the universe (like photons, matter, neutrinos, etc.) will be denoted by a subscript $b = \gamma, m, \nu, \cdots$. The conformal and physical Hubble parameters are $\mathcal{H} \equiv \dot{a}/a$ and $H \equiv a'/a$, respectively, with the scale factor $a$ normalized to unity today, $a_0 \equiv 1$. The dimensionfull and dimensionless power spectra $P_f(k)$ and $\mathcal{P}_f(k)$ of a Fourier mode $f(\pmb{k})$ are defined by

$$\langle f(\pmb{k}) \, f^*(\pmb{k}') \rangle = (2\pi)^3 \, P_f(k) \, \delta_D^{(3)}(\pmb{k} - \pmb{k}') = \frac{(2\pi)^3}{k^3} \, \mathcal{P}_f(k) \, \delta_D^{(3)}(\pmb{k} - \pmb{k}') \,, \qquad (1.2)$$

where $\delta_D^{(3)}$ is the three-dimensional Dirac delta function. Finally, statistical error bars are quoted as one Gaussian standard deviations ($1\sigma$, corresponding to about $68\%$ c.l.), unless stated otherwise.

## References

1. R. Essig et al., Working group report: new light weakly coupled particles. arXiv:1311.0029 [hep-ph]
2. R. Peccei, H. Quinn, CP conservation in the presence of pseudoparticles. Phys. Rev. Lett. **38**, 1440 (1977)
3. S. Weinberg, A new light boson? Phys. Rev. Lett. **40**, 223 (1978)
4. F. Wilczek, Problem of strong P and T invariance in the presence of instantons. Phys. Rev. Lett. **40**, 279 (1978)
5. A. Arvanitaki, S. Dimopoulos, S. Dubovsky, N. Kaloper, J. March-Russell, String axiverse. Phys. Rev. D **81**, 123530 (2010). arXiv:0905.4720 [hep-th]
6. B. Holdom, Two U(1)'s and epsilon charge shifts. Phys. Lett. B **166**, 196 (1986)
7. P. Galison, A. Manohar, Two Z's or not two Z's? Phys. Lett. B **136**, 279 (1984)
8. K. Abazajian et al., Light sterile neutrinos: a white paper. arXiv:1204.5379 [hep-ph]
9. C. Brust, D.E. Kaplan, M. Walters, New light species and the CMB. JHEP **12**, 058 (2013). arXiv:1303.5379 [hep-ph]
10. K. Abazajian et al. (CMB-S4 Collaboration), CMB-S4 Science Book, 1st edn. arXiv:1610.02743 [astro-ph.CO]
11. Z. Hou, R. Keisler, L. Knox, M. Millea, C. Reichardt, How massless neutrinos affect the cosmic microwave background damping tail. Phys. Rev. D **87**, 083008 (2013). arXiv:1104.2333 [astro-ph.CO]
12. S. Bashinsky, U. Seljak, Neutrino perturbations in CMB anisotropy and matter clustering. Phys. Rev. D **69**, 083002 (2004). arXiv:astro-ph/0310198 [astro-ph]

13. B. Follin, L. Knox, M. Millea, Z. Pan, First detection of the acoustic oscillation phase shift expected from the cosmic neutrino background. Phys. Rev. Lett. **115**, 091301 (2015). arXiv:1503.07863 [astro-ph.CO]
14. D. Baumann, D. Green, B. Wallisch, New target for cosmic axion searches. Phys. Rev. Lett. **117**, 171301 (2016). arXiv:1604.08614 [astro-ph.CO]
15. D. Baumann, D. Green, J. Meyers, B. Wallisch, Phases of new physics in the CMB. JCAP **01**, 007 (2016). arXiv:1508.06342 [astro-ph.CO]
16. D. Baumann, D. Green, B. Wallisch, Searching for light relics with large-scale structure. JCAP **08**, 029 (2018). arXiv:1712.08067 [astro-ph.CO]
17. D. Baumann, F. Beutler, R. Flauger, D. Green, A. Slosar, M. Vargas-Magaña, B. Wallisch, C. Yèche, First constraint on the neutrino-induced phase shift in the spectrum of baryon acoustic oscillations. Nat. Phys. **15**, 465 (2019). arXiv:1803.10741 [astro-ph.CO]

# Chapter 2
# Review of Modern Cosmology

Cosmology is the quantitative study of the structure and evolution of the universe. In the last few decades, it has emerged as a data-driven field of study which has revolutionized our understanding of the cosmos. The analysis of observations of type Ia supernovae [1, 2], measurements of the temperature anisotropies in the cosmic microwave background (especially by the satellite missions COBE [3, 4], WMAP [5, 6] and Planck [7, 8]) and maps of the large-scale structure [9–11] have contributed decisive insights. Together with important theoretical advances, this has led to the standard model of cosmology, which describes the roughly 13.8 billion years of cosmic expansion in terms of just six parameters.

In this chapter, we discuss both the theory and the observations underlying modern cosmology. In Sect. 2.1, we consider spatially homogeneous and isotropic spacetimes, collect the basic equations governing the universe on the largest scales, and introduce the $\Lambda$CDM model. In Sect. 2.2, we study the thermal history of the universe, including the cosmic neutrino and microwave backgrounds. In Sect. 2.3, we move beyond homogeneity and consider small fluctuations around the smooth universe. Apart from the mechanism underlying the growth and presence of structure, we will also encounter the sound waves in the primordial plasma which are of tremendous observational significance. In Sect. 2.4, we finally review the cosmic microwave background anisotropies, the large-scale structure of the universe and the baryon acoustic oscillations. These are the main cosmological observables employed in the rest of this thesis.

## 2.1 Homogeneous Cosmology

Cosmological observations indicate that the universe is both spatially homogeneous and isotropic on large scales and/or at early times. This is the basis of modern cosmology. In the following, we study the implications of these profound findings for

© Springer Nature Switzerland AG 2019
B. Wallisch, *Cosmological Probes of Light Relics*, Springer Theses,
https://doi.org/10.1007/978-3-030-31098-1_2

the shape and content of the universe (Sect. 2.1.1), and introduce the $\Lambda$CDM model which has emerged as the standard cosmological model (Sect. 2.1.2).

### *2.1.1 Geometry and Dynamics of the Universe*

A spatially homogeneous and isotropic spacetime can be described by the Friedmann–Lemaître–Robertson–Walker (FLRW) metric

$$ds^2 = \bar{g}_{\mu\nu}dx^\mu dx^\nu = -dt^2 + a^2(t)\gamma_{ij}dx^i dx^j , \tag{2.1}$$

where $\gamma_{ij}$ is the induced metric on the spatial hypersurfaces. The FLRW metric (2.1) depends on one time-dependent function, the scale factor $a(t)$. To see this, it is instructive to decompose $\bar{g}_{\mu\nu}$ into a scalar, $\bar{g}_{00}$, a three-vector, $\bar{g}_{i0}$, and a three-tensor, $\bar{g}_{ij}$. Homogeneity requires that the mean value of any scalar can only be a function of time. Absorbing this function into a redefinition of the time coordinate, we get $\bar{g}_{00} = -1$. Isotropy on the other hand implies that the mean value of any three-vector has to vanish, i.e. $\bar{g}_{i0} \equiv 0$. Finally, the mean value of the three-tensor $\bar{g}_{ij}$ has to be proportional to the three-metric $\gamma_{ij}$ based on isotropy. In turn, homogeneity restricts the three-curvature to be the same everywhere and the proportionality coefficient to be only a function of time. We therefore have $\bar{g}_{ij} = a^2(t)\gamma_{ij}$, with $\gamma_{ij}$ being restricted by a constant three-curvature $R^{(3)}$. Since there are three unique three-metrics that lead to $R^{(3)} = $ const, the spatial geometry of the universe can only be positively curved ($R^{(3)} > 0$), flat ($R^{(3)} = 0$) or negatively curved ($R^{(3)} < 0$).

Since we mostly rely on observations of photons to study the universe, we have to consider the propagation of light. It is therefore useful to work with a metric that is conformally invariant to Minkowski space. This can be achieved by introducing conformal time $d\tau \equiv dt/a$. As a consequence, we can express the FLRW metric as

$$ds^2 = a^2(\tau)\left(-d\tau^2 + d\boldsymbol{x}^2\right), \tag{2.2}$$

where we defined $d\boldsymbol{x}^2 \equiv \gamma_{ij}dx^i dx^j$. Observations of the CMB have constrained the three-curvature to be close to zero, which implies that the flat FLRW metric describes the global geometry of the universe to very good approximation. For the rest of this thesis, we will therefore neglect spatial curvature, $R^{(3)} = 0$, in which case $\gamma_{ij} = \delta_{ij}$ in Cartesian coordinates.

Of the four fundamental forces of Nature, only the gravitational force is relevant on cosmological scales because the other three are shielded by opposite charges or confined to subatomic scales. According to general relativity, the dynamics of the universe is therefore governed by the Einstein field equations [12],

$$G_{\mu\nu} = 8\pi G\, T_{\mu\nu} , \tag{2.3}$$

where $G$ is Newton's constant, and the Einstein tensor $G_{\mu\nu}$ depends on the metric $g_{\mu\nu}$ and its first two derivatives. The energy-momentum tensor $T_{\mu\nu}$ on the right-hand side captures the entire content of the universe. The fact that energy-momentum is conserved, $\nabla_\mu T^{\mu\nu} = 0$, is automatically ensured by the Bianchi identities.

The form of the energy-momentum tensor in a homogeneous and isotropic universe can be derived by following a similar argument as for the metric. It is again convenient to decompose $T_{\mu\nu}$ into a scalar, $T_{00}$, a three-vector, $T_{i0}$, and a three-tensor, $T_{ij}$. As above, homogeneity implies that the scalar can only depend on time, i.e. $\bar{T}_{00} = \bar{\rho}(t)$, with an arbitrary function $\bar{\rho}(t)$. Isotropy again requires $\bar{T}_{i0} \equiv 0$ since there would otherwise be a non-zero energy flux. By isotropy we also have $T_{ij} \propto g_{ij}$, with homogeneity restricting the proportionality coefficient to be a function of time alone. The energy-momentum tensor of a homogeneous and isotropic background can therefore only take the following form:

$$\bar{T}_{00} = \bar{\rho}(t)\,, \qquad \bar{T}_{i0} = \bar{T}_{0j} = 0\,, \qquad \bar{T}_{ij} = \bar{P}(t)\,g_{ij}\,. \tag{2.4}$$

This is the energy-momentum tensor of a perfect fluid as seen in the reference frame of a comoving observer. The content of a spatially homogeneous and isotropic universe can therefore be characterised by the energy density $\bar{\rho} = \bar{\rho}(t)$ and the pressure $\bar{P} = \bar{P}(t)$ in the rest frame of the cosmic fluid.

We obtain the evolution equations of the cosmic fluid by plugging the conformal FLRW metric (2.2) and the energy-momentum tensor (2.4) into the Einstein equations (2.3). Given the imposed symmetries, they simplify dramatically to the first and second Friedmann equations [13],

$$3\mathcal{H}^2 = 8\pi G a^2\,\bar{\rho}\,, \qquad 2\dot{\mathcal{H}} + \mathcal{H}^2 = -8\pi G a^2\,\bar{P}\,, \tag{2.5}$$

where we introduced the (conformal) Hubble rate $\mathcal{H} = \dot{a}/a$.[1] By combining these equations, or by directly employing energy-momentum conservation, we obtain the continuity equation,

$$\dot{\bar{\rho}} + 3\mathcal{H}\,(\bar{\rho} + \bar{P}) = 0\,, \tag{2.6}$$

from which we can read off the evolution of the cosmic fluid with a given equation of state $w \equiv \bar{P}/\bar{\rho}$. In this thesis, we will mostly be concerned with two types of cosmological content: radiation with $w_r = 1/3$, i.e. $\bar{\rho}_r \propto a^{-4}$, and (pressureless) matter with $w_m \approx 0$, i.e. $\bar{\rho}_m \propto a^{-3}$. It is easy to see that radiation dominates at early times in a universe with these two components, but matter takes over as the main constituent after matter-radiation equality at $a_{\text{eq}} \equiv \bar{\rho}_{r,0}/\bar{\rho}_{m,0}$, with $\rho_{b,0} = \rho_b(\tau = \tau_0)$. The third component, which is important in our universe and dominates at late times, is dark energy. Many observational clues point towards dark energy being described by an equation of state of $w_\Lambda = -1$ with $\rho_\Lambda = \text{const}$ and, hence, the cosmological constant $\Lambda$ (see e.g. [14]).

---

[1]For convenience, we invert the otherwise standard notation and use overdots (primes) to denote derivatives with respect to conformal (physical) time $\tau$ ($t$), i.e. $\dot{a} \equiv \mathrm{d}a/\mathrm{d}\tau$ and $a' \equiv \mathrm{d}a/\mathrm{d}t$. The physical Hubble rate is therefore given by $H = a'/a$.

### 2.1.2  Standard Model of Cosmology

It is rather remarkable that all current cosmological data (e.g. [6, 8, 11, 15]) is
fit by a simple six-parameter model—the $\Lambda$CDM model. As the name suggests,
at late times, the universe is dominated by the cosmological constant $\Lambda$ and cold
dark matter (CDM). At the same time, these are the components that we know least
about. The radiation energy density is very small today, but photons and neutrinos
still vastly dominate the entropy of the universe due to their large number densities
of about $411\,\text{cm}^{-3}$ and $112\,\text{cm}^{-3}$, respectively (cf. Sects. 2.2.3 and 2.2.4). Finally,
cosmologists refer to the visible matter comprised of the known Standard Model
particles as baryonic matter, which, for most of the universe's history, consists mainly
of electrons and protons.

The $\Lambda$CDM model includes two parameters characterising the initial conditions,
namely the amplitude $A_s$ and the tilt $n_s$ of the almost scale-invariant spectrum of
primordial curvature perturbations,

$$\mathcal{P}_\zeta(k) = A_s \left( \frac{k}{k_0} \right)^{n_s-1} , \tag{2.7}$$

where the arbitrary pivot scale is commonly set to $k_0 = 0.05\,\text{Mpc}^{-1}$. Such a spectrum
is notably predicted by generic inflationary models [16].

The remaining four parameters are associated with the geometry and composition of the universe. The matter content of the universe is described by the physical
baryon and cold dark matter densities, $\omega_b \equiv \Omega_b h^2$ and $\omega_c \equiv \Omega_c h^2$, where $\Omega_a \equiv
8\pi G/(3H_0^2)\,\rho_a$, with reduced Hubble constant $h \equiv H_0/\left(100\,\text{km s}^{-1}\text{Mpc}^{-1}\right)$. Some-
times the dark matter density $\Omega_c$ is traded for the total matter density $\Omega_m = \Omega_b +
\Omega_c + \Omega_\nu$, where $\Omega_\nu \approx \sum_i m_{\nu_i}/94.1\,\text{eV}$ is the (small) contribution of massive neutri-
nos to the matter density.[2] Instead of the Hubble constant $H_0$, we often use the angular
size of the sound horizon at photon decoupling (see below), $\theta_s \equiv r_s(z_\text{rec})/D_A(z_\text{rec})$,
where $r_s$ is the physical sound horizon and $D_A$ is the angular diameter distance, both
evaluated at the redshift of decoupling, $z_\text{rec}$. The parameter $\theta_s$ receives a contribution
from the dark energy density $\Omega_\Lambda \equiv \Lambda/(3H_0^2)$. We note that $\Lambda$CDM assumes the
universe to be exactly flat, $R^{(3)} \equiv 0$, i.e. $\Omega_r + \Omega_m + \Omega_\Lambda \equiv 1$, where the radiation
density is comprised of photons at a temperature of $T_0 = 2.7255\,\text{K} = 0.235\,\text{meV}$ [18]
as well as three relativistic neutrino species. We refer to Sect. 3.3 for further details
on the radiation density and neutrinos in particular, but note that neutrinos carry a
sizeable fraction of the energy density in the early universe. As illustrated in Fig. 2.1,
the history of the universe can therefore be divided into three large epochs: the

---

[2]The minimal sum of masses from neutrino oscillation experiments is about 58 meV [17] and in
particular cosmological measurements are closing in on this value with a current upper bound
of $\sum_i m_{\nu_i} < 0.23\,\text{eV}$ (95 % c.l.) [8]. As the masses are so small that neutrinos have been
(ultra-)relativistic for a large part of cosmological history, especially around the time of photon
decoupling ($T_\text{rec} \approx 0.26\,\text{eV}$), their effect on the aspects of interest in this thesis is small. For simplicity, we will therefore treat neutrinos as massless particles throughout, except in the BOSS
cosmology of Chap. 7.

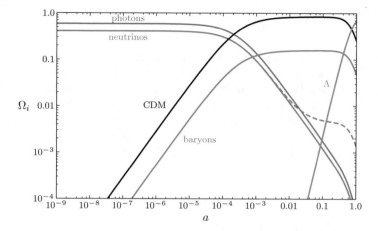

**Fig. 2.1** Evolution of the fractional energy densities $\Omega_i$ for photons, neutrinos, cold dark matter, baryons and dark energy described by a cosmological constant within the standard $\Lambda$CDM model. The dashed red line assumes massive neutrinos with $\sum_i m_{\nu_i} \approx 58\,\text{meV}$, whereas the solid red line takes these particles to be massless

**Table 2.1** Parameters of our reference cosmological model and their fiducial values based on [8]

| Parameter | Fiducial value | Description |
|---|---|---|
| $\omega_b$ | 0.02230 | Physical baryon density $\omega_b \equiv \Omega_b h^2$ |
| $\omega_c$ | 0.1188 | Physical dark matter density $\omega_c \equiv \Omega_c h^2$ |
| $100\,\theta_s$ | 1.04112 | $100\times$ angular size of the sound horizon at decoupling |
| $\tau$ | 0.066 | Optical depth due to reionization |
| $\ln(10^{10} A_s)$ | 3.064 | Log of scalar amplitude (at pivot scale $k_0 = 0.05\,\text{Mpc}^{-1}$) |
| $n_s$ | 0.9667 | Scalar spectral index (at pivot scale $k_0 = 0.05\,\text{Mpc}^{-1}$) |
| $N_{\text{eff}}$ | 3.046 | Effective number of (free-streaming) relativistic species |
| $Y_p$ | 0.2478 | Primordial helium fraction |

radiation-dominated, matter-dominated and dark energy-dominated eras. The standard six-parameter model is completed by the optical depth $\tau$ due to reionization in the late universe.[3] In Table 2.1, we list our fiducial values of the $\Lambda$CDM parameters, based on the Planck best-fit cosmology [8]. Unless otherwise stated, these values will be used throughout this thesis.

---

[3] Although we use the same symbol to denote the optical depth and conformal time, its meaning will always be clear from context.

## 2.2    Thermal History

We now turn our attention to the precise evolution of the different species in the universe. This is mainly a story of (local) thermal equilibrium, production and decoupling of particles. Before we provide any details, we give a brief history of the main events in the thermal history of the universe (Sect. 2.2.1). We then discuss its thermodynamic description at early times and the notion of particles freezing out (Sect. 2.2.2). We conclude the section by reviewing the formation of the cosmic neutrino and microwave backgrounds (Sects. 2.2.3 and 2.2.4). They are of particular importance in this thesis.

### 2.2.1    Brief History of the Hot Big Bang

We take the beginning of the hot big bang to be the end of inflation when the particles of the Standard Model (and possibly its extensions) were produced during reheating. The associated energy scale might be as large as $10^{16}$ GeV and all (known) particles were massless. When the temperature of the universe dropped to about 100 GeV, the electroweak symmetry of the SM became spontaneously broken. As a consequence, the SM particles acquired their mass through the Higgs mechanism, but most particles were still relativistic. During the quark-gluon transition, quarks and gluons became confined in composite hadronic states. This event occurred around the temperature of the non-perturbative QCD scale, $T \sim 150$ MeV, and is usually denoted as the QCD phase transition although it might be a cross-over. When the universe was about one second old, corresponding to a temperature of $T \sim 1$ MeV, neutrinos decoupled from the rest of the primordial plasma and the cosmic neutrino background was released because the weak interactions were no longer efficient enough to maintain local thermal equilibrium (cf. Sect. 2.2.3).

Around the same time, the interactions between neutrons and protons became inefficient leading to a relic abundance of neutrons. After about three minutes, the light elements, in particular hydrogen and helium, were synthesized from these neutrons and protons during big bang nucleosynthesis. By numerically solving coupled Boltzmann equations, the primordial helium fraction is predicted to be $Y_p = 4n_{He}/n_b \sim 0.25$, with the precise value depending on the baryon density and the amount of radiation in the universe. This estimate agrees well with both the observations of primordial abundances (see e.g. [19]) and the value inferred from CMB measurements (see e.g. [8]). In fact, BBN has become one of the main tools to constrain the evolution of the universe above the MeV scale. Moreover, the theoretical predictions for the primordial abundances as a function of $\omega_b$ and $\omega_r$ can be used to infer the value of $Y_p$ that is consistent within the $\Lambda$CDM model. We refer to this procedure as imposing consistency with BBN.

The last major event in the thermal history of our universe was the formation of the first hydrogen atoms, which is referred to as recombination (cf. Sect. 2.2.4). Since the number density of free electrons dropped sharply as a consequence, photons decoupled from matter at $T_{\text{rec}} \approx 0.26\,\text{eV}$ about 373 000 years after the beginning of the universe. They have been free-streaming ever since and we observe these relics of the big bang today as the cosmic microwave background.

## 2.2.2 In and Out of Equilibrium

Throughout the thermal evolution of the universe, the interaction rate of particles, $\Gamma \sim n\sigma$, with number density $n$ and thermally-averaged cross section $\sigma$, competes with the expansion rate $H$. As long as $\Gamma \gg H$, thermal equilibrium can be maintained locally because there are many particle interactions per Hubble time and the evolution is quasi-stationary. On the other hand, if the universe at some point expands faster than these particles can interact with each other, $\Gamma \ll H$, they are no longer in equilibrium and evolve separately. In the following, we will discuss both regimes, including the transition period when $\Gamma \sim H$.

### Thermal Equilibrium

At early times, all SM particles were in local thermal equilibrium, i.e. the interactions between them were efficient and kept these particles locally in close thermal contact. In phase space, every particle species $a$ can be described by its distribution function $f_a(t, x, p)$, the number of particles per unit phase space volume, with momentum $p$. Homogeneity and isotropy dictate that the distribution function can neither depend on the position $x$ nor the direction of the momentum $\hat{p}$, which implies that $f_a(t, x, p) \rightarrow \bar{f}_a(t, p)$. The thermal Bose–Einstein and Fermi–Dirac distribution functions are given by

$$\bar{f}_a(p) = \left[ e^{(E_a(p) - \mu_a)/T_a} \mp 1 \right]^{-1},$$  (2.8)

for bosons ($-$) and fermions ($+$), respectively, where $E_a$ is the relativistic energy which includes the mass $m_a$. The chemical potentials $\mu_a$ are likely small for all SM species and, in particular, vanish for photons (since the number of photons is not conserved). For electrons, for instance, we can estimate $\mu_e/T \sim 10^{-9}$ because the universe is electrically neutral, i.e. the proton number density is equal to the difference in the number densities of electrons and positrons, $n_p = n_e - \bar{n}_e$, the baryon-to-photon ratio is $\eta = n_b/n_\gamma \approx n_p/n_\gamma \sim 10^{-9}$ and $\mu_e/T \sim (n_e - \bar{n}_e)/n_\gamma$. For simplicity, we therefore set the chemical potentials to zero, $\mu_a \equiv 0$, from now on. All species which are in thermal equilibrium with one another of course share the same temperature $T_a = T$.

Often, we are only interested in the momentum-integrated quantities. Integrating the distribution function yields the number density

$$n_a = g_a \int \frac{\mathrm{d}^3 p}{(2\pi)^3} f_a(x, p),$$  (2.9)

where $g_a$ is the number of internal degrees of freedom of species $a$. Similarly, we obtain the energy density and pressure from the distribution function via the weighted integrals

$$\rho_a = g_a \int \frac{d^3 p}{(2\pi)^3} E_a(p) f_a(\boldsymbol{x}, \boldsymbol{p}), \qquad P_a = g_a \int \frac{d^3 p}{(2\pi)^3} \frac{p^2}{3 E_a(p)} f_a(\boldsymbol{x}, \boldsymbol{p}). \quad (2.10)$$

In the non-relativistic limit, $m_a \gg T_a$, it is easy to see that $\rho_a \approx m_a n_a$ and $P_a \ll \rho_a$, i.e. a non-relativistic gas of particles behaves like pressureless matter. If they become non-relativistic while being in thermal contact with other species, their number and energy densities get exponentially suppressed, $f_a \to e^{-m_a/T}$. Physically speaking, this arises because particles and anti-particles annihilate while the reverse process is kinematically forbidden below the mass threshold. As a consequence, the primordial plasma is dominated by relativistic species. If particles freeze out, on the other hand, they retain their equilibrium distribution function and do not get further depleted which results in a finite relic abundance. This explains why dark matter may be cold and comprised of thermal relics.

In the relativistic limit, $m_a \ll T_a$, we recover the equation of state of radiation, $P_a = \frac{1}{3}\rho_a$. Taking the thermal distributions (2.8) and defining the temperature of the universe as the photon temperature, $T \equiv T_\gamma$, the total radiation energy density can be written as

$$\bar{\rho}_r = \sum_a \bar{\rho}_a = \frac{\pi^2}{30} g_*(T) T^4, \qquad (2.11)$$

where we summed over all relativistic species and introduced the effective number of relativistic degrees of freedom

$$g_*(T) = \sum_{a=b} g_{*,a}(T) + \frac{7}{8} \sum_{a=f} g_{*,a}(T) = \sum_a g_a \left(\frac{T_a}{T}\right)^4 \left(1 - \frac{1}{8}\delta_{af}\right). \quad (2.12)$$

Here, the Kronecker delta $\delta_{af}$ vanishes for bosons, $a = b$, and equals unity for fermionic species, $a = f$, to account for their relative Fermi–Dirac suppression factor of $7/8$. We reiterate that all particles which are in thermal equilibrium with photons have the same temperature $T_a = T$, but the temperature of decoupled species may be different, $T_a \neq T$. In the Standard Model, this is only relevant for neutrinos after electron-positron annihilation (cf. Sect. 2.2.3). However, this difference will play a prominent role in Sect. 3.3 when we discuss additional light relics which might appear in BSM models. We also note that the effective number of relativistic degrees of freedom $g_*$ is approximately constant away from mass thresholds, $T \sim m_a$, but decreases when a species becomes non-relativistic and its contribution to the energy density becomes negligible.

The number of internal degrees of freedom $g_a$ of a particle species depends on some of its properties (cf. Table 2.2). For example, real scalar particles carry one degree of freedom, $g_s = 1$, while Weyl fermions have two spin states, $g_f = 2$. Since

**Table 2.2** Number of internal degrees of freedom, $g_a$, for scalar particles, spin-1/2 fermions and vector bosons

| Spin | 0 | | 1/2 | | 1 | |
|---|---|---|---|---|---|---|
| Type | Real | Complex | Weyl | Dirac | Massless | Massive |
| $g_a$ | 1 | 2 | 2 | 4 | 2 | 3 |

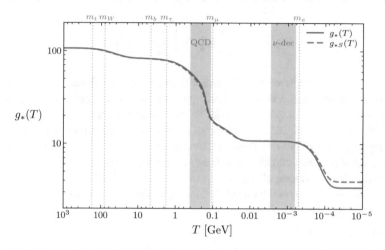

**Fig. 2.2** Evolution of the effective number of relativistic degrees of freedom and those in entropy, $g_*(T)$ and $g_{*S}(T)$, assuming the SM particle content. We used the state-of-the-art lattice QCD calculations of [20] for $20\,\mathrm{MeV} \lesssim T \lesssim 150\,\mathrm{GeV}$ and numerically evaluated the left-hand sides of (2.11) and (2.13) otherwise. The small differences between $g_*(T)$ and $g_{*S}(T)$ for $T \gtrsim 1\,\mathrm{MeV}$ arise from non-perturbative QCD effects which are not captured in (2.14). The gray bands indicate the QCD phase transition and neutrino decoupling, and the dotted lines denote some of the mass scales at which SM particles and anti-particles annihilate

massless vector bosons are transversely polarized, they provide $g_v = 2$ degrees of freedom, while their massive counterparts have an additional longitudinal polarization, resulting in $g_v = 3$. Accounting for all SM particles, there are effectively $g_{*,\mathrm{SM}} = 106.75$ relativistic degrees of freedom at high temperatures. After electron-positron annihilation around $T \sim 0.5\,\mathrm{MeV}$, the only relativistic SM particles are photons and neutrinos with $g_\gamma = 2$ and $g_v = 3 \cdot 2 = 6$. The entire evolution of $g_*(T)$ within the Standard Model is displayed in Fig. 2.2. Since the quarks and gluons are confined into hadrons during the QCD phase transition, and pions are the only relativistic composite particles afterwards, the value of $g_*(T)$ is reduced by about one order of magnitude around $T \sim 150\,\mathrm{MeV}$. This will be of further importance in subsequent chapters and dramatically impact the detectability of light BSM relics which might have decoupled at very early times.

When describing the thermal history of the universe, it is convenient to track conserved thermodynamic quantities. According to the second law of thermodynamics, the entropy of a system can only be constant or increase. For relativistic species, the entropy density is given by

$$s = \sum_a s_a = \sum_a \frac{\bar{\rho}_a + \bar{P}_a}{T_a} = \frac{2\pi^2}{45} g_{*S}(T)\, T^3 \,, \tag{2.13}$$

where we defined the effective number of relativistic degrees of freedom in entropy

$$g_{*S}(T) = \sum_a g_a \left(\frac{T_a}{T}\right)^3 \left(1 - \frac{1}{8}\delta_{af}\right), \tag{2.14}$$

in analogy to the definition of $g_*(T)$ in (2.12). In equilibrium, the entropy in a comoving volume is, in fact, conserved which implies $sa^3 = \text{const.}$ Moreover, the entropy is approximately constant even out of equilibrium because any non-equilibrium entropy production, e.g. from decaying particles, is usually small compared to the large entropy in photons. We can therefore treat the expansion of the universe as basically adiabatic. When particles and anti-particles annihilate, the released entropy is then redistributed among all species in thermal equilibrium. This implies that the temperature of the thermal bath redshifts slightly less than without the annihilation events, $T \propto g_{*S}^{-1/3} a^{-1}$.

**Beyond Equilibrium**

If the SM particles had remained in thermal equilibrium throughout the history of the universe, $\Gamma \gg H$, any past events would be irrelevant. In this case, our universe would almost entirely consist of photons and would not be an interesting place. Deviations from equilibrium, $\Gamma \lesssim H$, are therefore crucial. The full evolution of a species $a$ is determined by the Boltzmann equation, which is schematically given by

$$\frac{df_a}{dt} = C[f_a, \{f_b\}]. \tag{2.15}$$

This provides the time evolution of the distribution function $f_a(t, x, p)$ of each particle species $a$ as a function of its interactions with all other particles in the system. These interactions are captured by the collision term $C$ on the right-hand side. Solving the time evolution of an entire system may therefore involve a set of Boltzmann equations which can become computationally involved. If an exact treatment is not necessary, it is therefore advantageous to follow an approximate scheme.

At early times, the interactions are frequent enough that they keep the SM particles in thermal equilibrium. At some point, however, the interaction rate of a species may become of equal size to the Hubble rate,

$$\Gamma(T) \sim H(T), \tag{2.16}$$

and these particles freeze out, i.e. they loose their thermal contact with other species and decouple. This is, of course, not an instantaneous phenomenon at a specific freeze-out temperature $T_F$ defined by $\Gamma(T_F) = H(T_F)$. Having said that, freeze-out usually happens faster than a few Hubble times and the instantaneous decoupling limit often provides rather accurate estimates for the relic abundances at temperatures $T \ll T_F$ when $\Gamma \ll H$. We will therefore usually work within this approximation and avoid solving a set of Boltzmann equations.

Given the instantaneous decoupling limit, we can get qualitative insights into the competition between the interaction and the expansion rate in the early universe. By the first Friedmann equation (2.5), the Hubble rate in the radiation-dominated epoch is given by

$$H(T) = \sqrt{\frac{\bar{\rho}_r(T)}{3M_{\mathrm{pl}}^2}} = \sqrt{\frac{\pi^2}{90}g_*(T)}\,\frac{T^2}{M_{\mathrm{pl}}}, \tag{2.17}$$

where we inserted (2.11) and defined the reduced Planck mass $M_{\mathrm{pl}} \equiv (8\pi G)^{-1/2} \approx 2.4 \times 10^{18}\,\mathrm{GeV}$. Away from mass thresholds, in particular for $T \gtrsim 100\,\mathrm{GeV}$, the Hubble rate therefore has a quadratic temperature dependence, $H \propto T^2$. This means that any interaction with rate $\Gamma \propto T^m$ and $m > 2$ has the chance to be in thermal equilibrium at high enough temperatures and freeze out at some later point. In the SM, the particles generally follow $\Gamma \propto n \propto T^3$ above the electroweak symmetry breaking (EWSB) scale which implies that all particles (except gravitons which only have Planck-suppressed couplings) are in thermal equilibrium in the early universe. On the other hand, particles whose interactions were governed by rates $\Gamma \propto T^m$ with $m < 2$ would not be in thermal equilibrium at early times, but might have the possibility to come into thermal equilibrium later. This phenomenon is often referred to as freeze-in.

### 2.2.3 Cosmic Neutrino Background

We are now in the position to discuss the cosmic neutrino background which is one of the main subjects of this thesis. Neutrinos interact with the rest of the Standard Model only through the weak force. Because neutrinos are, in fact, the most weakly interacting SM particles, they are the first to decouple from the primordial plasma. There are in particular two processes, which keep these particles in thermal equilibrium at high temperatures: pair conversion of neutrinos into leptons (especially electrons and positrons), $\nu + \bar{\nu} \leftrightarrow l + \bar{l}$, and neutrino scattering off of leptons, $\nu + l \leftrightarrow \nu + l$. For energies far below the mass of the weak gauge bosons, $T \ll 80\,\mathrm{GeV}$, the corresponding cross section is $\sigma \sim G_F^2 T^2$, with Fermi's constant $G_F \approx 1.2 \times 10^{-5}\,\mathrm{GeV}^{-2}$. The weak interaction rate for neutrinos is therefore given by $\Gamma_\nu \sim G_F^2 T^2 \bar{n}_e \propto G_F^2 T^5$, where $\bar{n}_e$ is the electron number density in equilibrium. Neutrinos thus freeze out around a temperature of

$$T_{F,\nu} \sim \left(\frac{\sqrt{g_*}}{G_F^2 M_{\mathrm{pl}}}\right)^{-1/3} \sim 1\,\mathrm{MeV}, \tag{2.18}$$

when photons, electrons/positrons and neutrinos were the only relativistic particles left in the primordial plasma. After neutrinos decouple, they maintain their relativistic Fermi–Dirac distribution function $f_\nu(p) \approx [\exp(p/T_\nu) + 1]^{-1}$. Since the momentum $p$ redshifts according to $p \propto a^{-1}$, the neutrino temperature has to have the same scaling and evolve as $T_\nu \propto a^{-1}$.

Since we have measured the temperature of photons $T$ very well from the blackbody spectrum of the CMB (see below), we want to relate the neutrino temperature $T_\nu$ to this quantity. Until electrons and positrons annihilate shortly after neutrinos decouple, the temperatures are of course the same, $T_\nu = T$. As noted above, however, the annihilation process releases the entropy of electrons and positrons. As a consequence, over some short period, the photon bath cools more slowly than $T_\nu \propto a^{-1}$, which implies $T > T_\nu$ henceforth. In the limit of perfect and instantaneous neutrino decoupling,[4] electrons/positrons are the only relativistic particles remaining in thermal equilibrium with photons before the annihilation. The effective numbers of the relativistic degrees of freedom in entropy, which are in thermal equilibrium at the temperatures $T_>$ and $T_<$ before and after the annihilation, are therefore given by

$$g_{*S}(T_>) = g_\gamma + \frac{7}{8}g_e = 2 + \frac{7}{8} \cdot 4 = \frac{11}{2}, \qquad g_{*S}(T_<) = g_\gamma = 2, \qquad (2.19)$$

respectively. Since the entropy (2.13) of the thermal bath is conserved in a comoving volume and $T_\nu \propto a^{-1}$, we can easily relate the photon and neutrino temperatures:

$$g_{*S}\frac{T^3}{T_\nu^3} = \text{const} \quad \Rightarrow \quad \frac{T_\nu}{T_\gamma} = \left(\frac{g_{*S}(T_<)}{g_{*S}(T_>)}\right)^{1/3} = \left(\frac{4}{11}\right)^{1/3}. \qquad (2.20)$$

The measured photon temperature, $T_0 = 2.7255$ K, implies that the C$\nu$B has a thermal spectrum with a temperature of $T_{\nu,0} \approx 1.95$ K today. In consequence, the number density of cosmic neutrinos is very large, $\bar{n}_\nu \approx 112\,\text{cm}^{-3}$, and, in fact, exceeds the flux from astrophysical neutrino sources, such as our Sun. Nevertheless, the direct detection of the C$\nu$B is extremely challenging because the neutrino distribution peaks at the very small energy of $T_{\nu,0} \approx 0.17$ meV. Having said that, as we will also show in this thesis, there is more and more indirect evidence for its existence.

### 2.2.4  Cosmic Microwave Background

A very important event in the thermal history of the universe is the formation of the first atoms and the decoupling of photons. The observation of these photons, which make up the cosmic microwave background, has ultimately led to a number of major breakthroughs in modern cosmology and is still one of the main sources of information about our universe.

---

[4]We will come back to this assumption and correct for it in Sect. 3.3.

From the formation of the C$\nu$B, we fast forward about 250 000 years in cosmic history. In the meantime, the light elements, in particular helium, were synthesized during big bang nucleosynthesis and the universe entered into the matter-dominated epoch. The primordial plasma at that time consisted of (many) photons, free electrons and ionized nuclei (mostly protons). Compton scattering, $e^- + \gamma \leftrightarrow e^- + \gamma$, tightly coupled photons and electrons resulting in a small mean free path for photons. In turn, electrons strongly interacted with protons via Coulomb scattering, $e^- + p^+ \leftrightarrow e^- + p^+$. Finally, electromagnetic reactions such as those forming/ionizing neutral hydrogen, $e^- + p^+ \leftrightarrow H + \gamma$, kept the baryons and photons in equilibrium. However, once the universe cooled to a temperature of about $0.4\,\mathrm{eV}$, the ionization of neutral hydrogen became less and less efficient. This is usually referred to as recombination although it is the first time that electrons and protons combined without being ionized again immediately.

By this time, photon-electron scattering is governed by Thomson scattering, which is the low-energy limit of Compton scattering and has an interaction rate of $\Gamma_\gamma \sim n_e \sigma_T$, with constant Thomson cross section $\sigma_T$. However, the number density of free electrons $n_e$ got reduced dramatically by the increasing amount of neutral hydrogen and the Thomson scattering rate $\Gamma_\gamma$ dropped. When $\Gamma_\gamma \lesssim H$, the mean free path of photons became longer than the horizon size and the photons decoupled from matter. This happened at a temperature of about $T_{\mathrm{rec}} \approx 0.26\,\mathrm{eV}$ corresponding to a redshift of $z \approx 1090$. These photons have since freely streamed through the universe effectively unimpeded and are what we observe as the CMB today.

Recombination is not instantaneous, but requires a finite time, i.e. some photons encountered their last scattering event earlier than others while the plasma was still hotter.[5] This results in the so-called last-scattering surface to have a finite width. However, although these photons last-scattered at slightly different temperatures, we observe the CMB with an almost perfect black-body spectrum at a single temperature of $T_0 = 2.7255\,\mathrm{K}$ today. We do not measure a spectrum comprised of a set of black-body spectra because the photons which decoupled earlier redshifted according to $T \propto a^{-1}$, which is exactly the decrease in temperature that the photons which last-scattered later experienced as well. It was this uniform background that Penzias and Wilson discovered in 1965 [21] and whose black-body spectrum was measured exquisitely by the FIRAS instrument on the COBE satellite in the early 1990s [22]. In the main chapters of this thesis, we will get a glimpse of the wealth of information about our universe that is transmitted by these CMB photons.

---

[5]Recombination and (photon) decoupling are often used synonymously, in particular when referring to the time of decoupling, but actually are different processes. We will keep these two notions distinct in those cases where it is important and it will be apparent in all other cases.

## 2.3   Inhomogeneous Cosmology

So far, we have discussed the perfectly homogeneous and isotropic universe. This is an extremely good approximation on large scales, but breaks down at smaller distances. On the cosmological scales of interest to us, however, the spacetime is well described by perturbation theory around FLRW. For our applications, it will, in fact, be sufficient to work at linear order in perturbations as the departure from spatial homogeneity and isotropy is small. For instance, the CMB temperature varies across the sky at the level of one part in $10^4$ reflecting small spatial variations in the density of the primordial plasma (see below). Of course, this approach breaks down when high-density regions, such as galaxies, form through the gravitational instability of these small fluctuations. In the following, we will introduce first-order cosmological perturbation theory, and show how the large-scale structure of the universe formed and evolved (Sect. 2.3.1). Studying the perturbations in the primordial photon-baryon fluid will reveal the presence of sound waves which have since been imprinted in several cosmological observables (Sect. 2.3.2).

### 2.3.1   Structure Formation

Metric and matter fluctuations are coupled by the Einstein equations and, therefore, have to be treated simultaneously. We write the perturbations of the FLRW metric (2.2) and of the energy-momentum tensor of a perfect fluid (2.4) as

$$g_{\mu\nu}(\tau, \boldsymbol{x}) = \bar{g}_{\mu\nu}(\tau) + \delta g_{\mu\nu}(\tau, \boldsymbol{x}), \qquad T_{\mu\nu}(\tau, \boldsymbol{x}) = \bar{T}_{\mu\nu}(\tau) + \delta T_{\mu\nu}(\tau, \boldsymbol{x}). \quad (2.21)$$

Due to the coordinate independence of general relativity, these perturbations are not uniquely defined. For example, the metric perturbations can be non-zero, $\delta g_{\mu\nu} \neq 0$, even though the spacetime is described by a perfect FLRW universe, just in a different set of coordinates, $g_{\mu\nu}(\tau, \boldsymbol{x}) = \bar{g}_{\mu\nu}(\tau', \boldsymbol{x}')$. We therefore choose a particular coordinate system (or 'fix the gauge') when defining the metric perturbations. It is useful to decompose the perturbations in purely scalar, vector and tensor components which, at linear order, evolve separately under the Einstein equations. In this thesis, we will focus on the scalar degrees of freedom and, therefore, neglect vector and tensor perturbations from now on. We refer to the seminal papers [23, 24] (see also [25–29], for instance) for further details on these points and general treatments of cosmological perturbation theory. We choose to work in (conformal) Newtonian gauge where the scalar part of the metric is given by

$$ds^2 = a^2(\tau) \left[ -(1 + 2\Phi)d\tau^2 + (1 - 2\Psi)\delta_{ij}dx^i dx^j \right]. \quad (2.22)$$

The name of this gauge stems from the fact that the perturbations $\Phi$ and $\Psi$ are related to the (Newtonian) gravitational potential with $\Phi$ controlling the motion of non-relativistic particles and $\Psi$ being determined by the Poisson equation on small scales.

Every energy-momentum tensor has four scalar degrees of freedom which are related to the density $\rho$, the pressure $P$, the bulk velocity $v_i$ and the anisotropic stress $\Sigma_{ij}$. Since the contributions of different species are simply added, we define the perturbed energy-momentum tensor separately for each species $a$ as

$$T^0{}_{0,a} = -(\bar{\rho}_a + \delta\rho_a), \quad T^0{}_{i,a} = (\bar{\rho}_a + \bar{P}_a)v_{i,a}, \quad T^i{}_{j,a} = (\bar{P}_a + \delta P_a)\delta^i_j + (\bar{\rho}_a + \bar{P}_a)\Sigma^i{}_{j,a}.$$
(2.23)

The scalar part of the velocity can be written as $v_{i,a} = -\nabla_i u_a$, with the velocity potential $u_a$. Similarly, the anisotropic stress tensor $\Sigma_{ij,a}$ can be expressed as $\Sigma_{ij,a} = \frac{3}{2}(\nabla_i\nabla_j - \frac{1}{3}\delta_{ij}\nabla^2)\sigma_a$, where $\sigma_a$ is the scalar potential of the anisotropic stress and we introduced the factor of $\frac{3}{2}$ for later convenience. Instead of the density perturbation $\delta\rho_a$ we often employ the dimensionless overdensity

$$\delta_a \equiv \frac{\delta\rho_a}{\bar{\rho}_a}.$$
(2.24)

The previously introduced equation of state $w_a \equiv \bar{P}_a/\bar{\rho}_a$ and the speed of sound $c_a^2 \equiv \delta P_a/\delta\rho_a$ relate the (adiabatic) pressure $P_a$ to the density $\rho_a$, which effectively removes the pressure as a free variable for adiabatic fluctuations.

The evolution of the remaining three matter and two metric perturbations can be derived using the conservation of the energy-momentum tensor and the (linearised) Einstein equations. Energy-momentum conservation for each decoupled species, i.e. those without energy and momentum transfer, implies the continuity and Euler equations,

$$\dot{\delta}_a = (1 + w_a)\left(\nabla^2 u_a + 3\dot{\Psi}\right) - 3\mathcal{H}\left(\frac{\delta P_a}{\delta\rho_a} - w_a\right)\delta_a$$
(2.25)

$$\dot{u}_a = -\left[\mathcal{H}(1 - 3w_a) + \frac{\dot{w}_a}{1 + w_a}\right]u_a + \frac{1}{1 + w_a}\frac{\delta P_a}{\delta\rho_a}\delta_a + \nabla^2\sigma_a + \Phi.$$
(2.26)

These two equations can be combined into a second-order differential equation for the density contrast with a source term comprised of the other three perturbations. For adiabatic fluctuations, this evolution equation simplifies to

$$\ddot{\delta}_a + \chi_a\dot{\delta}_a - c_a^2\nabla^2\delta_a = (1 + w_a)\left(\nabla^4\sigma_a + \nabla^2\Phi + 3\ddot{\Psi} + 3\chi_a\dot{\Psi}\right),$$
(2.27)

where $\chi_a \equiv \mathcal{H}(1 - 3c_a^2)$ is the Hubble drag rate. The two metric potentials $\Phi$ and $\Psi$ are determined by the following first-order Einstein equations:

$$\nabla^2\Psi - 3\mathcal{H}(\dot{\Psi} + \mathcal{H}\Phi) = 4\pi G a^2 \delta\rho,$$
(2.28)

$$\ddot{\Psi} + \mathcal{H}(2\dot{\Psi} + \dot{\Phi}) + (2\dot{\mathcal{H}} + \mathcal{H}^2)\Phi + \frac{1}{3}\nabla^2(\Phi - \Psi) = 4\pi G a^2 \delta P,$$
(2.29)

where $\delta\rho \equiv \sum_a \delta\rho_a$ and $\delta P \equiv \sum_a \delta P_a$ are the total density and pressure perturbations, respectively. Equation (2.28) is known as the relativistic Poisson equation with the density perturbation sourcing the metric potentials. Finally, the spatial trace-free part of the Einstein equations results in the constraint equation

$$\Phi - \Psi = -12\pi G a^2 (\bar{\rho} + \bar{P})\sigma, \tag{2.30}$$

where $(\bar{\rho} + \bar{P})\sigma \equiv \sum_a (\bar{\rho}_a + \bar{P}_a)\sigma_a$. This implies that the metric potentials are equal for vanishing $\sigma_a$. If a finite anisotropic stress potential $\sigma_a$ is present, its evolution equation can be obtained from the corresponding (linearised) Boltzmann equation (cf. Sect. 5.2.3). In the standard cosmological model, free-streaming neutrinos notably induce a small anisotropic stress, but are essentially the only such source. This closes the system of equations and we can solve the entire evolution at first order in cosmological perturbation theory. We commonly decompose each variable into Fourier modes denoted by the same symbol, e.g.

$$\delta_k(\tau) = \int d^3 x \, e^{-ik \cdot x} \delta(\tau, x), \tag{2.31}$$

and often suppress the mode index, $\delta(\tau) = \delta_k(\tau)$, for convenience. This decomposition is particularly helpful as each Fourier mode evolves separately under the linear evolution equations.

When applying these evolution equations to our universe, we attempt to solve an initial value problem. It is convenient to set the initial conditions at sufficiently early times when all scales of interest in current observations were outside the Hubble radius, $k \ll \mathcal{H}$. As current observations strongly suggest adiabatic initial conditions, we will usually assume these in this thesis.[6] Adiabatic fluctuations are characterised by the fact that the initial overdensities of all species are related according to $(1 + w_b)\,\delta_{a,\text{in}} = (1 + w_a)\,\delta_{b,\text{in}}$, i.e. for example $\delta_{r,\text{in}} = 4\,\delta_{m,\text{in}}/3$. They can equivalently be described as perturbations induced by a common local shift in time $\delta\tau(x)$ of all background quantities, $\delta\rho_a(\tau, x) = \bar{\rho}_a(\tau + \delta\tau(x)) - \bar{\rho}_a(\tau)$. From this point of view, it might not be too surprising that generic single-field slow-roll models of inflation provide adiabatic initial conditions for the hot big bang given by

$$\delta_{a,k}(\tau_{\text{in}}) = -3(1 + w_a)\zeta_k, \tag{2.32}$$

where $\tau_{\text{in}}$ is the initial time and $\zeta$ is the primordial curvature perturbation. The latter is conserved on super-Hubble scales and predicted to follow the almost scale-invariant power spectrum $\mathcal{P}_\zeta(k)$ defined in (2.7). Importantly, the equation of state and the speed of sound in an adiabatically perturbed fluid are approximately equal, $c_a^2 \approx w_a$.

---

[6]There may also exist isocurvature perturbations for which the density fluctuations of one species do not necessarily correspond to density fluctuations in other species. These are disfavoured by current observations, in particular those of the CMB anisotropies.

The growth of these primordial density fluctuations is determined by a competition between gravity and pressure. While gravity attracts matter into overdense regions in the universe, pressure pushes matter out of these regions. This means that gravity leads to a growth of the initial inhomogeneities, whereas pressure will inhibit this growth. Specializing to matter perturbations with $c_m^2 \approx w_m \approx 0$, the evolution equation (2.27) implies in the subhorizon limit, $k \gg \mathcal{H}$, after time-averaging the gravitational potentials over a Hubble time[7] that the density perturbations only grow logarithmically during the radiation-dominated era (due to the large photon pressure), $\delta_m \propto \ln a$, but linearly while the universe is dominated by matter, $\delta_m \propto a$.[8] The sub-horizon gravitational potential $\Phi_k$, on the other hand, oscillates with a decaying amplitude $\propto a^{-2}$ during radiation domination, but approaches a constant after matter-radiation equality. In contrast, the super-horizon modes of the gravitational potential do not evolve in either epoch. The combination of these power laws gives rise to the characteristic shape of the matter power spectrum (cf. Sect. 2.4.2). Before we provide further details on the growth of structure, we will discuss photon perturbations because they also leave a small, but distinct imprint which is a key observable.

## 2.3.2 Cosmic Sound Waves

In the following, we study perturbations in the photon-baryon fluid of the early universe. It will turn out that the initial fluctuations excited sound waves in the primordial plasma which are observed today as the so-called baryon acoustic oscillations in both the anisotropies of the cosmic microwave background (see Sect. 2.4.1) and the clustering of galaxies (see Sect. 2.4.2). We will give an approximate description of the main features of these observables and refer to the excellent reviews [31–33] for a more detailed treatment.

Prior to recombination, photons, electrons and protons were tightly coupled through Thomson and Coulomb scattering in the photon-baryon fluid. The evolution equation for the density perturbations in this fluid can be obtained from (2.27) and is given by

$$\ddot{\delta}_\gamma + \frac{\mathcal{H}R}{1+R}\dot{\delta}_\gamma - c_s^2\nabla^2\delta_\gamma = \frac{4}{3}\nabla^2\Phi + 4\ddot{\Psi} + \frac{4\mathcal{H}R}{1+R}\dot{\Psi}, \qquad (2.33)$$

where we introduced the momentum density ratio of baryons to photons $R \equiv 3\bar{\rho}_b/(4\bar{\rho}_\gamma)$. We also defined the sound speed in the fluid $c_s^2 \equiv 1/[3(1 + R)]$ which is smaller than the standard value for a relativistic fluid, $c_s^2 = 1/3$, because the presence

---

[7]The radiation perturbations oscillate on small scales (cf. Sect. 2.3.2). After time-averaging over a Hubble time, these perturbations can however be neglected and the potentials are only sourced by the matter fluctuations [30]. We can therefore neglect the time derivatives of the potentials on subhorizon scales, $k^2\Phi \gg \ddot{\Psi}, \mathcal{H}\dot{\Psi}$.

[8]Once dark energy takes over as the main component of the universe, the clustering of matter stops and the growth of structures is halted by the accelerated expansion of the universe.

of baryons adds inertia to the fluid. The forced harmonic oscillator equation (2.33) essentially governs the entire BAO phenomenology. The metric potentials on the right-hand side evolve as determined by the matter in the universe (including neutrinos and dark matter). They source the fluctuations in the photon-baryon fluid on the left-hand side which are in turn supported by photon pressure and damped by Hubble friction. Note that the anisotropic stress of photons vanishes as it can only develop effectively after decoupling when the photons begin to stream freely.

To extract the general phenomenology of solutions to the master equation (2.33), we make a few simplifying assumptions.[9] However, we will revisit these considerations in Sect. 2.4.1 and especially in Chap. 5. First, we neglect the small anisotropic stress due to neutrinos, i.e. we set $\Phi = \Psi$ according to (2.30). Moreover, we also ignore the time dependence of $\Phi$ for now.[10] Defining $\Theta \equiv \frac{1}{4}\delta_\gamma + \Phi$, Eq. (2.33) can then be written as

$$\ddot{\Theta} + \frac{\mathcal{H}R}{1+R}\dot{\Theta} - c_s^2 \nabla^2(\Theta + R\Phi) = 0 \,. \tag{2.34}$$

Since the baryon-photon ratio evolves as $R \propto a$, the damping term is proportional to $\mathcal{H}R = \dot{R}$. Similar to the time evolution of the gravitational potentials, the baryon-photon ratio $R$ also changes on much larger time scales than $\Theta$. Treating $R$ as approximately constant, the evolution equation is therefore simply given by the differential equation of a harmonic oscillator with mode-dependent frequency $c_s k$,

$$\ddot{\vartheta}_k + c_s^2 k^2 \vartheta_k = 0 \,, \tag{2.35}$$

where we introduced $\vartheta \equiv \Theta + R\Phi = \frac{1}{4}\delta_\gamma + (1 + R)\Phi$. The solutions to this equation are of course sound waves[11]:

$$\vartheta_k(\tau) = A_k \cos(c_s k\tau) + B_k \sin(c_s k\tau) \,. \tag{2.36}$$

Imposing adiabatic initial conditions on superhorizon scales sets $B_k \equiv 0$ and $A_k = 3\zeta_k$. The quantity $\Theta$ consequently evolves according to

$$\Theta(k, \tau) = 3\zeta_k \cos(c_s k\tau) - R\Phi_k \,. \tag{2.37}$$

---

[9]Accurately computing the evolution of all perturbations in the universe requires solving many coupled equations as the interactions between the various species have to be captured by a set of Boltzmann equations. This can only be done numerically which is achieved in the current state-of-the-art Boltzmann solvers CAMB [34] and CLASS [35]. Nevertheless, it is instructive to obtain approximate analytic solutions and get analytic insights in order to deepen our understanding of the underlying physics (cf. e.g. Chap. 5).

[10]This is a good approximation in the matter-dominated era. During radiation domination, the time evolution of the gravitational potential around sound-horizon crossing leads to the radiation-driving effect which we will discuss below.

[11]Note that we could of course rewrite this solution in terms of an amplitude $A_k$ and a non-zero phase $\phi_k$, i.e. $\cos(c_s k\tau) \to \cos(c_s k\tau + \phi_k)$, with $B_k = 0$ implying $\phi_k = 0$.

The presence of baryons, $R \neq 0$, therefore not only changes the sound speed $c_s$ of the photon-baryon fluid, but also moves the equilibrium point of the oscillations from 0 to $-R\Psi$. This effect is often referred to as 'baryon loading' because baryons change the balance of pressure and gravity. Since photons decouple during recombination, we should evaluate these solutions at $\tau = \tau_{\mathrm{rec}}$.[12] At this time, modes with wavenumbers $k_n = n\pi/r_s$ had their extrema. This implies that the sound horizon at decoupling, $r_s \equiv c_s \tau_{\mathrm{rec}} \approx \tau_{\mathrm{rec}}/\sqrt{3}$,[13] is imprinted as a fundamental scale in the photon fluctuations that we can still observe today.

Up to now, we have considered the background to be essentially fixed. However, both the baryon-photon ratio and the gravitational potentials actually evolve. Including the evolution of the background densities in $R$ leads to the photon fluctuations being damped over time. While the gravitational potentials remain constant in the matter era, their oscillating amplitude decays proportional to $a^{-2}$ inside the horizon during the radiation-dominated epoch. Because the decay is due to photon pressure and happens when the photon-baryon fluid is in its most compressed state, the fluid bounces back without a counterbalancing effect from the gravitational potential. Since those fluctuation modes that entered the horizon during matter domination do not experience this radiation-driving effect, the amplitude of the photon perturbations in the subsequent rarefaction stage is enhanced in comparison.

Finally, we have to take the finite mean free path of photons into account. In the master equation (2.33), we assumed that photons and baryons are so tightly coupled that we can treat them as a single fluid. In reality, however, the mean free path of photons, which is given by the inverse Thomson scattering rate of photons with electrons, $\lambda_{\mathrm{mfp}} = 1/(a\sigma_T n_e)$ (in comoving coordinates), is small but finite even before decoupling. This diffusion process results in an additional damping of fluctuation modes. (Note that this damping effect is completely separate from the damping related to $\dot{R} \neq 0$ which we have just discussed.) Intuitively, this is caused by photons washing out inhomogeneities in the primordial plasma that are smaller than their mean free path. Small scales (large $k$) are therefore exponentially more damped than large scales according to $\exp[-(k/k_d)^2]$, where $k_d$ is the wavenumber associated with the mean squared diffusion distance at decoupling. A careful treatment, which includes corrections from the polarization of these photons, gives [36]

$$k_d^{-2} \equiv \int_0^{a_{\mathrm{rec}}} \frac{da}{a^3 \sigma_T n_e H} \frac{R^2 + \frac{16}{15}(1+R)}{6(1+R)^2}, \qquad (2.38)$$

where $a_{\mathrm{rec}}$ is the scale factor at decoupling. Since the diffusion scale depends on an integral over the Hubble parameter $H$, we see that the damping of the fluctuations in the photon-baryon fluid is sensitive to the expansion history of the early universe.

---

[12]As decoupling happens during the epoch of recombination, we use $\tau_{\mathrm{rec}}$ instead of $\tau_F$ or $\tau_{\mathrm{dec}}$ to specify the time of photon decoupling. This will allow easier discrimination of other freeze-out events in later parts of this thesis.

[13]To be precise, the sound horizon is given by $r_s(\tau) = \int_0^\tau d\tau\, c_s(\tau)$, which also captures the small time dependence of $c_s$ that we however neglect in our analytic treatment.

So far, we have focussed on photon fluctuations, but baryons inherit the same variations since they are tightly coupled to photons before the CMB is released. It is instructive to discuss the evolution of the baryon perturbations in real space instead of Fourier space and consider a single initial overdensity (following [37, 38]) because, for adiabatic fluctuations, the primordial density field is a superposition of such point-like overdensities. As illustrated in Fig. 2.3, the overdensities of photons and baryons spread out as spherical shells, while the dark matter perturbation does not move much and is left behind at the centre. After photon decoupling around $z \sim 1100$, the sound speed drops dramatically and the pressure wave slows down, producing a shell of gas at about 150 Mpc from the point of the initial overdensity. Subsequently, baryons fall into the dark matter potential well. At the same time, the baryonic shell also attracts the dark matter which therefore develops the same density profile with a peak at the same radius.

Going back to Fourier space, the presence of the baryonic shell corresponds to oscillations whose frequency is determined by the distance of propagation of the primordial sound waves, i.e. the sound horizon $r_s$.[14] At late times, galaxies formed preferentially in the regions of enhanced dark matter density. For the most part, these are located where the initial overdensities were, but there is a small (about 1 %) enhancement in the regions roughly $r_s \sim 150$ Mpc away from these positions. Consequently, there should be a small excess of galaxies 150 Mpc away from other galaxies. This is how the acoustic scale is imprinted in the two-point correlation function of galaxies (see below) as the so-called BAO peak. As a result, we can observe the remnants of cosmic sound waves today in both the CMB anisotropies and in the large-scale structure of the universe. We will discuss these cosmological observables in the next section.

Finally, let us briefly comment on the mathematical form of the BAO signal. From our discussion so far, one might expect the same pure cosine shape for the matter oscillations as for the photon perturbations in (2.37). However, we actually observe a pure sine solution, $\delta_m \sim \sin(kr_s)$. Heuristically, this can be understood as follows. Assuming decoupling to be instantaneous, we match the baryon perturbation and its time derivative at the last-scattering surface onto the growing and decaying mode of the matter fluctuations. Since $\dot{\delta}_b \sim c_s k \sin(kr_s)$, with $r_s = c_s \tau$, the time derivative dominates over $\delta_b \sim \cos(kr_s)$ for large wavenumbers $k$. In this way, the pure sine solution gets imprinted in the matter perturbations, with the initial conditions being fixed at the time of recombination (see e.g. [41] for an analytic treatment).

---

[14]The size of the sound horizon imprinted in the baryon perturbations is slightly larger than the size observed in the CMB anisotropies. The latter is set at the time when most photons had decoupled from the baryons. At this point, the baryons however still feel the drag of photons and essentially remain coupled to the photons because there are about $10^9$ times more photons than baryons. The end of the so-called drag epoch, $\tau_{\text{drag}} > \tau_{\text{rec}}$, marks the time when baryons finally loose this contact. The two sizes of the sound horizon inferred from the latest CMB measurements are $r_s(z_{\text{rec}} \approx 1090) \approx 145$ Mpc and $r_s(z_{\text{drag}} \approx 1060) \approx 148$ Mpc [8], i.e. they are relatively close, but different at a significance of more than $8\sigma$ at the current level of precision.

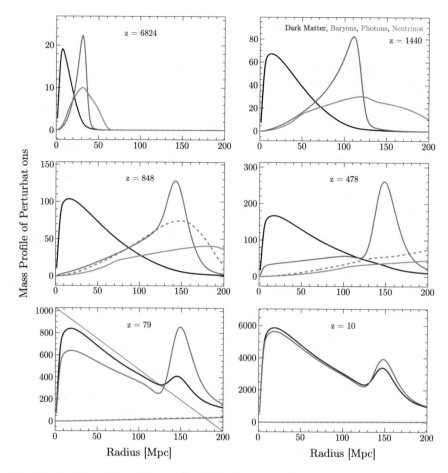

**Fig. 2.3** Evolution of the radial mass profile (density times radius squared) of initially point-like dark matter, baryon, photon and neutrino overdensities located at the origin as a function of the comoving radius (after [39]). All perturbations are fractional for that species and the photon and neutrino fluctuations were divided by 4/3 to put them on the same scale. The units of the mass profile are arbitrary, but are correctly scaled between the panels for synchronous gauge [40]. We observe photons and baryons travelling outwards due to radiation pressure. After the decoupling of photons, they stream freely at the speed of light, whereas the baryon perturbation is left behind in a shell. Dark matter remains concentrated at the origin, but partly falls into the gravitational potential created by the photons and baryons. The baryon and dark matter fluctuations grow as $\delta_m \propto a$ due to gravitational instability in the matter-dominated epoch, and finally trace each other. Today, we can observe the location of the matter shell at about 150 Mpc in the distribution of galaxies. Since neutrinos have been free-streaming close to the speed of light since they decoupled around $z \sim 10^{10}$, they may travel ahead of the sound horizon

## 2.4   Cosmological Observables

So far, we have explained how initially small fluctuations grow and evolve throughout the history of the universe. In particular, we have seen that cosmic sound waves are excited. In the following, we will relate the photon and matter perturbations to the quantities that we actually observe. This includes temperature anisotropies as well as polarization in the CMB sky and the locations of galaxies in the universe. These are the observables that we will use in the rest of the thesis to gain new insights into the early universe and particle physics. Although this is only a subset of the possible and currently employed observables, they dominate the cosmological information that has been inferred to date. We refer to the review literature [14, 42–45] for a comprehensive discussion of additional observables such as weak gravitational lensing, galaxy clusters, the Lyman-$\alpha$ forest and 21 cm tomography.

Understanding the physics behind these observables is important in the quest to uncover the laws of Nature. At the same time, we also have to know how the data can be best characterised and compared to theoretical expectations in a quantitative way. The most important statistic when studying both the cosmic microwave background and the large-scale structure is the two-point correlation function in real space or, equivalently, the power spectrum in Fourier space. These quantities contain all of the statistical information if the perturbations are drawn from a Gaussian distribution function. Since the initial conditions predicted by inflation are very nearly Gaussian, the primordial perturbations are almost entirely described by the primordial correlation function $\langle \zeta(x)\, \zeta(x') \rangle \equiv \xi_\zeta(x, x') = \xi_\zeta(|x - x'|)$, where we employed statistical homogeneity and isotropy in the second equality. The primordial power spectrum $\mathcal{P}_\zeta(k)$ is then defined as the Fourier transform of $\xi_\zeta(x)$ and given by

$$\langle \zeta_k\, \zeta_{k'}^* \rangle = \frac{(2\pi)^3}{k^3}\, \mathcal{P}_\zeta(k)\, \delta_D^{(3)}(k - k')\,, \tag{2.39}$$

where $\delta_D^{(3)}$ is the three-dimensional Dirac delta function. As long as density perturbations can be described in linear theory, their modes evolve independently and the power spectrum still captures most of the information. Having said that, the Einstein equations are inherently non-linear and the gravitational evolution will always couple different modes which introduces non-Gaussianities and, consequently, non-vanishing (connected) higher-point functions. Nevertheless, linear perturbation theory is reliable for many scales of interest, in particular for those modes measured in the CMB. As a consequence, the statistical analysis of CMB and LSS data heavily relies on the power spectrum at the moment.

### 2.4.1   CMB Anisotropies

Since the cosmic microwave background anisotropies were first observed by the DMR instrument on the COBE satellite in 1992 [3], they have proven to be a treasure-trove of information and have played a pivotal role in establishing the standard cosmological model introduced in Sect. 2.1.2. In the following, we will establish the relation between the photon fluctuations on the last-scattering surface and the measured CMB power spectra, and illustrate how general properties of our universe can be deduced from its characteristic shape. We will also discuss the generation of CMB polarization and mention gravitational lensing. More detailed derivations and an overview of current developments can be found in [31, 32, 46–48].

**Temperature Anisotropies**

Generally speaking, CMB experiments map the sky at microwave lengths[15] and mea-sure the intensity and polarization of the incident photons. As discussed in Sect. 2.2.4, these photons have a mean temperature of $\bar{T} = 2.7255$ K. We are now interested in the deviations $\delta \tilde{T} (\hat{n}) = T (\hat{n}) - \bar{T}$, where $\hat{n}$ indicates the line-of-sight direction in the sky. The dominant contribution to $\delta \tilde{T}$ comes from the motion of our solar sys-tem with respect to the CMB rest frame. The induced Doppler effect gives rise to an overall dipole anisotropy with $\delta \tilde{T} \approx 3.4$ mK $\sim 10^{-3} \bar{T}$ [49] which we generally subtract to get the primordial anisotropies $\delta T (\hat{n})$. The resulting CMB temperature map shows fluctuations of the order of $\delta T / \bar{T} \sim 10^{-4}$ as illustrated in Fig. 2.4. To relate the perturbations in the photon density at the last-scattering surface to the observed temperature inhomogeneities, we follow the free-streaming evolution from decoupling to today and project the acoustic oscillations onto the observer's celestial sphere. Both effects lead to additional modulations of the primordial density field that we have to take into account.

To simplify the discussion, we assume that recombination happened instanta-neously. This will capture most of the phenomenology, but the finite width of the last-scattering surface of course has to be taken into account when comparing to data. Integrating the Boltzmann equation of photons along their corresponding line-of-sight from decoupling, $\tau_{\text{rec}}$, to today, $\tau_0$, we find

$$\frac{\delta T}{\bar{T}} (\hat{n}) = \left( \frac{1}{4} \delta_\gamma + \Phi + \hat{n} \cdot v_e \right)_{\text{rec}} + \int_{\tau_{\text{rec}}}^{\tau_0} d\tau \left( \dot{\Phi} + \dot{\Psi} \right), \qquad (2.40)$$

where we dropped the term $\Phi (\tau_0)$ since it only affects the monopole perturbation. The first term, $\frac{1}{4} \delta_\gamma$, captures the intrinsic temperature variations $\theta$ from perturbing the ther-mal distribution (2.8) of photons, $f_\gamma = \bar{f}_\gamma + \delta f_\gamma = \left( \exp \left\{ p / [ \bar{T} (1 + \theta) ] \right\} - 1 \right)^{-1}$, whereas the presence of $\Phi$ accounts for the gravitational redshifting that occurs when photons climb out of a potential well at decoupling. The combination $\Theta = \frac{1}{4} \delta_\gamma + \Phi$ is called the Sachs–Wolfe (SW) term and can be thought of as the effective tempera-ture fluctuation of the primordial CMB. Note that an overdense region at decoupling, which has $\Phi < 0$, leads to a cold spot in the large-scale CMB sky because photons

---

[15] In this context, CMB experiments are surveys which map the CMB anisotropies in the sky. These measurements are usually not taken at a single, but at several frequencies. This is to reliably subtract galactic and astrophysical foregrounds, which are other sources of microwave emission and polarization originating, in particular, from galactic dust. We will generally assume that these foregrounds have been accounted for (or their effects can easily be marginalized over) so that we have direct access to the primordial signal.

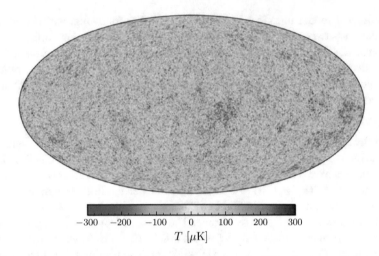

-300    -200    -100    0    100    200    300

$T\ [\mu\mathrm{K}]$

**Fig. 2.4** CMB intensity map at 5 arcmin resolution based on Planck observations using the SMICA component separation algorithm (after [50, 51]). A small strip of the Galactic plane was masked and subsequently filled in by a constrained realization with the same statistical properties as the rest of the sky. The characteristic spot size is about 1°

climbing out of this potential loose more energy than they had at the bottom of the well. Analogously, hot spots are observed at the locations of underdense regions. Since the electrons in the photon-baryon fluid are not at rest when the photons scatter off of them, the third term, $\hat{\boldsymbol{n}} \cdot \boldsymbol{v}_e$, is induced and describes the associated Doppler effect. The integral over the evolution of the gravitational potentials is denoted the integrated Sachs–Wolfe (ISW) term and has both an early and a late component because $\Phi$, $\Psi \neq$ const in the presence of either radiation or dark energy. Overall, the dominating component of $\delta T$ is the SW term $\Theta$, in particular on scales below the sound horizon.

To project the primordial sound waves onto the two-dimensional sky, it is useful to work in Fourier space and extract the multipole moments of the temperature anisotropies $\delta T$, which are defined by

$$\delta T_\ell = \frac{1}{(-\mathrm{i})^\ell} \int_{-1}^{1} \frac{\mathrm{d}\mu}{2} P_\ell(\mu)\, \delta T(\mu)\,, \tag{2.41}$$

with Legendre polynomials $P_\ell(\mu)$. We then find that the projection results in the SW and ISW terms being reweighted by Bessel functions $j_\ell(k\chi_{\mathrm{rec}})$, which arise via a Rayleigh plane-wave expansion. Similarly, the Doppler term is multiplied by the first derivative of these Bessel functions. Here, we introduced the comoving scale to the last-scattering surface $\chi_{\mathrm{rec}}$, which equals $\tau_0 - \tau_{\mathrm{rec}}$ in a flat universe. This means that each multipole moment $\ell$ in principle gets contributions from many different momentum modes $k$. Since the Bessel functions $j_\ell(x)$ are highly peaked near $x \approx \ell$ for large $\ell$, this effect is less pronounced on small scales than on large scales. In

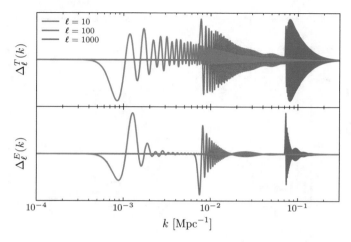

**Fig. 2.5** Transfer functions $\Delta_\ell^X(k)$ in temperature and polarization, $X = T, E$, for $\ell = 10, \ 100$ and $1000$ (normalized to the same maximum amplitude). Both functions peak around $\ell \sim k\chi_{\text{rec}}$ and decay towards larger wavenumbers. This decay is however much more pronounced in polarization and, in general, a smaller number of wavenumbers $k$ contribute to the same multipole $\ell$. We also note that the acoustic peaks in the two transfer functions are out of phase. Taken together, this explains why the polarization transfer is both cleaner than and complementary to the temperature transfer

multipole space, the acoustic peaks are therefore located at $\ell_n \sim k_n \chi_{\text{rec}} = n\pi \, \chi_{\text{rec}}/r_s$. Finally, it is convenient to introduce the transfer function $\Delta_\ell^T(k) \equiv \delta T_\ell(\tau_0, k)/\zeta_k$ which captures the entire linear evolution of the initial perturbations and includes these projection effects. The upper panel of Fig. 2.5 illustrates this function for three representative multipoles.

Finally, we turn to the two-point correlation function $\langle \delta T(\hat{n}) \, \delta T(\hat{n}') \rangle$ which is the quantity that we ultimately extract from CMB temperature maps. Assuming the initial conditions are statistically isotropic, we can expand this two-point function as

$$\langle \delta T(\hat{n}) \, \delta T(\hat{n}') \rangle = \sum_\ell \frac{2\ell + 1}{4\pi} \, C_\ell^{TT} \, P_\ell(\hat{n} \cdot \hat{n}'), \qquad (2.42)$$

where the Legendre polynomials $P_\ell(\hat{n} \cdot \hat{n}')$ only depend on the relative orientation of $\hat{n}$ and $\hat{n}'$. The expansion coefficients $C_\ell^{TT}$ in (2.42) are the famous angular (temperature) power spectrum and given by[16]

$$C_\ell^{TT} = \frac{4\pi}{(2\ell + 1)^2} \int d\ln k \, \left( \Delta_\ell^T(k) \right)^2 \mathcal{P}_\zeta(k). \qquad (2.43)$$

---

[16]We could have equivalently obtained the angular power spectrum by decomposing the temperature fluctuations $\delta T$ into spherical harmonics and computing the correlation function of the expansion coefficients. This is how measurements of $\delta T$ are commonly processed to obtain the power spectrum $C_\ell^{TT}$.

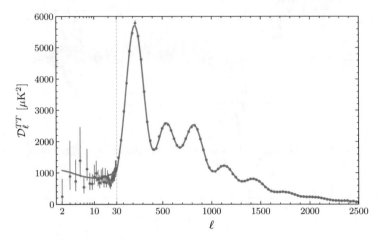

**Fig. 2.6** Planck 2015 temperature power spectrum $\mathcal{D}_\ell^{TT} \equiv \ell(\ell+1)/(2\pi)\,C_\ell^{TT}$ (based on data from [52]). The error bars of the binned data at high multipoles are smaller than the data points. The red line shows the best-fit theoretical spectrum of the six-parameter $\Lambda$CDM model inferred from the Planck TT likelihood

For convenience, we usually show the rescaled spectrum $\mathcal{D}_\ell^{TT} \equiv \ell(\ell+1)/(2\pi)\,C_\ell^{TT}$ which is displayed together with the latest measurement of the Planck satellite in Fig. 2.6. The shape of $\mathcal{D}_\ell^{TT}$ is very characteristic: The SW effect dominates on large scales (low multipoles), the acoustic oscillations are observed on intermediate scales (scales smaller than the projected sound horizon which corresponds to an angular scale of about $1°$ or $\ell \sim 200$) and the smallest scales (large multipoles) are exponentially damped. Because the initial power spectrum is almost scale-invariant, all features in the CMB power spectrum arise from the evolution of the cosmic sound waves being captured at the moment of last-scattering, i.e. the observed oscillations are a snapshot of these waves caught at different phases in their evolution, and subsequently projected onto the sky. The first acoustic peak was discovered by the Toco experiment [53]. The next peaks were tentatively detected by several experiments, but measured decisively by the WMAP satellite [54]. Nowadays, Planck has measured the temperature power spectrum to the cosmic variance limit (CVL)[17] for multipoles up to $\ell \approx 1600$ [52]. This measurement is complemented by the results of many ground-based experiments which mainly target large multipoles as a consequence of their better angular resolution.

Armed with these insights, we can infer the general dependence of the shape of the power spectrum on the cosmological parameters. The overall amplitude of the spectrum depends not only on the primordial amplitude $A_s$, but on the combination $A_s\,e^{-2\tau}$. The reason for this is that photons scattered off of electrons again after

---

[17]Cosmic variance refers to the statistical uncertainty inherent in cosmological measurements since we are only able to measure one realization of the true model underlying the universe. In a cosmic variance-limited measurement, the statistical error is dominated by this uncertainty, which is given by $\Delta C_\ell = \sqrt{2/(2\ell+1)}\,C_\ell$ for a CMB auto-spectrum.

the neutral hydrogen was reionized by the large amounts of ultraviolet radiation that were emitted after the first stars and galaxies had formed. As a result, the optical depth increased and scales smaller than the horizon at that time are suppressed by a factor of $e^{-\tau}$. This degeneracy between $A_s$ and $\tau$ can be broken to some extent by CMB lensing measurements (see below). From the measured value $A_s \approx 2.1 \times 10^{-9}$ we can infer the amplitude of the primordial perturbations (at $k_0 = 0.05\,\text{Mpc}^{-1}$): $\sqrt{\mathcal{P}_\zeta(k_0)} \approx 5 \times 10^{-5}$. Moreover, the spectrum is measured to be slightly red-tilted with a spectral index of $n_s = 0.968 \pm 0.004$, i.e. there is a bit more power on large scales than on small scales. This departure from scale invariance, $n_s - 1 < 0$, which is a natural consequence of many inflationary models, has now been measured at a significance of more than $7\sigma$ [8]. The spectral tilt and the optical depth are somewhat degenerate in the temperature power spectrum. This degeneracy can however be lifted by including information from polarization (see below) since reionization leads to a distinct bump at low multipoles in the polarization spectrum.

The positions of the peaks in the CMB spectrum, $\ell_n$, are particularly sensitive to the distance to last-scattering. In fact, the angular size of the sound horizon at decoupling, $\theta_s$, is a direct measure of the first peak location at an angular scale of about $1°$, which is the characteristic size of the spots in the CMB map of Fig. 2.4. Since curvature is exactly zero within $\Lambda$CDM,[18] the peak positions become a precise measure of the expansion history and, therefore, of the Hubble parameter $H_0$ and the physical matter density $\omega_m$. The overall peak heights relative to the large-scale plateau are however a much more sensitive probe of $\omega_m$ since the amplitude of the cosmic sound waves depends on the time of matter-radiation equality through the radiation-driving effect. As a consequence, the small-scale modes, which entered the horizon in the radiation era, are enhanced in comparison to the modes which started evolving only later during matter domination. The relative peak heights, on the other hand, are directly related to the baryon density $\omega_b$ as the odd peaks are larger than the even peaks due to baryon loading, cf. (2.37). Moreover, we can already anticipate from (2.38) that the damping tail of the spectrum is particularly sensitive to the early expansion history and, consequently, the radiation density (see Sect. 3.4 for a detailed discussion, including the related degeneracies). Finally, the late ISW effect on large scales is sensitive to dark energy, as we previously mentioned. This list of dependencies as derived from our relatively simple analytic treatment of the CMB phenomenology only indicates the potential of uncovering cosmological information encoded in the temperature anisotropies. In Sect. 5.2, we will see how much more information we can deduce when adopting a slightly more rigorous (but still analytic) treatment of cosmic sound waves.

---

[18]The location of the first peak is very sensitive to the curvature of the universe via the distance to last-scattering. The measurement of this peak famously led to the conclusion that our universe has a geometry that is very close to flat [55], which laid the groundwork for the $\Lambda$CDM model. Today, the famous $\Omega_m$-$\Omega_\Lambda$ plot shows that the confidence regions inferred from CMB, BAO and supernovae data, which all have different degeneracy lines, intersect in a single small region that is consistent with a flat universe, $\Omega_M + \Omega_\Lambda + \Omega_r = 1$.

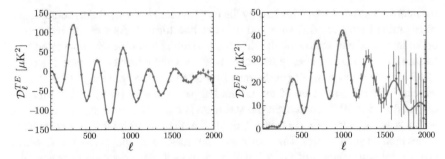

**Fig. 2.7** Planck 2015 high-$\ell$ TE and EE power spectra, $\mathcal{D}_\ell^{TE}$ and $\mathcal{D}_\ell^{EE}$ (based on data from [52]). The red line shows the best-fit $\Lambda$CDM theoretical spectrum inferred from the Planck TT likelihood. The smallness of the residuals with respect to this model indicates a very good fit providing a non-trivial confirmation of the standard cosmological model and CMB phenomenology

## Polarization

We do not only expect that the temperature varies across the CMB sky, but also that this ancient radiation is linearly polarized at the level of a few $\mu$K due to Thomson scattering at the time of decoupling (and reionization). As we will explain below, the scattering of photons with electrons is the only possibility to generate CMB polarization. The polarization signal therefore tracks free electrons and is a particularly clean probe of the physics at the last-scattering surface (and reionization). DASI detected CMB polarization about ten years after COBE announced their discovery of the temperature anisotropies [56]. Precise polarization measurements provide a non-trivial consistency check for the standard cosmological model because the temperature anisotropies and the polarization signal are directly related. As illustrated in Fig. 2.7, the Planck 2015 temperature-polarization cross- and polarization auto-spectra show this impressively. In addition, these spectra help to break degeneracies between cosmological parameters and provide complementary information. In the following, we give a lightning review of the main aspects of CMB polarization and refer to the seminal papers [57–61] and the pedagogical review [46] for further details.

The generation of CMB polarization is best described in the rest frame of a free electron in the primordial plasma. If the incident radiation field is isotropic, the Thomson-scattered radiation remains unpolarized since orthogonal polarization directions cancel out. The same statement holds if the incoming photons have a dipolar anisotropy. However, a net linear polarization arises if the radiation field around the electron has a non-zero quadrupole moment. Put differently, a non-vanishing quadrupole of the temperature anisotropy generates the linear polarization of the CMB. However, prior to decoupling, Thomson scattering keeps the CMB radiation very nearly isotropic in the rest frame of the electrons. A local temperature quadrupole can therefore only develop from a gradient in the velocity field once the photons have acquired an appreciable mean free path just before they decouple. CMB polarization is therefore only generated in the very last scattering events and results from the velocities of the electrons on scales smaller than the photon mean free path. Since both the temperature inhomogeneities and the velocity field of the photons originate from primordial density fluctuations and are out of phase, we expect the polarization peaks of the CMB to be both correlated and out of phase with the temperature peaks. This is exactly what we see in the data (see the left panel of Fig. 2.7). Moreover, cosmological parameters can be independently constrained from the temperature and polarization spectra because the CMB temperature primarily traces the density perturbations, whereas the polarization is effectively induced by the velocity fluctuations. In addition, the polarization spectrum should have less power than the temperature spectrum because the quadrupole moment is suppressed compared to the monopole and dipole moments. This prediction has also been confirmed quantitatively as displayed in the right panel of Fig. 2.7.

In general, a linearly-polarized radiation field can mathematically be described by three variables: the temperature $T$, and the Stokes parameters $Q$ and $U$. Whereas the temperature can be conveniently decomposed in terms of scalar spherical harmonics, the convenient complex combinations $Q \pm iU$ of the Stokes parameters are spin-2 quantities and have to be expanded in the more complicated tensor spherical harmonics. It is however possible to construct two scalar quantities which are invariant under coordinate transformations and commonly referred to as $E$ and $B$.[19] Importantly, scalar/density perturbations only create E-modes and no B-modes, while tensor perturbations (i.e. gravitational waves) induce both E- and B-modes. In this thesis, we therefore neglect B-modes and focus on E-modes, which we will generally refer to as polarization,[20] i.e. we consider the following temperature and polarization auto- and cross-spectra: TT, TE and EE (as displayed in Figs. 2.6 and 2.7).

Although the generation mechanism of polarization is somewhat more involved and the power spectrum is suppressed compared to the temperature anisotropies, the transfer function $\Delta_\ell^E(k)$ is simpler since there are no SW or ISW effects, for instance. Moreover, the mapping between wavenumbers $k$ and multipoles $\ell$ is much sharper for polarization as illustrated in Fig. 2.5. The underlying reasons are a slightly different projection onto the celestial sphere and the fact that the polarization signal is only generated effectively in the very last scattering events. The acoustic peaks in the polarization spectrum are therefore a more direct snapshot of the primordial sound waves and the peaks themselves are slightly sharper than in the temperature spectrum.

---

[19]In analogy with the properties of the electric and magnetic fields in electrodynamics, E-mode and B-mode polarization is curl- and divergence-free, respectively.

[20]A fraction of the E-modes is converted to B-modes in the late universe through gravitational lensing (see below). In contrast to the primary B-modes, these induced B-modes have been detected and can in principle be used to revert the effects of lensing on the temperature and E-mode spectra in a process referred to as delensing.

**Lensing**

There are a number of secondary effects that impact the observed CMB on small scales. For the purpose of this thesis, only one of them will be of some relevance: the weak gravitational lensing of the CMB (see [62] for a comprehensive review). On the long way from the last-scattering surface to our detectors, the CMB photons pass through the increasingly inhomogeneous matter distribution which acts as gravitational lenses. In effect, the CMB photons are deflected by gradients in the gravitational potential along the line-of-sight. This not only generates B-mode polarization on small scales, but also affects the temperature and E-mode power spectra. Since the photon paths are slightly perturbed, the location in the sky where we observe the photons is slightly offset from the location where they actually decoupled. The primordial hot and cold spots are therefore distorted and the acoustic peaks in the lensed power spectra are slightly smeared since power is transferred between multipoles. Because we will be interested in precisely measuring the acoustic peaks, this effect is a nuisance (although it is possible to 'delens', i.e. revert the effects of lensing, to some extent [63–68]). At the same time, the lensing power spectrum has now been extracted at high significance from Planck data [69] and CMB lensing allows to infer the integrated matter distribution between us and the last-scattering surface.

### 2.4.2  Large-Scale Structure

Having discussed the imprints of photon perturbations in the cosmic microwave background, we now return to the matter fluctuations, which grew under the influence of gravity. Eventually, galaxies formed which we can now observe in cosmological surveys (see Fig. 2.8). The principle LSS observable, inferred from both theory and data, is the two-point correlation function, $\xi(r) \equiv \xi_m(r)$, or the power spectrum of matter perturbations, $P(k) \equiv P_m(k)$, which is defined by

$$\langle \delta_m(\boldsymbol{k}) \, \delta_m(\boldsymbol{k}') \rangle = (2\pi)^3 \, P(k) \, \delta_D^{(3)}(\boldsymbol{k} - \boldsymbol{k}') \,. \tag{2.44}$$

The characteristic shape of the spectrum is easily derived by combining the Poisson equation on subhorizon scales, $\delta_m \propto k^2 \Phi$, which implies $P(k) \propto k^4 P_\Phi(k)$, and the evolution of the gravitational potential $\Phi$ as discussed at the end of Sect. 2.3.1. Modes which entered the horizon after matter-radiation equality, $k < k_{\mathrm{eq}}$, remain constant as $\Phi = \mathrm{const}$, i.e. $P_\Phi(k)$ is scale-invariant and $P(k) \propto k$.[21] On the other hand, the sub-horizon potential decays as $\Phi \propto a^{-2} \propto \tau^{-2}$ during radiation domination. Its power spectrum is therefore suppressed according to $P_\Phi(k) \propto k^{-3}(k_{\mathrm{eq}}/k)^4$ because a mode $k$ crosses the horizon at $\tau = 1/k$. This implies that the matter power spectrum peaks around $k \sim k_{\mathrm{eq}}$ and scales as $P(k) \propto k^{-3} \log^2(k/k_{\mathrm{eq}})$ for $k > k_{\mathrm{eq}}$,

---

[21]For simplicity, we set the primordial spectral tilt to unity, $n_s = 1$, in this discussion, i.e. assume a perfectly scale-invariant primordial power spectrum.

**Fig. 2.8** Map of galaxies
from the 14th data release of
the Sloan Digital Sky
Survey (SDSS; after [70])

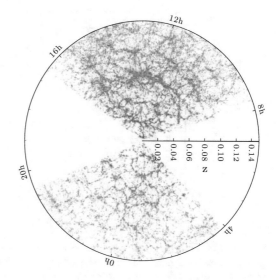

where we included the logarithmic growth of matter perturbations in the radiation
era. In addition, the cosmic sound waves are imprinted on top of this power law as we
discussed in Sect. 2.3.2. Figure 2.9 illustrates that we observe exactly this behaviour
on linear scales. Finally, the power increases in time proportional to $a^2$ since the
matter perturbations grow linearly during matter domination. In order to capture the
entire time evolution including the dark-energy era, we usually introduce the linear
growth function $D_1(z)$ such that $P(k, z) = \left[ D_1^2(0)/D_1^2(z) \right] P(k, z = 0)$.

As the density contrast grows, first linear perturbation theory and eventually all
perturbative treatments break down with small scales being affected earlier than
large scales. Understanding the non-linear evolution is one of the main challenges
when trying to connect LSS observables to fundamental physics. A description of
non-linear effects is however important because they can mimic or distort primordial
signals, but complicated by the fact that these effects are hard to characterise from first
principles. Fortunately, scales corresponding to $k \lesssim 0.1\,h\,\mathrm{Mpc}^{-1}$ (at $z = 0$ and larger
for earlier times) can be treated well in linear perturbation theory throughout cosmic
history. A lot of effort is currently being put into pushing the scale up to which we
trust perturbative computations into the mildly non-linear regime (see e.g. [79–84]).
For smaller scales, we have to resort to numerical simulations. In general, however,
the power on smaller scales is enhanced, in particular for $k \gtrsim 0.1\,h\,\mathrm{Mpc}^{-1}$. On the
bright side, we are able to model and account for these non-linearities to some extent.
Nevertheless, they clearly impose a limitation on how well we can use the small-scale
information to infer properties of the primordial plasma at the present time.

Up to now, we have assumed that we can directly measure the matter density field
in cosmological surveys. However, this is usually not the case because we generally
observe tracers of the matter density which may be highly non-linear objects. Galaxy
surveys for example measure the three-dimensional spatial distribution of galaxies

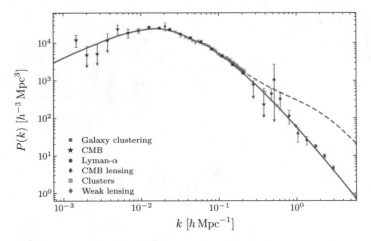

**Fig. 2.9** Linear matter power spectrum reconstructed from CMB temperature, CMB lensing, galaxy, cluster and Lyman-$\alpha$ forest measurements (after [71] with data from [72–78]). The solid and dashed lines display the linear and non-linear power spectra of the best-fit $\Lambda$CDM model inferred from the shown CMB temperature data. The observed agreement highlights the consistency of the measurements conducted by an array of different cosmological probes over a large range of scales

which we subsequently have to relate to the underlying distribution of matter. This relation is described, in a statistical sense, by the galaxy bias $b = b(z)$. In the limit of linear bias, we have $\delta_{g,k} = b\,\delta_{m,k}$, where the galaxy distribution is usually captured by its number density field, $\delta_g(\boldsymbol{x}) = (n(\boldsymbol{x}) - \bar{n})/\bar{n}$, with the mean density of galaxies $\bar{n} = \langle n(\boldsymbol{x}) \rangle$. The linear-bias approximation may be sufficient on large scales, but the bias also picks up a scale dependence, $b(k, z)$, on smaller scales. While there have been a number of advances in the recent past, this dependence is challenging to predict (cf. [85] for a comprehensive review). Although the wavenumbers at which this becomes important get smaller at higher redshifts, the observed objects tend to be more strongly biased since they are intrinsically brighter in order to be detected.

Further sources of uncertainty in the mapping between theory and observations arise, for instance, from redshift-space distortions due to the peculiar velocity of galaxies (the relative velocity with respect to the Hubble flow), and because we cannot measure the positions $\boldsymbol{x}$ of objects in the universe, but only their redshifts and angular positions on the sky. All these points (and more) have to be accounted for when trying to link LSS observables, such as the power spectrum of galaxies, to the physics in the early universe. The large number of potentially available modes is very encouraging in principle, but it seems as if Nature makes us work hard to harness this information.

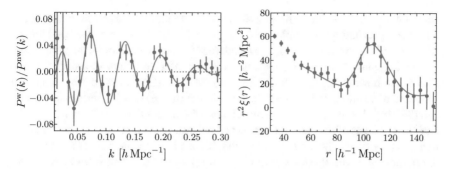

**Fig. 2.10** BAO spectrum (*left*) and BAO peak (*right*) measurements in the redshift bin $0.5 < z_3 < 0.75$ of the last data release of the Baryon Oscillation Spectroscopic Survey together with the respective best-fit model in the employed fitting range (after [88])

### 2.4.3 Baryon Acoustic Oscillations

As we have just discussed, several theoretical challenges in the galaxy power spectrum are related to its overall shape and amplitude. This is in particular the case for the issues of non-linear evolution and biasing. Subtracting this smooth ('no-wiggle') part, $P^{nw}(k)$, from the full spectrum, we are left with the oscillatory ('wiggle') part, $P^w(k) \equiv P(k) - P^{nw}(k)$. This contains the signal of the cosmic sound waves in the primordial plasma which is why we refer to the ratio of the oscillatory to the smooth spectrum,

$$O(k) = \frac{P^w(k)}{P^{nw}(k)} = \frac{P(k) - P^{nw}(k)}{P^{nw}(k)}, \qquad (2.45)$$

as the BAO spectrum.[22] In 2005, this BAO signal was first detected in both the two-point correlation function of the Sloan Digital Sky Survey (SDSS) [86] and the power spectrum measured by the 2dF Galaxy Redshift Survey [87]. The currently highest signal-to-noise measurements are provided by the Baryon Oscillation Spectroscopic Survey and displayed in Fig. 2.10 for one of their redshift bins.

As in the case of the matter power spectrum, gravity still non-linearly processed the BAO signal from its primordial form [38, 89, 90]. However, we have been able to understand and get certain aspects of this observable under better theoretical control. We can schematically express the BAO spectrum as

$$O(k) = A(k) \sin[\omega(k) \, k + \phi(k)], \qquad (2.46)$$

with the amplitude $A$, frequency $\omega$ and phase $\phi$ (see our discussion at the end of Sect. 2.3.2). The amplitude depends on the wavenumber $k$ because of the expo-

---

[22] Although we observe the BAO signal as acoustic peaks both in CMB and in LSS measurements, we will usually refer to the latter when we mention BAO observations. Strictly speaking, the spectrum of baryon acoustic oscillations is $P^w(k)$. For convenience, we will however also refer to $O(k)$ as the BAO spectrum.

nential damping both due to photon diffusion in the photon-baryon fluid and due to non-linear gravitational evolution (see e.g. [38, 91–93]). This is why the BAO amplitude is currently not employed in the standard inference of cosmological parameters.

The frequency $\omega$ of the BAO spectrum is also affected by non-linearities. These effects can however approximately be removed by BAO reconstruction, which is a well-tested technique to better estimate the initial (linear) density perturbations (see e.g. [94–98]). By reversing the displacements of galaxies due to their bulk flow, the non-linear effects of structure formation and redshift-space distortions can be undone to a certain extent. Thanks to this method, it has now become possible to measure the frequency of the BAO spectrum [99] and the location of the BAO peak [100, 101] at the sub-percent level. Using the BAO signal as a standard ruler then allows to break degeneracies between cosmological parameters in the CMB (e.g. between $\Omega_m$ and $\Omega_\Lambda$) resulting in tighter constraints.[23]

Although the phase is absent for adiabatic initial conditions, as we will explicitly show in Sect. 5.2, it can be induced by free-streaming neutrinos. In general, the phase is also affected by the non-linearities induced by gravity. However, it has been proven that a constant phase, $\phi = $ const, is immune to these effects [104]. This suggests that we can reliably extract the primordial phase from late-time observables. We will explore how to use the information encoded in the BAO spectrum, in particular the phase $\phi$, in Chap. 6 and establish a modified BAO analysis in Chap. 7, which takes some of these considerations into account.

# References

1.  A. Riess et al. (Supernova Search Team), Observational evidence from supernovae for an accelerating universe and a cosmological constant. Astron. J. **116**, 1009 (1998). arXiv:astro-ph/9805201 [astro-ph]
2.  S. Perlmutter et al. (Supernova Cosmology Project), Measurements of $\Omega$ and $\Lambda$ from 42 high-redshift supernovae. Astrophys. J. **517**, 565 (1999). arXiv:astro-ph/9812133 [astro-ph]
3.  G. Smoot et al. (COBE Collaboration), Structure in the COBE differential microwave radiometer first-year maps. Astrophys. J. **396**, L1 (1992)
4.  C. Bennett et al., Four-year COBE DMR cosmic microwave background observations maps and basic results. Astrophys. J. **464**, L1 (1996). arXiv:astro-ph/9601067 [astro-ph]
5.  D. Spergel et al. (WMAP Collaboration), First-year Wilkinson Microwave Anisotropy Probe (WMAP) observations: determination of cosmological parameters. Astrophys. J. Suppl. **148**, 175 (2003). arXiv:astro-ph/0302209 [astro-ph]
6.  G. Hinshaw et al. (WMAP Collaboration), Nine-year Wilkinson Microwave Anisotropy Probe (WMAP) observations: cosmological parameter results. Astrophys. J. Suppl. **208**, 19 (2013). arXiv:1212.5226 [astro-ph.CO]
7.  P.A.R. Ade et al. (Planck Collaboration), Planck 2013 results. XVI. Cosmological parameters. Astron. Astrophys. **571**, A16 (2014). arXiv:1303.5076 [astro-ph.CO]

---

[23] Another common dataset that is used to break degeneracies are the local $H_0$ measurements from supernovae. Having said that, with both global, such as those from the CMB and BAO, and local measurements improving, the inferred values of $H_0$ are currently statistically discrepant at the $3\sigma$ level [102]. It is however questionable whether we should be paying too much attention given the vast statistical power of the CMB in particular (see e.g. [103]).

8. P.A.R. Ade et al. (Planck Collaboration), Planck 2015 results. XIII. Cosmological parameters. Astron. Astrophys. **594**, A13 (2016). arXiv:1502.01589 [astro-ph.CO]

9. M. Colless et al. (2dFGRS Collaboration), The 2dF Galaxy Redshift Survey: spectra and redshifts. Mon. Not. Roy. Astron. Soc. **328**, 1039 (2001). arXiv:astro-ph/0106498 [astro-ph]

10. C. Stoughton et al. (SDSS Collaboration), The Sloan Digital Sky Survey: early data release. Astron. J. **123**, 485 (2002)

11. S. Alam et al. (BOSS Collaboration), The clustering of galaxies in the completed SDSS-III Baryon Oscillation Spectroscopic Survey: cosmological analysis of the DR12 galaxy sample. Mon. Not. Roy. Astron. Soc. **470**, 2617 (2017). arXiv:1607.03155 [astro-ph.CO]

12. A. Einstein, Die Grundlage der Allgemeinen Relativitätstheorie. Ann. Phys. **49**, 769 (1916)

13. A. Friedmann, Über die Krümmung des Raumes. Z. Phys. **10**, 377 (1922)

14. D. Weinberg, M. Mortonson, D. Eisenstein, C. Hirata, A. Riess, E. Rozo, Observational probes of cosmic acceleration. Phys. Rept. **530**, 87 (2013). arXiv:1201.2434 [astro-ph.CO]

15. T. Abbott et al. (DES Collaboration), Dark Energy Survey year 1 results: cosmological constraints from galaxy clustering and weak lensing. Phys. Rev. D **98**, 043526 (2018). arXiv:1708.01530 [astro-ph.CO]

16. D. Baumann, TASI Lectures on Inflation. arXiv:0907.5424 [hep-th]

17. C. Patrignani et al. (Particle Data Group), Review of particle physics. Chin. Phys. C **40**, 100001 (2016)

18. D. Fixsen, The temperature of the cosmic microwave background. Astrophys. J. **707**, 916 (2009). arXiv:0911.1955 [astro-ph.CO]

19. E. Aver, K. Olive, E. Skillman, The effects of He-I $\lambda$ 10830 on helium abundance determinations. JCAP **07**, 011 (2015). arXiv:1503.08146 [astro-ph.CO]

20. S. Borsanyi et al., Calculation of the axion mass based on high-temperature lattice quantum chromodynamics. Nature **539**, 69 (2016). arXiv:1606.07494 [hep-lat]

21. A. Penzias, R. Wilson, A measurement of excess antenna temperature at 4080 Mc/s. Astrophys. J. **142**, 419 (1965)

22. D. Fixsen, E. Cheng, J. Gales, J. Mather, R. Shafer, E. Wright, The cosmic microwave background spectrum from the full COBE FIRAS dataset. Astrophys. J. **473**, 576 (1996). arXiv:astro-ph/9605054 [astro-ph]

23. J. Bardeen, Gauge invariant cosmological perturbations. Phys. Rev. D **22**, 1882 (1980)

24. H. Kodama, M. Sasaki, Cosmological perturbation theory. Prog. Theor. Phys. Suppl. **78**, 1 (1984)

25. V. Mukhanov, H. Feldman, R. Brandenberger, Theory of cosmological perturbations. Phys. Rept. **215**, 203 (1992)

26. C.-P. Ma, E. Bertschinger, Cosmological perturbation theory in the synchronous and conformal Newtonian gauges. Astrophys. J. **455**, 7 (1995). arXiv:astro-ph/9506072 [astro-ph]

27. S. Bashinsky, U. Seljak, Neutrino perturbations in CMB anisotropy and matter clustering. Phys. Rev. D **69**, 083002 (2004). arXiv:astro-ph/0310198 [astro-ph]

28. V. Mukhanov, *Physical Foundations of Cosmology* (Cambridge University Press, Cambridge, 2005)

29. K. Malik, D. Wands, Cosmological perturbations. Phys. Rept. **475**, 1 (2009). arXiv:0809.4944 [astro-ph]

30. S. Weinberg, Cosmological fluctuations of short wavelength. Astrophys. J. **581**, 810 (2002). arXiv:astro-ph/0207375 [astro-ph]

31. W. Hu, Wandering in the background: a CMB explorer. Ph.D. Thesis, University of California, Berkeley (1995). arXiv:astro-ph/9508126 [astro-ph]

32. W. Hu, S. Dodelson, Cosmic microwave background anisotropies. Ann. Rev. Astron. Astrophys. **40**, 171 (2002). arXiv:astro-ph/0110414 [astro-ph]

33. A. Challinor, H. Peiris, Lecture notes on the physics of cosmic microwave background anisotropies. AIP Conf. Proc. **1132**, 86 (2009). arXiv:0903.5158 [astro-ph.CO]

34. A. Lewis, A. Challinor, A. Lasenby, Efficient computation of CMB anisotropies in closed FRW models. Astrophys. J. **538**, 473 (2000). arXiv:astro-ph/9911177 [astro-ph]

35. D. Blas, J. Lesgourgues, T. Tram, The Cosmic Linear Anisotropy Solving System (CLASS) II: approximation schemes. JCAP **07**, 034 (2011). arXiv:1104.2933 [astro-ph.CO]
36. M. Zaldarriaga, D. Harari, Analytic approach to the polarization of the cosmic microwave background in flat and open universes. Phys. Rev. D **52**, 3276 (1995). arXiv:astro-ph/9504085 [astro-ph]
37. S. Bashinsky, E. Bertschinger, Dynamics of cosmological perturbations in position space. Phys. Rev. D **65**, 123008 (2002). arXiv:astro-ph/0202215 [astro-ph]
38. D. Eisenstein, H.-J. Seo, M. White, On the robustness of the acoustic scale in the low-redshift clustering of matter. Astrophys. J. **664**, 660 (2007). arXiv:astro-ph/0604361 [astro-ph]
39. Adapted from [38] with permission. Copyright by the American Astronomical Society
40. S. Weinberg, *Cosmology* (Oxford University Press, Oxford, 2008)
41. Z. Slepian, D. Eisenstein, A simple analytic treatment of linear growth of structure with baryon acoustic oscillations. Mon. Not. Roy. Astron. Soc. **457**, 24 (2016). arXiv:1509.08199 [astro-ph.CO]
42. M. Bartelmann, P. Schneider, Weak gravitational lensing. Phys. Rept. **340**, 291 (2001). arXiv:astro-ph/9912508 [astro-ph]
43. J. Carlstrom, G. Holder, E. Reese, Cosmology with the Sunyaev-Zel'dovich effect. Ann. Rev. Astron. Astrophys. **40**, 643 (2002). arXiv:astro-ph/0208192 [astro-ph]
44. S. Furlanetto, S.P. Oh, F. Briggs, Cosmology at low frequencies: the 21 cm transition and the high-redshift universe. Phys. Rept. **433**, 181 (2006). arXiv:astro-ph/0608032 [astro-ph]
45. M. Kilbinger, Cosmology with cosmic shear observations: a review. Rept. Prog. Phys. **78**, 086901 (2015). arXiv:1411.0115 [astro-ph.CO]
46. W. Hu, M. White, A CMB polarization primer. New Astron. **2**, 323 (1997). arXiv:astro-ph/9706147 [astro-ph]
47. S. Dodelson, *Modern Cosmology* (Academic Press, San Diego, 2003)
48. S. Staggs, J. Dunkley, L. Page, Recent discoveries from the cosmic microwave background: a review of recent progress. Rept. Prog. Phys. **81**, 044901 (2018)
49. R. Adam et al. (Planck Collaboration), Planck 2015 results. VIII. High frequency instrument data processing: calibration and maps. Astron. Astrophys. **594**, A8 (2016). arXiv:1502.01587 [astro-ph.CO]
50. Map of CMB temperature from SMICA, http://www.cosmos.esa.int/documents/387566/425793/2015_SMICA_CMB/, Original image credit: ESA and the Planck collaboration
51. R. Adam et al. (Planck Collaboration), Planck 2015 results. I. Overview of products and scientific results. Astron. Astrophys. **594**, A1 (2016). arXiv:1502.01582 [astro-ph.CO]
52. N. Aghanim et al., (Planck Collaboration), Planck 2015 results. XI. CMB power spectra, likelihoods and robustness of parameters. Astron. Astrophys. **594**, A11 (2015). arXiv:1507.02704 [astro-ph.CO]
53. A. Miller et al., A Measurement of the Angular Power Spectrum of the CMB from $\ell = 100$ to 400. Astrophys. J. **524**, L1 (1999). arXiv:astro-ph/9906421 [astro-ph]
54. G. Hinshaw et al. (WMAP Collaboration), First year wilkinson microwave anisotropy probe (WMAP) observations: the angular power spectrum. Astrophys. J. Suppl. **148**, 135 (2003).arXiv:astro-ph/0302217 [astro-ph]
55. P. de Bernardis et al. (BOOMERanG Collaboration), A flat universe from high-resolution maps of the cosmic microwave background radiation. Nature **404**, 955 (2000). arXiv:astroph/0004404 [astro-ph]
56. J. Kovac, E. Leitch, C. Pryke, J. Carlstrom, N. Halverson, W. Holzapfel, Detection of polarization in the cosmic microwave background using DASI. Nature **420**, 772 (2002). arXiv:astro-ph/0209478 [astro-ph]
57. J. Bond, G. Efstathiou, Cosmic background radiation anisotropies in universes dominated by non-baryonic dark matter. Astrophys. J. **285**, L45 (1984)
58. A. Polnarev, Polarization and anisotropy induced in the microwave background by cosmological gravitational waves. Astron. Zh. **62**, 1041 (1985)
59. U. Seljak, Measuring polarization in the cosmic microwave background. Astrophys. J. **482**, 6 (1997). arXiv:astro-ph/9608131 [astro-ph]

60. M. Zaldarriaga, U. Seljak, An all-sky analysis of polarization in the microwave background. Phys. Rev. D **55**, 1830 (1997). arXiv:astro-ph/9609170 [astro-ph]
61. M. Kamionkowski, A. Kosowsky, A. Stebbins, Statistics of cosmic microwave background polarization. Phys. Rev. D **55**, 7368 (1997). arXiv:astro-ph/9611125 [astro-ph]
62. A. Lewis, A. Challinor, Weak gravitational lensing of the CMB. Phys. Rept. **429**, 1 (2006). arXiv:astro-ph/0601594 [astro-ph]
63. W. Hu, T. Okamoto, Mass reconstruction with cosmic microwave background polarization. Astrophys. J. **574**, 566 (2002). arXiv:astro-ph/0111606 [astro-ph]
64. L. Knox, Y.-S. Song, A limit on the detectability of the energy scale of inflation. Phys. Rev. Lett. **89**, 011303 (2002). arXiv:astro-ph/0202286 [astro-ph]
65. C. Hirata, U. Seljak, Reconstruction of lensing from the cosmic microwave background polarization. Phys. Rev. D **68**, 083002 (2003). arXiv:astro-ph/0306354 [astro-ph]
66. D. Green, J. Meyers, A. van Engelen, CMB delensing beyond the B-modes. JCAP **12**, 005 (2017). arXiv:1609.08143 [astro-ph.CO]
67. N. Sehgal, M. Madhavacheril, B. Sherwin, A. van Engelen, Internal delensing of cosmic microwave background acoustic peaks. Phys. Rev. D **95**, 103512 (2017). arXiv:1612.03898 [astro-ph.CO]
68. J. Carron, A. Lewis, A. Challinor, Internal delensing of Planck CMB temperature and polarization. JCAP **05**, 035 (2017). arXiv:1701.01712 [astro-ph.CO]
69. P.A.R. Ade et al. (Planck Collaboration), Planck 2015 results. XV. Gravitational lensing. Astron. Astrophys. **594**, A15 (2016). arXiv:1502.01591 [astro-ph.CO]
70. The SDSS's map of the universe, http://www.sdss.org/science/orangepie/, Original image credit: M. Blanton and SDSS
71. Adapted from [78] with permission. Copyright by the American Astronomical Society
72. P. McDonald et al. (SDSS Collaboration), The Lyman-$\alpha$ forest power spectrum from the Sloan Digital Sky Survey. Astrophys. J. Suppl. **163**, 80 (2006). arXiv:astro-ph/0405013 [astro-ph]
73. A. Vikhlinin et al., Chandra Cluster Cosmology Project III: cosmological parameter constraints. Astrophys. J. **692**, 1060 (2009). arXiv:0812.2720 [astro-ph]
74. B. Reid et al., Cosmological constraints from the clustering of the Sloan Digital Sky Survey DR7 luminous red galaxies. Mon. Not. Roy. Astron. Soc. **404**, 60 (2010). arXiv:0907.1659 [astro-ph.CO]
75. N. Sehgal et al., The Atacama Cosmology Telescope: cosmology from galaxy clusters detected via the Sunyaev-Zel'dovich effect. Astrophys. J. **732**, 44 (2011). arXiv:1010.1025 [astro-ph.CO]
76. S. Das et al., Detection of the power spectrum of cosmic microwave background lensing by the Atacama Cosmology Telescope. Phys. Rev. Lett. **107**, 021301 (2011). arXiv:1103.2124 [astro-ph.CO]
77. J. Tinker et al., Cosmological constraints from galaxy clustering and the mass-to-number ratio of galaxy clusters. Astrophys. J. **745**, 16 (2012). arXiv:1104.1635 [astro-ph.CO]
78. R. Hložek et al., The atacama cosmology telescope: a measurement of the primordial power spectrum. Astrophys. J. **749**, 90 (2012). arXiv:1105.4887 [astro-ph.CO]
79. F. Bernardeau, S. Colombi, E. Gaztanaga, R. Scoccimarro, Large-scale structure of the universe and cosmological perturbation theory. Phys. Rept. **367**, 1 (2002). arXiv:astroph/0112551 [astro-ph]
80. M. Crocce, R. Scoccimarro, Renormalized cosmological perturbation theory. Phys. Rev. D **73**, 063519 (2006). arXiv:astro-ph/0509418 [astro-ph]
81. D. Baumann, A. Nicolis, L. Senatore, M. Zaldarriaga, Cosmological non-linearities as an effective fluid. JCAP **07**, 051 (2012). arXiv:1004.2488 [astro-ph.CO]
82. J. Carrasco, M. Hertzberg, L. Senatore, The effective field theory of cosmological large-scale structures. JHEP **09**, 082 (2012). arXiv:1206.2926 [astro-ph.CO]
83. M. Bartelmann, F. Fabis, D. Berg, E. Kozlikin, R. Lilow, C. Viermann, A microscopic, nonequilibrium, statistical field theory for cosmic structure formation. New J. Phys. **18**, 043020 (2016). arXiv:1411.0806 [cond-mat.stat-mech]

84. D. Blas, M. Garny, M.M. Ivanov, S. Sibiryakov, Time-sliced perturbation theory for large-scale structure I: general formalism. JCAP **07**, 052 (2016). arXiv:1512.05807 [astro-ph.CO]
85. V. Desjacques, D. Jeong, F. Schmidt, Large-scale galaxy bias. Phys. Rept. **733**, 1 (2018). arXiv:1611.09787 [astro-ph.CO]
86. D. Eisenstein et al. (SDSS Collaboration), Detection of the baryon acoustic peak in the large-scale correlation function of SDSS luminous red galaxies. Astrophys. J. **633**, 560 (2005). arXiv:astro-ph/0501171 [astro-ph]
87. S. Cole et al. (2dFGRS Collaboration), The 2dF Galaxy Redshift Survey: power-spectrum analysis of the final dataset and cosmological implications. Mon. Not. Roy. Astron. Soc. **362**, 505 (2005). arXiv:astro-ph/0501174 [astro-ph]
88. Adapted from [99, 100] with permission of Oxford University Press on behalf of the Royal Astronomical Society
89. M. Crocce, R. Scoccimarro, Non-linear evolution of baryon acoustic oscillations. Phys. Rev. D **77**, 023533 (2008). arXiv:0704.2783 [astro-ph]
90. N. Sugiyama, D. Spergel, How does non-linear dynamics affect the baryon acoustic oscillation? JCAP **02**, 042 (2014). arXiv:1306.6660 [astro-ph.CO]
91. H.-J. Seo, D. Eisenstein, Improved forecasts for the baryon acoustic oscillations and cosmological distance scale. Astrophys. J. **665**, 14 (2007). arXiv:astro-ph/0701079 [astro-ph]
92. T. Baldauf, M. Mirbabayi, M. Simonović, M. Zaldarriaga, Equivalence Principle and the Baryon Acoustic Peak. Phys. Rev. D **92**, 043514 (2015). arXiv:1504.04366 [astro-ph.CO]
93. D. Blas, M. Garny, M. Ivanov, S. Sibiryakov, Time-sliced perturbation theory II: baryon acoustic oscillations and infrared resummation. JCAP **07**, 028 (2016). arXiv:1605.02149 [astro-ph.CO]
94. D. Eisenstein, H.-J. Seo, E. Sirko, D. Spergel, Improving cosmological distance measurements by reconstruction of the baryon acoustic peak. Astrophys. J. **664**, 675 (2007). arXiv:astro-ph/0604362 [astro-ph]
95. N. Padmanabhan, X. Xu, D. Eisenstein, R. Scalzo, A. Cuesta, K. Mehta, E. Kazin, A two-percent distance to z = 0.35 by reconstructing baryon acoustic oscillations – I. Methods and application to the Sloan Digital Sky Survey. Mon. Not. Roy. Astron. Soc. **427**, 2132 (2012). arXiv:1202.0090 [astro-ph.CO]
96. B. Sherwin, M. Zaldarriaga, The shift of the baryon acoustic oscillation scale: a simple physical picture. Phys. Rev. D **85**, 103523 (2012). arXiv:1202.3998 [astro-ph.CO]
97. M. Schmittfull, Y. Feng, F. Beutler, B. Sherwin, M.Y. Chu, Eulerian BAO reconstructions and N-point statistics. Phys. Rev. D **92**, 123522 (2015). arXiv:1508.06972 [astro-ph.CO]
98. H.-J. Seo, F. Beutler, A. Ross, S. Saito, Modelling the reconstructed BAO in Fourier space. Mon. Not. Roy. Astron. Soc. **460**, 2453 (2016). arXiv:1511.00663 [astro-ph.CO]
99. F. Beutler et al. (BOSS Collaboration), The clustering of galaxies in the completed SDSS-III Baryon Oscillation Spectroscopic Survey: baryon acoustic oscillations in Fourier space. Mon. Not. Roy. Astron. Soc. **464**, 3409 (2017). arXiv:1607.03149 [astro-ph.CO]
100. A. Ross et al. (BOSS Collaboration), The clustering of galaxies in the completed SDSS-III Baryon Oscillation Spectroscopic Survey: observational systematics and baryon acoustic oscillations in the correlation function. Mon. Not. Roy. Astron. Soc. **464**, 1168 (2017). arXiv:1607.03145 [astro-ph.CO]
101. M. Vargas-Magaña et al. (BOSS Collaboration), The clustering of galaxies in the completed SDSS-III Baryon Oscillation Spectroscopic Survey: theoretical systematics and baryon acoustic oscillations in the galaxy correlation function. Mon. Not. Roy. Astron. Soc. **477**, 1153 (2018). arXiv:1610.03506 [astro-ph.CO]
102. A. Riess et al., A 2.4 % determination of the local value of the Hubble constant. Astrophys. J. **826**, 56 (2016). arXiv:1604.01424 [astro-ph.CO]

103. D. Scott, The standard model of cosmology: a skeptic's guide, arXiv:1804.01318 [astro-ph.CO]
104. D. Baumann, D. Green, M. Zaldarriaga, Phases of new physics in the BAO spectrum. JCAP **11**, 007 (2017). arXiv:1703.00894 [astro-ph.CO]

# Chapter 3
# Light Species in Cosmology and Particle Physics

We saw in the last chapter that cosmological observables are very sensitive to the composition of the universe. We assumed that the Standard Model of particle physics accurately describes the baryonic matter content, which indeed it does to remarkable accuracy. In addition, we had to however invoke the presence of non-baryonic dark matter to explain some of the cosmological data, in particular the acoustic peaks of the CMB. This is only one of the many reasons to believe that the SM is incomplete. Other famous theoretical and experimental shortcomings of the SM are the strong CP problem, the origin of neutrino masses, the stability of the Higgs mass and the matter-antimatter asymmetry. Models of physics beyond the Standard Model usually invoke new degrees of freedom to address some or all of these problems. Many interesting SM extensions contain new light species [1], such as axions [2–4], axion-like particles [5], dark photons [6, 7] and light sterile neutrinos [8].[1] These particles are often so weakly coupled to the SM that they escape detection in terrestrial experiments. However, they may be efficiently produced in astrophysical systems and in the early universe, which therefore presents an alternative way of probing these elusive species.

In this chapter, we provide the connection between cosmological observables and the additional light relics predicted in some BSM models. In Sect. 3.1, we give order-of-magnitude estimates for the constraining power of astrophysical systems and cosmology on weakly-interacting light particles. In Sect. 3.2, we introduce the particle physics aspects of this thesis. We first review the current status of the Standard Model of particle physics and motivate its extensions with additional light particles. We will then present an effective field theory framework which offers a convenient parametrization of large classes of light species and their interactions with the SM. In Sect. 3.3, we introduce the effective number of relativistic species, $N_{\text{eff}}$, as the main cosmological parameter capturing neutrinos and any other light thermal relics, and

---

[1]In addition, BSM models often require new massive particles which can be too heavy to be produced at the energies available at colliders even in the distant future. Although it is also possible to constrain these types of particles using cosmology (see e.g. [9–23]), in this thesis, we will focus on light weakly-coupled species.

© Springer Nature Switzerland AG 2019
B. Wallisch, *Cosmological Probes of Light Relics*, Springer Theses,
https://doi.org/10.1007/978-3-030-31098-1_3

determine the relation between the observable relic abundance of these particles and their decoupling temperatures. We show that a single species in thermal equilibrium produces a minimal non-zero contribution to $N_{\mathrm{eff}}$ which provides an interesting target for future cosmological measurements. In Sect. 3.4, we finally examine the possible signatures that these particles might leave in cosmological observables such as the CMB and LSS power spectra. This will allow us to search for these signals and constrain the interactions of light relics with the Standard Model in the following chapters of this thesis.

## 3.1   The Power of Astrophysics and Cosmology

Detecting new light species is challenging because their couplings to the Standard Model degrees of freedom are necessarily small (since we would have already detected them otherwise). In particular, since the scattering cross sections are tiny, it is difficult to probe these SM extensions in terrestrial experiments on the intensity or energy frontier of particle physics, i.e. in the laboratory or at particle accelerators. In astrophysics and cosmology, however, we have access to high-density environments and/or the ability to follow the evolution over long time scales which can overcome the small cross sections and allow a significant production of the extra species. For example, new light particles can be produced in the interior of stars [24]. Since these species are weakly interacting, they carry energy away from the stellar core similar to neutrinos. The absence of an anomalous extra cooling over the long lifetime of stars puts some of the best current constraints on weakly-coupled particles.

To illustrate the origin of this sensitivity, we consider the fractional change in the number densities of the particles involved in the production process, which can schematically be written as

$$\frac{\Delta n}{n} \sim n\sigma \times \Delta t\,, \tag{3.1}$$

i.e. it is equal to the interaction rate, $\Gamma \sim n\sigma$, with thermally-averaged cross section $\sigma$, times the interaction time $\Delta t$. This highlights how small cross sections can be compensated for by the high densities in the stellar interior, $n \sim (1\,\mathrm{keV})^3$, and especially the very long lifetime of stars which is typically of the order of $\Delta t \sim 10^8$ yrs. We therefore find significant changes in the stellar evolution, $\Delta n/n \gtrsim 1$, if

$$\sigma > (n\Delta t)^{-1} \sim \left(10^{10}\,\mathrm{GeV}\right)^{-2}\,. \tag{3.2}$$

These particles may also be produced in extreme astrophysical events, such as supernovae explosions, which happen on much shorter time scales, $\Delta t \sim 10\,\mathrm{s}$. Because the densities in this case are much higher, $n \sim (10\,\mathrm{MeV})^3$, the constraints which can be derived from the observed energy loss are at an order of magnitude comparable to (3.2).

Since the early universe was dominated by radiation, constraints on light relics can also be inferred from cosmological measurements. In order to get a sense for the power of these possible bounds, a similar argument based on (3.1) can be applied to cosmology. The high densities of the early universe, $n \sim T^3 \gg (1 \, \text{MeV})^3$, allow these light particles to have been in thermal equilibrium with the SM (and therefore efficiently produced) for time scales of $\Delta t < 1$ s. They can therefore make a significant contribution to the total radiation density of the universe and, hence, be possibly detected in CMB and LSS observables. The estimate (3.2) suggests that cosmological constraints will improve over astrophysical bounds for temperatures above $10^4 \, \text{GeV}$. Moreover, cosmology may constrain all couplings to the Standard Model equally since thermal equilibrium in the early universe is democratic,[2] whereas astrophysical systems and laboratory experiments are often only sensitive to a subset of these interactions, e.g. the coupling to photons. This universality of cosmological constraints is one of the reasons why the search for light thermal relics has been adopted as one of the main science targets of the next generation of CMB experiments, such as the CMB-S4 mission [27]. We will further quantify the constraining power of cosmology below and especially in Chap. 4.

## 3.2 Physics Beyond the Standard Model

The Standard Model of particle physics is a great success. It provides a theoretically consistent description of all known particles and their interactions (except gravity) up to the Planck scale. On the experimental side, large improvements in detector technology and analysis techniques have led to all predicted particles being found, including the Higgs boson [28, 29], and their interactions being measured to exquisite precision. Having said that, the SM is clearly incomplete and cannot be a fundamental theory since it does not address a number of open problems and several pieces of evidence for new physics. In the following, we briefly review the main aspects of the SM (Sect. 3.2.1) and give a few pieces of evidence for dark BSM sectors (Sect. 3.2.2). Finally, we introduce an effective field theory of light species as a convenient way of parametrizing the new dark sector and its interactions with the SM for cosmological searches (Sect. 3.2.3).

### 3.2.1 Standard Model of Particle Physics

Theoretically speaking, the Standard Model of particle physics is a highly successful quantum field theory based on the local gauge group $SU(3)_C \otimes SU(2)_L \otimes U(1)_Y$.

---

[2] Any new light particle that was in thermal equilibrium in the past will have a number density which is comparable to that of photons. This is the reason why neutrinos have been detected with high significance in the CMB despite their weak coupling (cf. Chap. 5; see also [25, 26]).

The elementary fields associated with these groups are the eight gluons $G^a$, which mediate the strong force, and the four gauge bosons $W^a$ and $B$, which mediate the electroweak force. The matter fields can be divided into quarks and leptons of three generations each. Quarks are charged under the entire SM gauge group, whereas leptons only interact via the electroweak gauge group $SU(2)_L \otimes U(1)_Y$ as described by the Glashow–Weinberg–Salam model [30–32]. With the gauge bosons being spin-1 particles and matter consisting of elementary spin-1/2 fermions, the Higgs field is the only scalar (spin-0) quantity in the SM.

At high energies, all SM particles are exactly massless and the Standard Model Lagrangian can schematically be written as

$$\mathcal{L} = -\frac{1}{4} X_{\mu\nu} X^{\mu\nu} + i \bar{\psi} \slashed{D} \psi + \left( y_{ij} H \psi_i \psi_j + \text{h.c.} \right) + |D_\mu H|^2 - V(H). \quad (3.3)$$

The first term captures the kinetic terms of the electroweak and strong gauge bosons, where $X_{\mu\nu} \equiv \{B_{\mu\nu}, W^a_{\mu\nu}, G^a_{\mu\nu}\}$ are the relevant field strength tensors. The second term is the kinetic term of the SM Weyl fermions $\psi$ and anti-fermions $\bar{\psi} = \psi^\dagger \gamma^0$, with $\slashed{D} \equiv \gamma^\mu D_\mu$ and Dirac matrices $\gamma^\mu$. The covariant derivative $D_\mu$ encodes the interaction of these fermions with the force carriers. The third term are the Yukawa couplings of the matter fields to the Higgs doublet $H$ with Yukawa matrix $y_{ij}$. The last two terms finally characterise the kinetic term, the gauge boson interactions and the quartic potential of the Higgs field.

At low energies ($\lesssim 100 \,\text{GeV}$), the Higgs field develops a non-zero vacuum expectation value, $v = 246 \,\text{GeV}$, and the electroweak symmetry is spontaneously broken to $U(1)_{\text{em}}$ with the photon being the associated gauge boson. In the process, the Higgs boson itself, the weak gauge bosons ($W^\pm$ and $Z$) as well as the quarks and leptons[3] receive their mass by means of the Higgs mechanism [33–36]. The fermion masses are given by $m_i = y_{ii} v / \sqrt{2}$, for example. Moreover, the $B$ and $W^3$ bosons mix and become the photon and the $Z$ boson. The parameter governing this relationship is Weinberg's weak mixing angle $\theta_w \approx 30°$ which also sets the difference between the $W$ and $Z$ boson masses, $m_W = m_Z \cos \theta_w$.

The Standard Model has 19 free parameters which have to be measured in experiments. Most of these parameters have been determined to exquisite precision and the SM has passed most of its tests with flying colours [37]. For instance, the electromagnetic fine-structure constant, $\alpha \approx 1/137$, has now been measured with a relative uncertainty of less than one part in $10^9$ [37, 38]. Remarkably, the value obtained from a measurement of the electron magnetic moment together with a quantum electrodynamics calculation to tenth-order in perturbation theory achieves the same level of precision and agrees [39, 40].

---

[3] It is currently unclear whether neutrinos are coupled to the SM Higgs field as well. In the SM, they are exactly massless, i.e. the possible Yukawa couplings are taken to be zero.

### 3.2.2 Motivations for New Physics

Apart from the obvious fact that the Standard Model of particle physics only describes three of the four fundamental forces of Nature, there are several experimental and theoretical pieces of evidence pointing towards physics beyond the Standard Model. In the following, we describe some of these puzzles with a particular focus on those problems which might be solved by dark sectors containing new light and weakly-interacting particles. For more in-depth reviews, we refer the reader to [1, 37, 41–43].

It is well known that the weak interactions break charge-parity (CP) symmetry through a complex phase in the CKM matrix describing the mixing of quarks. A long-standing puzzle, however, is the non-observation of CP violation in the strong interactions although the QCD Lagrangian allows a CP-violating term,

$$\mathcal{L} \supset \frac{\theta}{32\pi^2} G_{\mu\nu,a} \tilde{G}^{\mu\nu,a} , \tag{3.4}$$

where $\theta$ is the CP-violating phase and $\tilde{G}^{\mu\nu,a} \equiv \frac{1}{2}\epsilon^{\mu\nu\rho\sigma} G^a_{\rho\sigma}$ is the dual gluon field strength tensor. Among other implications, this term induces a neutron electric dipole moment. Current experimental bounds on this quantity imply that $|\theta| < 10^{-10}$ [44, 45] and not $\theta \sim \mathcal{O}(1)$ as one might expect. The absence of CP violation in QCD is therefore a fine-tuning problem with the number of proposed solutions being limited. The most popular suggestion is to promote this parameter to a dynamical field which can naturally make $\theta$ small. This can be achieved by spontaneous breaking of the (approximate) Peccei–Quinn symmetry [2] at a scale $f_a$ which is known as the decay constant. The spontaneous symmetry breaking gives rise to a pseudo-Nambu-Goldstone boson, the QCD axion, with a specific relation between the axion mass and its decay constant, $m_a \propto 1/f_a$, and generic couplings not only to the gluon, but also to quarks. Interestingly, axions which are non-thermally produced for example via the misalignment mechanism could make up all or part of the dark matter in the universe [46–49] and thereby solve two of the major problems in particle physics at once. In principle, axions can also be thermally produced [50–55]. This is possible if the decay constant $f_a$ is small enough and/or the reheating temperature of the universe large enough, so that the axion production rate $\Gamma \sim T^3/f_a^2$ was larger than the Hubble rate $H$ at early times.

The neutrino sector is the least understood part of the SM. These very weakly interacting particles are famously predicted to be massless, but flavour oscillations[4] have been observed for solar, atmospheric, reactor and accelerator neutrinos. This implies that these particles must have non-zero masses which are however constrained to be at the sub-eV level and, therefore, much smaller than the mass of any other known particle. From these neutrino oscillation experiments, we know the mass-squared differences, $\Delta m^2$, of two pairs of neutrinos, but their ordering is still unknown, i.e. we do not know whether there are two light neutrinos and one slightly heavier neu-

---

[4] A neutrino which is produced or emitted in a well-defined flavour eigenstate has a non-zero probability of being detected in a different flavour state.

trino (normal hierarchy) or whether it is the other way around (inverse hierarchy). In fact, the overall mass scale and the mechanism by which neutrinos obtain their masses is still a mystery. It might be the standard Higgs mechanism with very small Yukawa couplings (if neutrinos are not their own anti-particle), it could be through a new Higgs-like field or the underlying process might be of an entirely different nature. Many attempts to incorporate neutrino masses in the Standard Model rely on new hidden sectors with new forces and/or particles, some of which might be light or even massless (see e.g. [56–58]). This also applies to solving additional problems related to neutrinos, such as the possibility of CP violation and the question whether they are Majorana or Dirac fermions, i.e. their own anti-particle or not.

The anomalous magnetic moment of the muon can be measured accurately in the laboratory by studying the precession of $\mu^+$ and $\mu^-$ in a constant external magnetic field. Precise SM calculations, however, are discrepant at the level of about $3.5\,\sigma$ [37]. One possibility for addressing this tension is the presence of a new force mediated by an extra Abelian $U(1)$ gauge boson $A'_\mu$ [59], which is usually referred to as a dark photon. This new particle may be very weakly coupled to electrically charged particles (including muons) through kinetic mixing with the photon [6] and thereby alleviate the $g_\mu - 2$ discrepancy. Further motivation for this type of BSM model [60] comes from the unexpected energy-dependent rise in the ratio of positrons to electrons in cosmic rays, as observed for example by PAMELA [61]. Moreover, a substantial number of dark matter models also employ new millicharged species, i.e. particles with small un-quantized electric charge, as dark matter candidates. These new particles usually arise naturally in many SM extensions which include dark photons or extra dimensions. As in the case of axions, light dark photons themselves may also constitute (part of) the dark matter.

One of the most famous theoretical problems of the SM is the hierarchy problem. At the core, it raises the question why the weak force is $10^{24}$ times stronger than gravity or, in other words, why the Higgs boson mass, $m_H \approx 125$ GeV, is so much smaller than the Planck mass, $M_{\rm pl} \approx 2.4 \times 10^{18}$ GeV, despite of quantum corrections. For a long time, the most popular solution has been to introduce a new spacetime symmetry known as supersymmetry. One of its consequences would be that every SM particle has a (heavier) superpartner. Apart from being theoretically appealing in a number of ways, supersymmetry could also, for instance, solve the problem of gauge coupling unification, provide a dark matter candidate and, imposed as a local symmetry, lead to a theory of supergravity. Having said that, no sign of any supersymmetric extension of the Standard Model has so far been seen at the Large Hadon Collider (LHC) or elsewhere. This has led particle physicists to further explore alternative ways of solving the hierarchy problem. Examples include composite Higgs models in which the Higgs boson may arise as a pseudo-Nambu-Goldstone boson (pNGB) [62–64], extra dimensions [65–67], a relaxation mechanism [68] or a large number of hidden sectors [69]. Instead of new heavy degrees of freedom, which have not been observed to date, light particles arise in these approaches for instance as mediators between the dark sector and the Standard Model or as a consequence of symmetry breaking patterns.

Ultimately, we would like to find a unified theory of the four fundamental forces or, in other words, of particle physics and gravity. Candidates for such a theory, e.g. string theory, generically predict (very rich) hidden sectors with a large number of moduli (scalar) fields, additional gauge bosons or even higher-spin particles. Some of these extra fields may have small or vanishing masses and may only be very weakly coupled to the SM degrees of freedom. In fact, most of the scenarios discussed above, in particular axions and hidden photons, generically occur in string compactifications [41, 49, 70, 71]. For example, axions arise as Kaluza–Klein modes of higher-dimensional form fields when compactifying the $d > 4$ spacetime dimensions to the usual four. These axions are exactly massless to all orders in perturbation theory, but receive their mass by non-perturbative effects, such as instantons. The presence of a large number of such axions with a wide range of masses is sometimes referred to as the string axiverse [5].

This comparably short and condensed list of current puzzles and their possible solutions only indicates the wealth of particle physics phenomenology these days. Having said that, many of the proposed SM extensions have in common that they contain a dark sector which is only (very) weakly coupled to the Standard Model. Since the SM symmetries restrict the kind of interactions or "portals" between the sectors, four of them are commonly used: the axion and vector portals, which we introduced above, as well as the Higgs and neutrino portals, which contain additional scalars or sterile neutrinos to mediate the interaction. There has been substantial experimental effort as well as progress in recent years at both the energy and intensity frontiers. This includes probing higher and higher energies especially at particle colliders, and using intense sources and ultra-sensitive detectors, for example in laser experiments [1, 41, 42]. In this thesis, we will contribute to the efforts on the cosmic frontier, i.e. the search for new physics in cosmology.

## *3.2.3 Effective Field Theory of Light Species*

We have just seen that additional light species arise in many well-motivated extensions of the Standard Model. In addition, cosmology tends to provide constraints on broad classes of models rather than very specific scenarios. Instead of working through these BSM models one by one, it is therefore more efficient to study the interactions between the new species with the SM degrees of freedom within the framework of effective field theory and thereby capture their main phenomenology. Generally speaking, this means parametrizing these interactions as

$$\mathcal{L} \supset \sum g \, \mathcal{O}_X \mathcal{O}_{\text{SM}}, \qquad (3.5)$$

where $\mathcal{O}_X$ and $\mathcal{O}_{\text{SM}}$ are operators of light and SM fields, respectively. Since the small masses of $X$ could receive large quantum corrections, we employ (approximate) symmetries to prevent this. The allowed couplings in (3.5) are then restricted by these protective symmetries. Ultimately, we will be able to put constraints on the interaction terms from cosmological measurements because the relic abundance of a new species is governed by its decoupling temperature, cf. Fig. 3.2, which in turn depends on the coupling parameters $g$.


**Introduction to Effective Field Theory**

The effective field theory framework builds upon the realization that Nature comes with many separated scales and that we can usually analyse natural phenomena by considering one relevant scale at a time. For example, we are able to describe the formation of hydrogen during recombination in terms of protons and electrons, and do not have to take the dynamics of quarks and gluons inside the proton into account. Similarly, we can treat the primordial plasma prior to decoupling as a fluid with certain properties without knowing the exact trajectories of the baryons and photons. In the same spirit, we were able to use Fermi theory instead of the full Standard Model to study neutrino decoupling. This is due to the fact that Fermi theory well describes the weak interactions of neutrinos with the other leptons at energy scales far below the mass of the $W$ boson and can therefore be seen as an effective description of this part of the Standard Model.

An effective field theory is a quantum field theory which takes advantage of scale separation and only includes the appropriate degrees of freedom to describe certain phenomena occurring at a particular energy scale. This way, we capture the important aspects at the scale of interest and do not have to worry about the potentially rich spectrum of states in the underlying microscopic theory. In the case of Fermi theory, the full description in terms of the weak force within the SM is known (and perturbative), but we can simplify the calculation of neutrino scattering by turning to the effective description. In the search for new physics, on the other hand, we usually do not have a full high-energy theory, but EFTs allow us to parametrize the unknown interactions, to estimate the magnitudes of these interactions and to classify their relative importance.

The guiding principle in the construction of EFTs are the symmetries obeyed by the relevant particle or field content. In this thesis, we are interested in EFTs which contain light fields in addition to the SM degrees of freedom. Following [72], we only consider models that are minimal and technically natural. Minimality here means that the additional particle content is as small as possible, i.e. it usually consists of only one additional elementary particle. The possible theory space is further reduced by imposing an (approximate) symmetry to protect the small mass from large quantum corrections.[5] We therefore require that our EFT not only obeys the SM symmetries, in particular Poincaré and gauge invariance, but also additional protective symmetries related to the specific scenario. Since the available symmetries depend on the spin

---

[5]The requirement of minimality essentially prevents us from employing strong dynamics in the new sector to generate large anomalous dimensions for the mass terms to render them small in the EFT because these models generically have a rich spectrum.

of the new particle, it is convenient to organize the EFT according to spin. In the following, we consequently study the effective field theories of light species with different spins $s \leq 2$. Since interacting particles with spin $s > 2$ necessarily have to be composites [73], we therefore exhaust all possibilities for elementary particles.

### Spin-0: Goldstone Bosons

A particularly well-motivated example of a new light particle are Goldstone bosons which generically appear when global symmetries are spontaneously broken. Goldstone bosons are either massless (if the broken symmetry was exact) or naturally light (if it was approximate). Examples of light pseudo-Nambu-Goldstone bosons are axions [2–4], familons [74–76], and majorons [77, 78], associated with spontaneously broken Peccei–Quinn, family and lepton-number symmetry, respectively. Below the scale of the spontaneous symmetry breaking, the couplings of the Goldstone boson $\phi$ to the SM particles can be characterised through a set of effective interactions

$$\mathcal{L} \supset \sum \frac{\mathcal{O}_\phi \mathcal{O}_{\text{SM}}}{\Lambda^\Delta} , \tag{3.6}$$

where $\Lambda$ is related to the symmetry breaking scale and the operators $\mathcal{O}_\phi$ are restricted by an approximate shift symmetry, $\phi \to \phi + \text{const}$.

Axion, familon and majoron models are characterised by different couplings in (3.6).[6] For example, below the electroweak symmetry breaking scale the axion couplings to the photon and gluon fields is given by

$$\mathcal{L} \supset -\frac{1}{4} \left( \frac{\phi}{\Lambda_\gamma} F_{\mu\nu} \tilde{F}^{\mu\nu} + \frac{\phi}{\Lambda_g} G_{\mu\nu,a} \tilde{G}^{\mu\nu,a} \right) , \tag{3.7}$$

with the field strength tensors $F_{\mu\nu}$ and $G^a_{\mu\nu}$, and their duals $\tilde{F}_{\mu\nu}$ and $\tilde{G}^a_{\mu\nu}$. The photon interaction term is somewhat model dependent, but typically arises together with the gluon coupling. Strictly speaking, only the latter has to be present in order to solve the strong CP problem. The interaction of familons with the charged SM fermions are generally governed by a current, e.g. the axial vector current $J_5^\mu$, derivatively coupled to the scalar field,

$$\mathcal{L} \supset -\frac{\partial_\mu \phi}{\Lambda_\psi} J_5^\mu = -\frac{\partial_\mu \phi}{\Lambda_\psi} \bar{\psi} \gamma^\mu \gamma^5 \psi , \tag{3.8}$$

where $\psi$ is any charged SM fermion. Majorons may arise as Goldstone bosons associated with the spontaneous breaking of the neutrino flavour symmetry which might be related to the existence of neutrino masses. Their interactions with the SM

---

[6]We will follow the common practice of reserving the name axion or axion-like particle for pNGBs that couple to the gauge bosons of the SM through operators like $\phi F_{\mu\nu} \tilde{F}^{\mu\nu}$. For simplicity, we will refer to all such particles simply as axions.

may be of a similar form as for familons. For a more in-depth discussion of light spin-0 fields, we refer to Chap. 4.


## Spin-$\frac{1}{2}$: Light Fermions

Light spin-$\frac{1}{2}$ particles can either arise as Weyl or Dirac fermions. Both are natural possibilities since their mass terms can be protected by a chiral and axial symmetry, respectively. A hidden Dirac fermion $\Psi$, which is invariant under $\Psi \rightarrow e^{i\alpha\gamma^5} \Psi$, with arbitrary phase $\alpha$ and $\gamma^5 \equiv i\gamma^0\gamma^1\gamma^2\gamma^3$, can couple to the hypercharge gauge boson $B_\mu$ through a dimension-5 dipole interaction,

$$\mathcal{L} \supset -\frac{1}{\Lambda_d} \bar{\Psi}\sigma^{\mu\nu}\Psi \, B_{\mu\nu} \,, \tag{3.9}$$

with $\sigma^{\mu\nu} \equiv \frac{i}{2}[\gamma^\mu, \gamma^\nu]$. This coupling may arise from loops in a high-energy theory that involve heavy charged particles. Another option is to couple the new particle $\Psi$ to any SM fermion $\psi$ via dimension-6 four-fermion interactions,

$$\mathcal{L} \supset \frac{1}{\Lambda_f^2} \left( d_s \, \bar{\Psi}\Psi \, \bar{\psi}\psi + d_p \, \bar{\Psi}\gamma^5\Psi \, \bar{\psi}\gamma^5\psi + d_a \, \bar{\Psi}\gamma^\mu\Psi \, \bar{\psi}\gamma_\mu\psi + d_v \, \bar{\Psi}\gamma^\mu\gamma^5\Psi \, \bar{\psi}\gamma_\mu\gamma^5\psi \right), \tag{3.10}$$

where the parameters $d_i$, $i = s, p, a, v$, are in principle $\mathcal{O}(1)$ numbers for the scalar, pseudo-scalar, axial and vector couplings, respectively. One possibility for inducing such an interaction is the exchange of a massive scalar or vector boson, such as those generated by spontaneously broken gauge symmetries. A well-studied example are light sterile neutrinos which may be coupled to the SM via a new massive vector boson $Z'$ from a spontaneously broken $U(1)$ symmetry [8, 79, 80].

A Weyl fermion $\chi$, which is protected by the chiral symmetry $\chi \rightarrow e^{i\alpha}\chi$, with arbitrary phase $\alpha$, can be coupled to SM fermions by four-fermion interactions as in (3.10). Alternatively, we can couple $\chi$ to the hypercharge gauge boson $B_\mu$ through a dimension-6 anapole moment,

$$\mathcal{L} \supset -\frac{1}{\Lambda_a^2} \chi^\dagger \bar{\sigma}^\mu \chi \, \partial^\nu B_{\mu\nu} \,, \tag{3.11}$$

where $(\bar{\sigma}^\mu) = (\mathbb{1}, -\sigma^i)$ with Pauli matrices $\sigma^i$.


## Spin-1: Vector Bosons

Massless spin-1 particles carry fewer degrees of freedom than their massive counterparts which means that a mass term cannot arise from perturbative quantum effects. This implies that massless vector bosons are technically natural. A small protected mass can nevertheless be generated, for example via the standard Higgs mechanism, a Stückelberg mechanism or in LARGE volume compactifications of string theory [41]. The most minimal scenario is to couple one hidden $U(1)$ gauge boson $A'_\mu$, a dark photon, to the SM fermions via a dipole interaction. Prior to electroweak symmetry breaking, this coupling can be written as

$$\mathcal{L} \supset -\frac{1}{\Lambda^2_{A'}} F'_{\mu\nu} H \bar{\psi} \sigma^{\mu\nu} \psi, \tag{3.12}$$

where $F'_{\mu\nu}$ is the field strength associated with the dark photon and the presence of the Higgs field $H$ is required by gauge invariance. Once the Higgs has acquired its non-zero vacuum expectation value $v$, this operator will be effectively of dimension-5 with an effective scale of $\tilde{\Lambda}_{A'} = \Lambda^2_{A'}/v$.

Another option to couple a dark photon to the Standard Model is by kinetically mixing the new boson $A'_\mu$ with the hypercharge gauge boson, $\mathcal{L} \supset -(\epsilon/2) F'^{\mu\nu} B_{\mu\nu}$ [6]. If the extra particle $A'_\mu$ is exactly massless, we can decouple it from the SM by a field redefinition. This scenario therefore requires to enlarge the particle content by a new Dirac fermion $\chi$ which is millicharged under the electroweak SM gauge group, e.g.

$$\mathcal{L} \supset -\epsilon g_{A'} \cos\theta_w \, \bar{\chi} \slashed{A} \chi - \epsilon g_{A'} \sin\theta_w \, \bar{\chi} \slashed{Z} \chi - g_{A'} \, \bar{\chi} \slashed{A}' \chi, \tag{3.13}$$

where $g_{A'}$ is the coupling between $\chi$ and the dark photon, and $\epsilon$ is the kinetic mixing parameter. As mentioned above, the additional fermion $\chi$ can be (a fraction of) the dark matter in many models.

## Spin-$\frac{3}{2}$: Gravitino

There is one unique elementary particle with spin-3/2, the gravitino. The existence of the gravitino as the superpartner to the graviton is a universal prediction of supergravity. Its mass is set by the supersymmetry breaking scale, $m_{3/2} \sim F/M_{\mathrm{pl}}$, and can be very small in low-scale supersymmetry breaking scenarios. Although the gravitino typically interacts with gravitational strength, its longitudinal component couples to the SM in the same way as the Goldstino,

$$\mathcal{L} \supset -\frac{1}{F^2} \chi^\dagger \sigma_\mu \partial_\nu \chi \, T^{\mu\nu}. \tag{3.14}$$

The coupling parameter of this component of the gravitino is therefore enhanced compared to the Planck scale.

**Spin-2: Graviton**

The graviton is the unique elementary spin-2 particle and the force carrier of gravity. It is massless and only interacts with gravitational strength, i.e. all its interactions with the SM fields are suppressed by $M_{pl} \approx 2.4 \times 10^{18}\,\mathrm{GeV}$.

This concludes our discussion of the effective field theory of light species. We will return to these EFT couplings in Sect. 3.3.3 where we discuss some of the current constraints from particle accelerators, laboratory experiments, astrophysics and cosmology.

## 3.3  Neutrinos and Dark Radiation

In this thesis, we are particularly interested in measurements of the radiation density of the universe to probe particle physics. The contribution from photons, $\rho_\gamma$, is fixed by the very well measured value of the CMB temperature,

$$\rho_{\gamma,0} = \frac{\pi^2}{15} T_0^4 \approx 2.0 \times 10^{-15} \mathrm{eV}^4 \,, \tag{3.15}$$

which corresponds to a physical photon density of $\omega_\gamma = \Omega_\gamma h^2 \approx 2.5 \times 10^{-5}$. As we discussed in Sect. 2.2.3, the standard models of cosmology and particle physics also predict a contribution from neutrinos. According to Eqs. (2.12) and (2.20), the expected radiation density from each neutrino species in the instantaneous decoupling limit is[7]

$$\rho_{\nu_i} = \frac{7}{8}\left(\frac{T_{\nu,0}}{T_0}\right)^{4/3}\rho_\gamma = \frac{7}{8}\left(\frac{4}{11}\right)^{4/3}\rho_\gamma \equiv a_\nu^{-1}\rho_\gamma \,, \tag{3.16}$$

where we defined $a_\nu \approx 4.40$ instead of its inverse for later convenience. The three neutrino species of the Standard Model consequently contribute a significant amount to the total radiation density in the early universe: $\rho_\nu/\rho_r = \sum_i \rho_{\nu_i}/\rho_r \approx 41\%$. The gravitational effects of neutrinos are therefore significant at early times which is why we can observe their imprints in the CMB and BAO spectra (cf. Chaps. 5 and 7; see also [26]) although their contribution to the total energy density today is very small.

---

[7]In principle, the decoupling temperatures of the three different neutrino species slightly differ from each other (see e.g. [81, 82]). We neglect this and effectively take $T_\nu$ to be the average neutrino temperature.

### 3.3.1  Effective Number of Relativistic Species

We have assumed so far that neutrinos decoupled instantaneously and, therefore, did not receive any of the entropy that was released when electrons and positrons annihilated. However, a small amount of the entropy is actually transferred to the neutrino sector as well. Instead of changing the prefactor $a_\nu$ in (3.16), it is common to introduce the effective number of neutrinos $N_\nu$ as

$$\rho_r = \rho_\gamma + \rho_\nu = \left[1 + a_\nu^{-1} N_\nu\right]\rho_\gamma . \tag{3.17}$$

In the instantaneous decoupling limit, we have $N_\nu = 3$. Accounting for plasma corrections of quantum electrodynamics, flavour oscillations and, in particular, the fact that neutrinos have not fully decoupled when electrons and positrons annihilated, one finds $N_\nu = 3.046$ in the SM [83].[8] In this sense, measurements of $N_\nu > 0$ probe the energy density of the C$\nu$B and $N_\nu \neq 3.046$ could be a sign for non-standard properties of neutrinos or changes to the standard thermal history.

The introduction of this new parameter also provides a convenient way to include the possibility of radiation in excess of the SM expectations. BSM physics may add extra radiation density $\rho_X$ to the early universe which is often referred to as dark radiation. It is conventional to measure this radiation density relative to the density $\rho_{\nu_i}$ of a single SM neutrino,

$$\Delta N_{\text{eff}} \equiv \frac{\rho_X}{\rho_{\nu_i}} = a_\nu \frac{\rho_X}{\rho_\gamma} , \tag{3.18}$$

and define the *effective number of relativistic species*

$$N_{\text{eff}} \equiv N_\nu + \Delta N_{\text{eff}} = 3.046 + \Delta N_{\text{eff}} . \tag{3.19}$$

The parameter $N_{\text{eff}}$ therefore captures the difference between the total radiation density of the universe, $\rho_r$, and the CMB photon energy density, $\rho_\gamma$, normalized to the energy density of a single neutrino species, $\rho_{\nu_i}$. This is the central cosmological parameter under investigation in this thesis. Slightly generalizing the statement from above, measurements of $N_{\text{eff}} > 0$ therefore probe the energy density of the C$\nu$B, and $N_{\text{eff}} \neq 3.046$ would be a signature of physics beyond the standard models of particle physics and/or cosmology.

---

[8]Recently, a more accurate calculation including neutrino oscillations with the present values of the mixing parameters found $N_\nu = 3.045$ [84]. At the level of precision anticipated in the near future and under consideration in this thesis, the difference is irrelevant which is why we keep the standard value of $N_\nu = 3.046$ as our baseline assumption.

Current observations of the CMB anisotropies and the light element abundances find [25, 85][9, 10]

$$N_{\text{eff}} = 3.04 \pm 0.18 \quad (\text{CMB}),\tag{3.20}$$

$$N_{\text{eff}} = 2.85 \pm 0.28 \quad (\text{BBN}),\tag{3.21}$$

which are consistent with the SM prediction of $N_{\text{eff}} = N_\nu = 3.046$. These measurements represent a highly significant detection of the energy density associated with the cosmic neutrino background and a non-trivial confirmation of the thermal history back to about one second after the big bang when neutrinos decoupled. The consistency of the measurements is remarkable, although the interpretation is somewhat sensitive to assumptions about the cosmological model and constraints weaken considerably in some extensions of the $\Lambda$CDM model (cf. Sect. 5.3). Furthermore, these measurements put interesting limits on many extensions of the SM containing additional light fields and/or thermal histories that enhance or dilute the radiation density (see e.g. [72, 87–95]).

Strictly speaking, the parameter $N_{\text{eff}}$ is usually taken to capture neutrinos and neutrino-like species, i.e. it refers to free-streaming radiation. As we illustrate in Fig. 3.1, not all relativistic BSM particles have to fall into this category. We will hence also allow for a contribution from non-free-streaming radiation and capture this by the following parameter[11, 12]

$$N_{\text{fluid}} \equiv a_\nu \frac{\rho_Y}{\rho_\gamma},\tag{3.22}$$

with non-free-streaming radiation density $\rho_Y$. In Chap. 5, we will characterise the different effects of $N_{\text{eff}}$ and $N_{\text{fluid}}$ on the photon-baryon fluid, and study their distinct cosmological imprints in detail. In particular, we will keep $N_{\text{fluid}} \neq 0$ when

---

[9]It is sometimes stated imprecisely that particle colliders also constrain the number of neutrino species through precision measurements of the width of the $Z$ decay. Strictly speaking, collider measurements however only tell us how many fermions with mass below $\frac{1}{2}m_Z$ couple to the $Z$ boson [86]. These experiments find very close agreement with three families of active neutrinos.

[10]The quoted CMB constraint includes high-$\ell$ polarization data which has been labelled as preliminary by the Planck collaboration. Only considering temperature and low-$\ell$ polarization data results in $N_{\text{eff}} = 3.13^{+0.30}_{-0.34}$ [25].

[11]The introduction of a viscosity parameter $c_{\text{vis}}$ [98], which has recently been detected by Planck, $c_{\text{vis}}^2 = 0.331 \pm 0.037$ [25] (see also [99, 100]), constitutes another attempt to parametrize (non-)free-streaming radiation. However, as detailed in [101], the fiducial value of $c_{\text{vis}}^2 = \frac{1}{3}$ is not equivalent to free-streaming radiation and differs from $\Lambda$CDM by $\Delta\chi^2 = 20$. The advantage of our parametrization is that it exactly reproduces $\Lambda$CDM when $N_{\text{eff}} = N_\nu = 3.046$ and $N_{\text{fluid}} = 0$. Previously, a similar parametrization in terms of these effective numbers was discussed in [95, 102–104] and analysed with WMAP data in [102, 103]. Having said this, distinguishing these parameters with high significance has only recently become possible (see Sect. 5.3).

[12]The use of simplified models to search for new physics at particle accelerators, such as the Large Hadron Collider [105], is analogous to this approach. In both cases, large numbers of models are reduced to a single model, which captures their essential features.

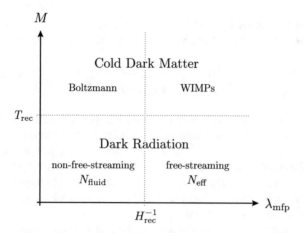

**Fig. 3.1** Classification of particles beyond the Standard Model according to their masses $M$ and their mean free paths $\lambda_{\text{mfp}}$, normalized relative to the temperature and Hubble rate at recombination, $T_{\text{rec}}$ and $H_{\text{rec}}$, which are relevant for CMB observations (after [96, 97]). While particles with $M > T_{\text{rec}}$ contribute to the cold dark matter of the universe at the time of recombination, lighter particles with $M < T_{\text{rec}}$ are relativistic and behave as radiation. Since massive, strongly interacting particles are Boltzmann-suppressed, they do not contribute a cosmologically interesting radiation density. Depending on their mean free path, we separate dark radiation into free-streaming and non-free-streaming particles. We note that non-thermal relics, such as non-thermally produced axions, are not captured by the simple characterisation presented in this figure

analysing Planck data and forecasting future constraints. In Chaps. 6 and 7, we will restrict ourselves to $N_{\text{eff}}$ and implicitly set $N_{\text{fluid}} \equiv 0$ because many BSM models, in particular simple extensions of the SM, only predict free-streaming species.

## 3.3.2 Thermal History with Additional Species

A natural source for $\Delta N_{\text{eff}} \neq 0$ are extra relativistic particles.[13] Let us therefore consider a light species $X$ as the only additional particle in some BSM theory. Assuming this species was in thermal equilibrium with the SM particles at some point in the history of the universe, we can compute its contribution to $\Delta N_{\text{eff}}$ in a similar way to our calculation of the relic density of neutrinos. Furthermore, assuming that the

---

[13]For simplicity, we set $N_{\text{fluid}} = 0$ in this section, but note that the thermal history is unchanged if an additional species contributes to $N_{\text{fluid}}$ instead of $\Delta N_{\text{eff}}$ because we will only be concerned with the relic energy density. As we will see later, distinguishing $N_{\text{eff}}$ and $N_{\text{fluid}}$ relies on more subtle effects.

species $X$ freezes out well before neutrino decoupling,[14] $T_{F,X} \gtrsim 10 \, \text{MeV}$, the temperature associated with $X$ in relation to the neutrino temperature $T_\nu$ is

$$T_X = \left( \frac{g_*(T_{F,\nu})}{g_*(T_{F,X})} \right)^{1/3} T_\nu = \left( \frac{10.75}{g_{*,SM}} \right)^{1/3} \left( \frac{g_{*,SM}}{g_*(T_{F,X})} \right)^{1/3} T_\nu \approx 0.465 \left( \frac{g_{*,SM}}{g_*(T_{F,X})} \right)^{1/3} T_\nu \,,$$
(3.23)

where we employed entropy conservation among the particles in thermal equilibrium, i.e. we assumed no significant entropy production after the decoupling of $X$. We also inserted $g_{*,SM} = 106.75$, and $g_*(T_{F,\nu}) = 10.75$ due to photons, electrons and neutrinos. (We omitted the additional subscript for entropy since $g_*(T) = g_{*S}(T)$ in thermal equilibrium.) Note that neither $g_*(T_{F,\nu})$ nor $g_*(T_{F,X})$ include a contribution from the particle $X$ as it has decoupled and does not receive any of the released entropy of later annihilation processes.

After neutrinos decouple, the evolution of $T_X$ and $T_\nu$ is the same since both scale as $T_i \propto a^{-1}$. Provided that the extra species are relativistic when they freeze out and remain decoupled, their energy ratio stays constant and we get

$$\Delta N_{\text{eff}} = \frac{\rho_X}{\rho_{\nu_i}} = \frac{g_{*,X}}{g_{*,\nu_i}} \frac{T_X^4}{T_\nu^4} \approx 0.027 \, g_{*,X} \left( \frac{g_{*,SM}}{g_*(T_{F,X})} \right)^{4/3} , \qquad (3.24)$$

where we used $g_{*,\nu_i} = 7/4$ and $g_{*,X}$ depends on the spin of the particle (cf. Sect. 2.2.2). Figure 3.2 shows the contribution to $\Delta N_{\text{eff}}$ from a single thermally-decoupled species as a function of the decoupling temperature $T_F$ and the spin of the particle based on (3.24). We see that decoupling after the QCD phase transition produces a contribution to $N_{\text{eff}}$ that is comparable to that of a single neutrino species, which is ruled out (or at least strongly disfavoured) by current observations. On the other hand, decoupling before the QCD phase transition creates an abundance that is smaller by an order of magnitude due to the much larger number of available degrees of freedom $g_*$. Since Planck is blind to these particles, such scenarios are still consistent with current limits. Future observations will therefore give us access to particles that are more weakly coupled than neutrinos and decoupled before the end of the QCD phase transition.

Employing $g_*(T_{F,X}) \leq g_{*,SM} = 106.75$ in (3.24), we get the asymptotic values displayed in Fig. 3.2. We find that any additional species $X$ which has been in thermal equilibrium with the Standard Model at any point in the history of the universe contributes the following minimal amount to the radiation density of the universe:

---

[14]If the particle $X$ froze out after electron-positron annihilation, its temperature would be the same as for photons, $T_X = T$, for all times. On the other hand, we would have $T > T_X > T_\nu$ if the new species decoupled in the relatively small window between the onset of neutrino decoupling and the conclusion of $e^+e^-$ annihilation. As illustrated in Fig. 3.2, both scenarios are ruled out by current measurements at more than $5\sigma$ ($3\sigma$) for new particles with (without) spin. We therefore do not discuss them further.

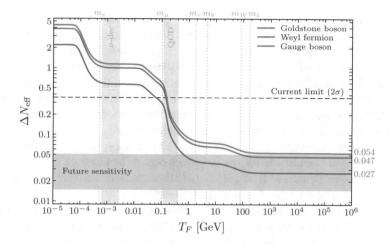

**Fig. 3.2** Contributions of a single thermally-decoupled Goldstone boson, Weyl fermion or massless gauge boson to the effective number of relativistic species, $\Delta N_{\text{eff}}$, as a function of its decoupling temperature $T_F$. The current Planck limit at 95% c.l. from [25] and the possible future sensitivity (cf. Chap. 6) illustrate the current and future power of cosmological surveys to constrain light thermal relics. The drop in $\Delta N_{\text{eff}}$ by about one order of magnitude around $T_F \sim 150\,\text{MeV}$ is due to the QCD phase transition, which is denoted by a vertical gray band as is neutrino decoupling. The dotted lines indicate some of the mass scales at which SM particles and anti-particles annihilate

$$\Delta N_{\text{eff}} \geq 0.027 g_{*,X} = \begin{cases} 0.027 & \text{Goldstone boson (spin-0),} \\ 0.047 & \text{Weyl fermion (spin-1/2),} \\ 0.054 & \text{massless gauge boson (spin-1).} \end{cases} \tag{3.25}$$

This is an important result. Provided that future cosmological surveys are sensitive to $\Delta N_{\text{eff}} = 0.027$, we can detect any particle which has ever been in thermal equilibrium with the SM. Reaching this minimal thermal abundance is therefore a very interesting target for upcoming measurements with important consequences for BSM physics (see Chap. 4 and [72, 95], for instance). We find it intriguing that this threshold seems to be within reach of future observations. In Chaps. 5 and 6, we will quantify this expectation and see that future CMB and LSS experiments indeed have the potential to achieve such measurements. In this case, these cosmological surveys either have to detect extra light relics or we can put strong constraints on their couplings to the SM, as we will show in Chap. 4 for the case of Goldstone bosons.

When deriving equation (3.24), we assumed an extension of the SM in which there is no significant entropy production and the species $X$ is the only addition to the SM particle content. We briefly revisit these assumptions in the following. Arguably, many BSM models come with additional massive particles which increase the effective number of relativistic degrees of freedom $g_*$ at high temperatures, $T \gg 100\,\text{GeV}$. Moreover, their annihilation might subsequently reduce the relic abundance of $X$ according to (3.24), since $g_*(T_{F,X}) > g_{*,SM}$, allowing for $\Delta N_{\rm eff} < 0.027 g_{*,X}$. This is however degenerate with the uncertainty on the reheating temperature, i.e. the question whether such an additional particle has ever been in thermal equilibrium, because the energy density is only diluted for decoupling temperatures above the masses of these new particles. Furthermore, a dilution of the minimal contribution to $N_{\rm eff}$ by a factor of two requires $g_{*,BSM} \approx 1.7 g_{*,SM}$, i.e. almost a doubling of the SM degrees of freedom. Although this is possible in SM extensions, these models commonly contain many additional light particles as well (see e.g. [5, 69]) which in turn enhance the expected value of $\Delta N_{\rm eff}$. For instance, just three light degrees of freedom can compensate for the large increase in field content in the Minimal Supersymmetric Standard Model (MSSM). In this sense, our assumption of one new particle $X$ is to be considered as a conservative choice.

In certain BSM models, the minimal contribution to $\Delta N_{\rm eff}$ might also be diluted because entropy is produced, e.g. by out-of-equilibrium decays of massive BSM particles or a phase transition. Introducing the parameter $\gamma = s(T_{F,X})/s(T_{F,\nu})$ to capture the amount of entropy production after the species $X$ decoupled, Eq. (3.24) becomes

$$\Delta N_{\rm eff} = 0.027\, g_{*,X} \left( \frac{g_{*,SM}}{g_*(T_{F,X})} \right)^{4/3} \times \gamma^{-4/3} . \tag{3.26}$$

To illustrate the possible size of $\gamma$, we consider a massive particle $\chi$ which decays out of equilibrium, i.e. after it has decoupled and acquired an abundance $Y_i = n_\chi/s$. If we assume that this particle dominates the energy density of the universe when it decays, the entropy ratio is given by $\gamma \sim g_{*,d}^{1/4} Y_i \, (m_\chi^2 \tau_\chi / M_{\rm pl})^{1/2}$ [106], with particle mass $m_\chi$ and lifetime $\tau_\chi$. This might potentially be large and, for example, heavy gravitinos are constrained to have never been in thermal equilibrium during the hot big bang evolution in this way. For concreteness, let us consider a light scalar particle, which froze out at high temperatures, and an additional massive species with mass $m_\chi = 1\,{\rm TeV}$, which decoupled at $T > m_\chi$ and has a lifetime corresponding to a decay temperature $T_D \sim 10\,{\rm GeV}$. Such a particle would dominate the energy density of the universe at $T \sim 1\,{\rm TeV}$ and easily dilute the contribution of the light particle to $\Delta N_{\rm eff}$. Having said that, this scenario is no different than assuming the reheating temperature to be $T_R \sim 10\,{\rm GeV}$, in which case the scalar particle would have never been in thermal equilibrium. In a sense, we are therefore expanding the definition of the reheating temperature $T_R$ to include a possible second reheating phase caused by an out-of-equilibrium decay process after inflationary reheating. This is therefore degenerate with our general uncertainty on the reheating temperature. Depending on the involved time scales, the combined constraints on $N_{\rm eff}$ from BBN and the CMB/LSS, together with other cosmological signatures of particle decays, might be able to put some constraints on the entropy factor $\gamma$ and thereby limit the possible dilution of $\Delta N_{\rm eff}$ due to entropy production at correspondingly late times (see e.g. [11, 14, 107]).

So far, we have implicitly assumed that $\Delta N_{\rm eff} > 0$. The annihilation or decay of particles into neutrinos after their decoupling could however produce $\Delta N_{\rm eff} < 0$, for instance. This type of signature would therefore hint towards certain types of BSM models. One such example may be a light particle which is unstable and may decay to neutrinos. Interestingly, its out-of-equilibrium decay, i.e. after it decoupled at early times, would lead to a larger suppression of $N_{\text{eff}}$ than its equilibrium decay and would therefore be easier to detect. Alternatively, non-standard neutrino properties or particles decaying to photons could effectively lead to $\Delta N_{\text{eff}} < 0$ as well.

### 3.3.3 Current Constraints on EFT Parameters

As we have already argued in Sect. 3.1, astrophysics and cosmology can put strong constraints on light and weakly-coupled particles. In the following, we reconsider the EFT of light species and discuss some of the constraints on the effective interactions from current measurements. We will also connect these couplings to the thresholds (3.25) for $\Delta N_{\text{eff}}$. As in Sect. 3.2.3, we will study the new species separately according to their spin.

#### Spin-0: Goldstone Bosons

The Goldstone couplings (3.7) and (3.8) are constrained by laboratory experiments [1, 108], by astrophysics [24, 109] and by cosmology [49, 72] (see [110] for a recent review).[15] While laboratory constraints, such as those from light-shining-through-walls experiments, helioscopes or haloscopes, have the advantage of being direct measurements, their main drawback is that they are usually rather model-specific and sensitive only to narrow windows of pNGB masses. Astrophysical and cosmological constraints are complementary since they are relatively insensitive to the detailed form of the couplings to the SM and span a wide range of masses. The main astrophysical constraints on these new light particles come from stellar cooling [24]. In order not to disrupt successful models of stellar evolution, any new light particles must be more weakly coupled than neutrinos. The axion-photon coupling, for example, is bounded by $\Lambda_\gamma \gtrsim 1.5 \times 10^{10}\,\text{GeV}$ (95% c.l.) in this way [116], in agreement with our estimate (3.2). Moreover, since neutrinos couple to the rest of the SM through a dimension-six operator (suppressed by the electroweak scale), the constraints on extra particles are particularly severe for dimension-four and dimension-five couplings to the SM.

We refer to Chap. 4 for constraints on the SM interactions of Goldstone bosons from precise cosmological measurements. We will see that current bounds may be improved by many orders of magnitude if the threshold value $\Delta N_{\text{eff}} = 0.027$ can be observationally excluded. In addition, most bounds on familon couplings can already be substantially improved by cosmological experiments that are only sensitive to $\Delta N_{\text{eff}} \sim 0.05$.

---

[15]With the advent of gravitational-wave astronomy, very light pNGBs could potentially also be probed around rapidly-rotating black holes (see e.g. [111–115]).

**Spin-$\frac{1}{2}$: Light Fermions**

The best constraint on the dipole interaction (3.9) between new light Dirac fermions and the Standard Model comes from stellar cooling with $\Lambda_d \gtrsim 10^9$ GeV [109]. On the other hand, collider searches provide the most competitive bounds on the four-fermion couplings of (3.10), $\Lambda_f \gtrsim \mathcal{O}(1\,\text{TeV})$, with the precise numbers depending somewhat on the type of SM fermion [72]. The anapole coupling (3.11) of extra Weyl fermions has similar constraints from particle accelerators.

Figure 3.2 shows that any dark Dirac or Weyl fermion must have decoupled before the end of the QCD phase transition in order to be compatible with the current Planck limits on the radiation density. Cosmological measurements will be able to put strong constraints on these couplings since Dirac fermions minimally contribute $\Delta N_{\text{eff}} = 0.094$ to the radiation density, which will be well within the sensitivity of future surveys. For interactions including light Weyl fermions, the cosmological threshold is smaller by a factor of 2, but the combination of upcoming CMB and LSS experiments still has the potential to improve these bounds by many orders of magnitude.

**Spin-1: Vector Bosons**

The constraints on the dipole coupling (3.12) of dark photons from stellar and supernova cooling depend on the type of SM fermions that are involved. For instance, the coupling to electrons is bounded by $\Lambda_{A'} \gtrsim 10^7$ GeV and the muon interaction by $\Lambda_{A'} \gtrsim 10^4$ GeV [72]. Models with extra millicharged particles, such as those of (3.13), also have strong stellar and supernova bounds. These constraints are complemented by a number of collider searches and laboratory experiments, for instance in beam dump experiments and using helioscopes (see e.g. [1, 41]). For hidden millicharged fermion masses of $m_\chi \lesssim 100\,\text{keV}$ for example, the current bound on the kinetic mixing parameter is $\epsilon \lesssim 10^{-13}$ [41]. These constraints are again supplemented by current measurements of $N_{\text{eff}}$ from BBN and the CMB as well [72].

Since massless gauge bosons in thermal equilibrium contribute at least 0.054 to $N_{\text{eff}}$, future cosmological measurements will be able to severely restrict the allowed parameter space of these models, including scenarios where millicharged fermions make up the dark matter. It goes without saying that bounds on non-Abelian gauge fields will be much stronger because they would carry a radiation density equivalent to $N_{\text{fluid}} \sim 0.07(N^2 - 1)$ for a dark $SU(N)$ gauge group [117].

**Spin-$\frac{3}{2}$: Gravitino**

Since the gravitino coupling (3.14) is suppressed by the supersymmetry breaking scale and not the Planck scale, the gravitino may have been in thermal equilibrium with the Standard Model at early times. It minimally contributes $\Delta N_{\text{eff}} = 0.047$ to the radiation density in the early universe because it behaves like a Weyl fermion as far as its contribution to $\Delta N_{\text{eff}}$ is concerned. In fact, current cosmological constraints on $m_{3/2}$ could indicate that the gravitino may be so light (and hence $F$ so small) that it only decouples at $T \lesssim 100\,\text{GeV}$ [27], i.e. the remaining minimal contribution to $N_{\text{eff}}$ gets enhanced compared to the value for a Weyl fermion, $\Delta N_{\text{eff}} \gtrsim 0.057$, and therefore more easily detected in upcoming observations.

**Spin-2: Graviton**

It is unlikely that the graviton has ever been in thermal equilibrium because it only interacts with gravitational strength. Its thermal abundance is therefore negligible. Having said that, since gravitational waves are massless and free-streaming, the energy density of a stochastic background in the early universe may contribute a small amount to the radiation density and therefore to the effective number of relativistic species $N_{\text{eff}}$ [118–121].

## 3.4 Cosmological Signatures of Light Relics

Changing the radiation density by adding additional light species [16] to the universe will impact a number of cosmological observables in more or less subtle ways. In the following, we will describe the influence of light relics on the CMB anisotropies, the matter power spectrum, the BAO spectrum and big bang nucleosynthesis. We will see that precise measurements of the various observables and reliable extraction of several signatures turns cosmology into an accurate tool for probing cosmic neutrinos as well as other light thermal relics and, hence, this type of SM and BSM physics.

### 3.4.1 Diffusion Damping

A change in the radiation density impacts the CMB power spectrum in several ways (see [125] for an excellent discussion). At the level of the homogeneous cosmology, the largest effect of relativistic particles is a change in the expansion rate during radiation domination, which according to the first Friedmann equation (2.5) is given by

$$3M_{\text{pl}}^2 H^2 = \rho_\gamma \left( 1 + \frac{N_{\text{eff}} + N_{\text{fluid}}}{a_\nu} \right), \tag{3.27}$$

where free-streaming particles ($N_{\text{eff}}$) and non-free-streaming particles ($N_{\text{fluid}}$) contribute equally, and $a_\nu \approx 4.40$ was introduced in (3.16). As we anticipated in

---

[16] A change in $N_{\text{eff}}$ might not be related to new light, but new heavy particles which could decay or otherwise mimic the effects of light degrees of freedom [89, 122–124]. However, these scenarios are usually accompanied by additional signatures. The late decay or annihilation to photons led to $\Delta N_{\text{eff}} > 0$, for example, but could also change BBN, prompt spectral distortions of the CMB frequency spectrum or leave distinct signals in the CMB anisotropy spectrum, depending on when this additional energy injection into the thermal bath of photons occurred (see e.g. [10, 13, 17] and references therein).

Sect. 2.3.2, the change in the Hubble rate manifests itself in a modification of the damping tail of the CMB (see Fig. 3.3). However, understanding the precise impact of a change in the radiation density is non-trivial [125], since changing $H$ will also affect the location of the first acoustic peak, which is extremely well measured. To study the effects of $N_{eff} + N_{fluid}$, it is instructive to consider the ratio of the angular sizes of the damping scale and the sound horizon, $\theta_d = 1/(k_d D_A)$ and $\theta_s = r_s/D_A$, in order to eliminate the a priori unknown angular diameter distance to the last-scattering surface, $D_A$, which also depends on the Hubble rate. Around the time of recombination, the wavelength associated with the mean squared diffusion distance is proportional to the Hubble rate, $k_d^2 \propto H_{rec}$, according to (2.38). Since the sound horizon is inversely proportional to $H_{rec}$, the ratio scales as

$$\frac{\theta_d}{\theta_s} = \frac{1}{k_d \, r_s} \propto \frac{1}{H_{rec}^{1/2} \, H_{rec}^{-1}} = H_{rec}^{1/2} \,. \tag{3.28}$$

Increasing $N_{eff}$ (and hence $H_{rec}$) therefore leads to a larger $\theta_d$ if we keep the angular scale of the first acoustic peak, $\theta_s$, fixed, e.g. by simultaneously varying the Hubble constant $H_0$. This implies that the damping kicks in at larger angular scales (smaller multipoles) and reduces the power in the damping tail when increasing the radiation density. This is exactly the behaviour that we observe for the CMB power spectrum in Fig. 3.3. The constraint in (3.20) is, in fact, mostly derived from measurements of the CMB damping tail [25, 126].

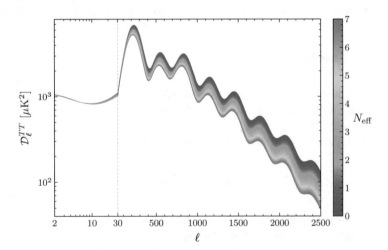

**Fig. 3.3** Variation of the CMB temperature power spectrum $\mathcal{D}_\ell^{TT}$ as a function of $N_{eff}$ for fixed angular size of the sound horizon $\theta_s$

It is useful to anticipate the possible degeneracies between the effects of extra relativistic species and changes in the cosmological parameters as these may limit the constraining power. The physical origin of the effect on the damping tail is given by (2.38), so we can understand the most severe degeneracy analytically. As pointed out in [127], there is an important degeneracy between the expansion rate $H$ and the primordial helium fraction $Y_p$. Since helium has a much larger binding energy than hydrogen, increasing the helium fraction for fixed $\omega_b$ (which is well determined by the relative peak heights) will decrease the number of free electrons at the time of hydrogen recombination, $n_e \propto (1 - Y_p)$, resulting in a larger photon mean free path and, in consequence, more damping. It is then possible to change $Y_p$ and $H$ simultaneously in a way that keeps the damping scale fixed, cf. $k_d^{-2} \propto (n_e H)^{-1}$. Including the dependence on $n_e$ in (3.28), we therefore have $\theta_d/\theta_s \propto (H_{\mathrm{rec}}/n_e)^{1/2}$. For fixed $\theta_s$, the damping scale consequently remains unchanged if $Y_p$ is reduced while simultaneously increasing $N_{\mathrm{eff}}$, i.e. the parameters $N_{\mathrm{eff}}$ and $Y_p$ are anti-correlated. If we want to measure $Y_p$ independently from BBN, this degeneracy will be one of the main limiting factors for constraints on extra relativistic species in the future.

Given that the physics underlying the acoustic peaks in the CMB and BAO spectra is the same, we expect to see similar effects of relativistic species. In particular, the baryon acoustic oscillations are also damped due to photon diffusion and hence sensitive to $N_{\mathrm{eff}}$ in the same way as the CMB anisotropies. However, there is an additional exponential damping in the BAO spectrum because it is non-linearly processed by gravitational evolution. Since a change in diffusion damping is therefore degenerate with theoretical uncertainties in the amount of non-linear damping, the constraining power of the damping tail is restricted in this observable.

### 3.4.2  Phase Shift

Before Planck, the CMB constraints on $N_{\mathrm{eff}}$ were mainly provided by the damping tail and therefore just probed the homogeneous radiation density. Recently, the experimental sensitivity has however improved to such a level that the measurements have started to become susceptible to neutrino perturbations (and those of other free-streaming relics). These affect the photon-baryon fluid through their gravitational influence and lead to a distinct imprint in the acoustic oscillations [127]. In the following, we present a heuristic description of this effect and refer to Sect. 5.2 for the proper treatment.

For this purpose, let us reconsider the single point-like overdensities that we discussed in Sect. 2.3.2 and whose evolution is visualized in Fig. 2.3. Since neutrinos freely stream through the universe close to the speed of light after their decoupling, they induce metric perturbations ahead of the sound horizon. The baryons and photons experience this change in the gravitational potential and the peak of these perturbations is slightly displaced to a larger radius. In Fourier space, this results in a

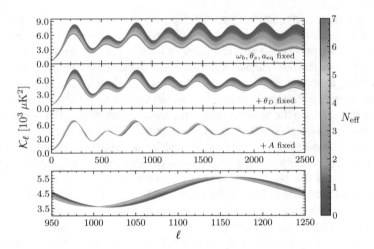

**Fig. 3.4** Variation of the CMB temperature power spectrum as a function of $N_{\text{eff}}$ (from [128]). The spectra have been rescaled, so that the fiducial spectrum for $N_{\text{eff}} = 3.046$ is undamped, $\mathcal{K}_\ell = \mathcal{D}_\ell^{TT} \exp\{a(\ell\theta_d)^\kappa\}$, with $\theta_d \approx 1.6 \times 10^{-3}$ and the fitting parameters $a \approx 0.68$, $\kappa \approx 1.3$, i.e. the exponential diffusion damping was removed. Following [26], the physical baryon density $\omega_b$, the scale factor at matter-radiation equality $a_{\text{eq}} \equiv \omega_m/\omega_r$ and the angular size of the sound horizon $\theta_s$ are held fixed in all panels. The dominant effect in the first panel is the variation of the damping scale $\theta_D$. In the second panel, we fixed $\theta_D$ by adjusting the helium fraction $Y_p$. The dominant variation is now the amplitude perturbation $\delta A$. In the third panel, the spectra are normalized at the fourth peak. The remaining variation is the phase shift $\phi$ (see the zoom-in in the fourth panel)

small change in the temporal phase of the acoustic oscillations which is imprinted in both the temperature and polarization spectra of the CMB as a coherent shift in the peak locations (see Fig. 3.4). A precise determination of the acoustic peaks therefore allows a measurement of free-streaming radiation independent of the Hubble rate. As a consequence, the degeneracy between $N_{\text{eff}}$ and $N_{\text{fluid}}$, as well as between $N_{\text{eff}}$ and $Y_p$ can be broken by this subtle effect. We will study this *phase shift* in the acoustic oscillations and its measurable implications in great detail throughout this thesis.

Since the baryon acoustic oscillations in the matter power spectrum originate from the same mechanism as those in the CMB spectra, the BAO spectrum exhibits the same phase shift induced by the supersonic propagation of free-streaming species, cf. Fig. 3.5. Importantly, it was recently pointed out that this shift should also be robust to non-linear gravitational evolution in the late universe [129]. This means that we can circumvent the usual LSS complications related to non-linearities and use linear perturbation theory to predict and measure this shift in late-time observables such as the BAO spectrum of galaxy clustering. This will be the basis of the modified BAO analysis that we propose in Chap. 6 and will ultimately lead to the first measurement of this imprint of neutrinos in the distribution of galaxies in Chap. 7.

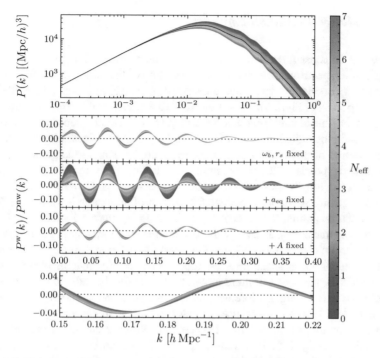

**Fig. 3.5** Variation of the matter power spectrum $P(k)$ (*top*) and the BAO spectrum $P^{\mathrm{w}}(k)/P^{\mathrm{nw}}(k)$ (*bottom*) as a function of $N_{\mathrm{eff}}$ (from [128]). The physical baryon density $\omega_b$ and the physical sound horizon at the drag epoch, $r_s$, are held fixed in all panels of the BAO spectrum. In the second BAO panel, we fixed the scale factor at matter-radiation equality, $a_{\mathrm{eq}} \equiv \omega_m/\omega_r$. Note that we effectively fix $r_s$ by appropriately rescaling the wavenumbers for each spectrum, whereas we can vary $\omega_c$ to fix $a_{\mathrm{eq}}$. The variation in the BAO amplitude $\delta A$ is then the dominant contribution. In the third BAO panel, the spectra are normalized at the fourth peak and the bottom panel shows a zoom-in illustrating the remaining phase shift

The largest impact on the BAO spectrum is actually a change in the sound horizon coming from a difference in the expansion history. By increasing the expansion rate during the radiation era and, hence, reducing the time over which the sound waves can propagate and diffuse, the acoustic scale decreases, $r_s \propto H^{-1}$. This of course appears as a variation in the frequency of the acoustic oscillations that BAO surveys are very sensitive to. The influence of light species on the BAO amplitude should be taken with care when constraining $N_{\mathrm{eff}}$ because it is degenerate with theoretical uncertainties on the non-linear damping.

### 3.4.3  Matter-Radiation Equality

The two main consequences of an enhanced (diminished) radiation density on the matter power spectrum are a change in the location of the turn-over of the spectrum towards larger (smaller) scales and a decrease (increase) in power on small scales (see the top panel of Fig. 3.5). These two effects are related as they are both linked to a change in the time of matter-radiation equality. For fixed matter density, $\omega_m = \text{const}$, an increase in the radiation density leads to a longer epoch of radiation domination. The maximum of the matter power spectrum therefore shifts to larger scales as its location corresponds to the wavenumber which enters the horizon at $a_{\text{eq}}$. At the same time, the amplitude of those modes that crossed the horizon before matter-radiation equality is suppressed since the growth of structures is only logarithmic during radiation domination (compared to linear during matter domination). Larger scales remain unchanged as they evolve deep in the matter-dominated era. Although these effects are clearly visible in the linear matter power spectrum, non-linearities make the matter power spectrum a less robust probe of $N_{\text{eff}}$ than the phase shift.

In the CMB, the power of more modes gets increased by radiation driving if the radiation density is larger. Moreover, a later matter-radiation equality also results in an enhanced ISW effect because the gravitational potentials evolve by a slightly larger amount after recombination. This effect is however sub-dominant, unless we marginalize over $Y_p$ [125]. In this case, the ISW effect contributes some constraining power as the sensitivity of the damping tail to $N_{\text{eff}}$ is greatly reduced.

### 3.4.4  Light Element Abundances

Before the advent of precision cosmology, in particular before the leap in CMB detector sensitivity, the most important model-independent constraints on light relics came from big bang nucleosynthesis. Still today, the measurements of the primordial relic abundance of light elements are an important and independent probe of additional light species. The sensitivity of the predicted abundances from BBN on $N_{\text{eff}}$ is mainly through its impact on the expansion rate between about one second and a few minutes after the big bang. This is due to the fact that a larger expansion rate implies an earlier freeze-out of neutrons which leads to an increase in the neutron-to-proton ratio. As a consequence, more neutrons are available for the synthesis of helium and the other elements which results in an enhancement of the primordial abundances (see e.g. [14, 37, 85, 130–132] for reviews of BBN and its implications on BSM physics).

With the sensitivity of the CMB and BBN to $N_{\text{eff}}$ and $Y_p$ reaching similar levels, there are a number of exciting possibilities to use these measurements in conjunction. First, we can independently check the predictions of BBN by measuring the two BBN input parameters $N_{\text{eff}}$ and $\omega_b$ as well as the output $Y_p$. Moreover, since the CMB and BBN provide snapshots of the universe at different times, we can also compare the measured values and investigate whether the radiation density might have evolved over time, for example due to the decay or presence of additional non-relativistic particles [90, 122]. Having said that, for most parts of this thesis, we will assume that the physics underlying the cosmological observables and BBN are consistent which will result in the tightest possible constraints on the radiation density in the early universe.

# References

1. R. Essig et al., Working group report: new light weakly coupled particles. arXiv:1311.0029 [hep-ph]
2. R. Peccei, H. Quinn, CP conservation in the presence of pseudoparticles. Phys. Rev. Lett. **38**, 1440 (1977)
3. S. Weinberg, A new light boson? Phys. Rev. Lett. **40**, 223 (1978)
4. F. Wilczek, Problem of strong P and T invariance in the presence of instantons. Phys. Rev. Lett. **40**, 279 (1978)
5. A. Arvanitaki, S. Dimopoulos, S. Dubovsky, N. Kaloper, J. March-Russell, String axiverse. Phys. Rev. D **81**, 123530 (2010). arXiv:0905.4720 [hep-th]
6. B. Holdom, Two U(1)'s and epsilon charge shifts. Phys. Lett. B **166**, 196 (1986)
7. P. Galison, A. Manohar, Two Z's or not two Z's? Phys. Lett. B **136**, 279 (1984)
8. K. Abazajian et al., Light sterile neutrinos: a white paper. arXiv:1204.5379 [hep-ph]
9. G. Steigman, M. Turner, Cosmological constraints on the properties of weakly interacting massive particles. Nucl. Phys. B **53**, 375 (1985)
10. W. Hu, J. Silk, Thermalization constraints and spectral distortions for massive unstable relic particles. Phys. Rev. Lett. **70**, 2661 (1993)
11. M. Kawasaki, K. Kohri, N. Sugiyama, Cosmological constraints on late-time entropy production. Phys. Rev. Lett. **82**, 4168 (1999). arXiv:astro-ph/9811437 [astro-ph]
12. N. Padmanabhan, D. Finkbeiner, Detecting dark matter annihilation with CMB polarization: signatures and experimental prospects. Phys. Rev. D **72**, 023508 (2005). arXiv:astro-ph/0503486 [astro-ph]
13. J. Chluba, R. Sunyaev, Pre-recombinational energy release and narrow features in the CMB spectrum. Astron. Astrophys. **501**, 29 (2009). arXiv:0803.3584 [astro-ph]
14. F. Iocco, G. Mangano, G. Miele, O. Pisanti, P. Serpico, Primordial nucleosynthesis: from precision cosmology to fundamental physics. Phys. Rept. **472**, 1 (2009). arXiv:0809.0631 [astro-ph]
15. S. Galli, F. Iocco, G. Bertone, A. Melchiorri, CMB constraints on dark matter models with large annihilation cross section. Phys. Rev. D **80**, 023505 (2009). arXiv:0905.0003 [astro-ph.CO]
16. D. Baumann, D. Green, Signatures of supersymmetry from the early universe. Phys. Rev. D **85**, 103520 (2012). arXiv:1109.0292 [hep-th]
17. J. Chluba, Distinguishing different scenarios of early energy release with spectral distortions of the cosmic microwave background. Mon. Not. R. Astron. Soc. **436**, 2232 (2013). arXiv:1304.6121 [astro-ph.CO]
18. E. Dimastrogiovanni, M. Fasiello, M. Kamionkowski, Imprints of massive primordial fields on large-scale structure. JCAP **02**, 017 (2016). arXiv:1504.05993 [astro-ph.CO]
19. J. Lesgourgues, G. Marques-Tavares, M. Schmaltz, Evidence for dark matter interactions in cosmological precision data? JCAP **02**, 037 (2016). arXiv:1507.04351 [astro-ph.CO]
20. N. Arkani-Hamed, J. Maldacena, Cosmological collider physics. arXiv:1503.08043 [hep-th]
21. R. Flauger, M. Mirbabayi, L. Senatore, E. Silverstein, Productive interactions: heavy particles and non-Gaussianity. JCAP **10**, 058 (2017). arXiv:1606.00513 [hep-th]

22. H. Lee, D. Baumann, G. Pimentel, Non-Gaussianity as a particle detector. JHEP **12**, 040 (2016). arXiv:1607.03735 [hep-th]
23. D. Green, P.D. Meerburg, J. Meyers, Aspects of dark matter annihilation in cosmology. JCAP **04**, 025 (2019). arXiv:1804.01055 [astro-ph.CO]
24. G. Raffelt, *Stars as Laboratories for Fundamental Physics* (University of Chicago Press, Chicago, 1996)
25. P.A.R. Ade et al., (Planck Collaboration), Planck 2015 results. XIII. Cosmological parameters. Astron. Astrophys. **594**, A13 (2016). arXiv:1502.01589 [astro-ph.CO]
26. B. Follin, L. Knox, M. Millea, Z. Pan, First detection of the acoustic oscillation phase shift expected from the cosmic neutrino background. Phys. Rev. Lett. **115**, 091301 (2015). arXiv:1503.07863 [astro-ph.CO]
27. K. Abazajian et al. (CMB-S4 Collaboration), CMB-S4 Science Book, 1st edn., arXiv:1610.02743 [astro-ph.CO]
28. G. Aad et al. (ATLAS Collaboration), Observation of a new particle in the search for the standard model higgs boson with the ATLAS detector at the LHC. Phys. Lett. B **716**, 1 (2012). arXiv:1207.7214 [hep-ex]
29. S. Chatrchyan et al. (CMS Collaboration), Observation of a new boson at a mass of 125 GeV with the CMS experiment at the LHC. Phys. Lett. B **716**, 30 (2012). arXiv:1207.7235 [hep-ex]
30. S. Glashow, Partial symmetries of weak interactions. Nucl. Phys. **22**, 579 (1961)
31. S. Weinberg, A model of leptons. Phys. Rev. Lett. **19**, 1264 (1967)
32. A. Salam, Weak and electromagnetic interactions. Conf. Proc. **C680519**, 367 (1968)
33. P. Higgs, Broken symmetries, massless particles and gauge fields. Phys. Lett. **12**, 132 (1964)
34. F. Englert, R. Brout, Broken symmetry and the mass of gauge vector mesons. Phys. Rev. Lett. **13**, 321 (1964)
35. P. Higgs, Broken symmetries and the masses of gauge bosons. Phys. Rev. Lett. **13**, 508 (1964)
36. G. Guralnik, C. Hagen, T. Kibble, Global conservation laws and massless particles. Phys. Rev. Lett. **13**, 585 (1964)
37. C. Patrignani et al. (Particle Data Group), Review of particle physics. Chin. Phys. C **40**, 100001 (2016)
38. R. Bouchendira, P. Clade, S. Guellati-Khelifa, F. Nez, F. Biraben, New determination of the fine-structure constant and test of quantum electrodynamics. Phys. Rev. Lett. **106**, 080801 (2011). arXiv:1012.3627 [physics.atom-ph]
39. T. Aoyama, M. Hayakawa, T. Kinoshita, M. Nio, Tenth-order QED contribution to the electron $g - 2$ and an improved value of the fine-structure constant. Phys. Rev. Lett. **109**, 111807 (2012). arXiv:1205.5368 [hep-ph]
40. T. Aoyama, M. Hayakawa, T. Kinoshita, M. Nio, Tenth-order electron anomalous magnetic moment: contribution of diagrams without closed lepton loops. Phys. Rev. D **91**, 033006 (2015) [Erratum: Phys. Rev. D **96**, 019901 (2017)]. arXiv:1412.8284 [hep-ph]
41. J. Jaeckel, A. Ringwald, The low-energy frontier of particle physics. Ann. Rev. Nucl. Part. Sci. **60**, 405 (2010). arXiv:1002.0329 [hep-ph]
42. A. de Gouvea et al., Working group report: neutrinos. arXiv:1310.4340 [hep-ex]
43. J. Alexander et al., Dark sectors 2016 workshop: community report. arXiv:1608.08632 [hep-ph]
44. C. Baker et al., Improved experimental limit on the electric dipole moment of the neutron. Phys. Rev. Lett. **97**, 131801 (2006). arXiv:hep-ex/0602020 [hep-ex]
45. J. Pendlebury et al., Revised experimental upper limit on the electric dipole moment of the neutron. Phys. Rev. D **92**, 092003 (2015). arXiv:1509.04411 [hep-ex]
46. J. Preskill, M. Wise, F. Wilczek, Cosmology of the invisible axion. Phys. Lett. B **120**, 127 (1983)
47. L. Abbott, P. Sikivie, A cosmological bound on the invisible axion. Phys. Lett. B **120**, 133 (1983)
48. M. Dine, W. Fischler, The not so harmless axion. Phys. Lett. B **120**, 137 (1983)
49. D.J.E. Marsh, Axion cosmology. Phys. Rep. **643**, 1 (2016). arXiv:1510.07633 [astro-ph.CO]

50. M. Turner, Early-universe thermal production of not-so-invisible axions. Phys. Rev. Lett. **59**, 2489 (1987) [Erratum: Phys. Rev. Lett. **60**, 1101 (1988)]

51. E. Braaten, T. Yuan, Calculation of screening in a hot plasma. Phys. Rev. Lett. **66**, 2183 (1991)

52. M. Bolz, A. Brandenburg, W. Buchmüller, Thermal production of gravitinos. Nucl. Phys. B **606**, 518 (2001) [Erratum: Nucl. Phys. B **790**, 336 (2008)]. arXiv:hep-ph/0012052 [hep-ph]

53. E. Masso, F. Rota, G. Zsembinszki, On axion thermalization in the early universe. Phys. Rev. D **66**, 023004 (2002). arXiv:hep-ph/0203221 [hep-ph]

54. P. Graf, F. Steffen, Thermal axion production in the primordial quark-gluon plasma. Phys. Rev. D **83**, 075011 (2011). arXiv:1008.4528 [hep-ph]

55. A. Salvio, A. Strumia, W. Xue, Thermal axion production. JCAP **01**, 011 (2014). arXiv:1310.6982 [hep-ph]

56. R. Mohapatra et al., Theory of neutrinos: a white paper. Rep. Prog. Phys. **70**, 1757 (2007). arXiv:hep-ph/0510213 [hep-ph]

57. S. King, C. Luhn, Neutrino mass and mixing with discrete symmetry. Rep. Prog. Phys. **76**, 056201 (2013). arXiv:1301.1340 [hep-ph]

58. S. King, Models of neutrino mass, mixing and CP violation. J. Phys. G **42**, 123001 (2015). arXiv:1510.02091 [hep-ph]

59. F. Jegerlehner, A. Nyffeler, The muon g − 2. Phys. Rep. **477**, 1 (2009). arXiv:0902.3360 [hep-ph]

60. N. Arkani-Hamed, D. Finkbeiner, T. Slatyer, N. Weiner, A theory of dark matter. Phys. Rev. D **79**, 015014 (2009). arXiv:0810.0713 [hep-ph]

61. O. Adriani et al. (PAMELA Collaboration), An anomalous positron abundance in cosmic rays with energies 1.5–100 GeV. Nature **458**, 607 (2009). arXiv:0810.4995 [astro-ph]

62. D.B. Kaplan, H. Georgi, SU(2)×U(1) breaking by vacuum misalignment. Phys. Lett. B **136**, 183 (1984)

63. D.B. Kaplan, H. Georgi, S. Dimopoulos, Composite Higgs scalars. Phys. Lett. B **136**, 187 (1984)

64. K. Agashe, R. Contino, A. Pomarol, The minimal composite Higgs model. Nucl. Phys. B **719**, 165 (2005). arXiv:hep-ph/0412089 [hep-ph]

65. N. Arkani-Hamed, S. Dimopoulos, G. Dvali, The hierarchy problem and new dimensions at a millimeter. Phys. Lett. B **429**, 263 (1998). arXiv:hep-ph/9803315 [hep-ph]

66. I. Antoniadis, N. Arkani-Hamed, S. Dimopoulos, G. Dvali, New dimensions at a millimeter to a fermi and superstrings at a TeV. Phys. Lett. B **436**, 257 (1998). arXiv:hep-ph/9804398 [hep-ph]

67. L. Randall, R. Sundrum, A large mass hierarchy from a small extra dimension. Phys. Rev. Lett. **83**, 3370 (1999). arXiv:hep-ph/9905221 [hep-ph]

68. P. Graham, D.E. Kaplan, S. Rajendran, Cosmological relaxation of the electroweak scale. Phys. Rev. Lett. **115**, 221801 (2015). arXiv:1504.07551 [hep-ph]

69. N. Arkani-Hamed, T. Cohen, R. D'Agnolo, A. Hook, H. Kim, D. Pinner, Solving the hierarchy problem at reheating with a large number of degrees of freedom. Phys. Rev. Lett. **117**, 251801 (2016). arXiv:1607.06821 [hep-ph]

70. P. Svrcek, E. Witten, Axions in string theory. JHEP **06**, 051 (2006). arXiv:hep-th/0605206 [hep-th]

71. D. Baumann, L. McAllister, *Inflation and String Theory* (Cambridge University Press, Cambridge, 2015). arXiv:1404.2601 [hep-th]

72. C. Brust, D.E. Kaplan, M. Walters, New light species and the CMB. JHEP **12**, 058 (2013). arXiv:1303.5379 [hep-ph]

73. N. Arkani-Hamed, T.-C. Huang, Y.-T. Huang, Scattering amplitudes for all masses and spins. arXiv:1709.04891 [hep-th]

74. F. Wilczek, Axions and family symmetry breaking. Phys. Rev. Lett. **49**, 1549 (1982)

75. D. Reiss, Can the family group be a global symmetry? Phys. Lett. B **115**, 217 (1982)

76. J. Kim, Light pseudoscalars, particle physics and cosmology. Phys. Rep. **150**, 1 (1987)

77. Y. Chikashige, R. Mohapatra, R. Peccei, Are there real Goldstone bosons associated with broken lepton number? Phys. Lett. B **98**, 265 (1981)

78. Y. Chikashige, R. Mohapatra, R. Peccei, Spontaneously broken lepton number and cosmological constraints on the neutrino mass spectrum. Phys. Rev. Lett. **45**, 1926 (1980)
79. L.A. Anchordoqui, H. Goldberg, G. Steigman, Right-handed neutrinos as the dark radiation: status and forecasts for the LHC. Phys. Lett. B **718**, 1162 (2013). arXiv:1211.0186 [hep-ph]
80. K. Abazajian, Sterile neutrinos in cosmology. Phys. Rep. **711**, 1 (2017). arXiv:1705.01837 [hep-ph]
81. A. Dolgov, Neutrinos in cosmology. Phys. Rept. **370**, 333 (2002). arXiv:hep-ph/0202122 [hep-ph]
82. J. Lesgourgues, G. Mangano, G. Miele, S. Pastor, *Neutrino Cosmology* (Cambridge University Press, Cambridge, 2013)
83. G. Mangano, G. Miele, S. Pastor, T. Pinto, O. Pisanti, P. Serpico, Relic neutrino decoupling including flavour oscillations. Nucl. Phys. B **729**, 221 (2005). arXiv:hepph/0506164 [hep-ph]
84. P. de Salas, S. Pastor, Relic neutrino decoupling with flavour oscillations revisited. JCAP **07**, 051 (2016). arXiv:1606.06986 [hep-ph]
85. R. Cyburt, B. Fields, K. Olive, T.-H. Yeh, Big bang nucleosynthesis: 2015. Rev. Mod. Phys. **88**, 015004 (2016). arXiv:1505.01076 [astro-ph.CO]
86. S. Schael et al., Precision electroweak measurements on the Z resonance. Phys. Rept. **427**, 257 (2006). arXiv:hep-ex/0509008 [hep-ex]
87. G. Jungman, M. Kamionkowski, A. Kosowsky, D. Spergel, Cosmological parameter determination with microwave background maps. Phys. Rev. D **54**, 1332 (1996). arXiv:astro-ph/9512139 [astro-ph]
88. D. Cadamuro, S. Hannestad, G. Raffelt, J. Redondo, Cosmological bounds on sub-MeV mass axions. JCAP **02**, 003 (2011). arXiv:1011.3694 [hep-ph]
89. J. Menestrina, R. Scherrer, Dark radiation from particle decays during big bang nucleosynthesis. Phys. Rev. D **85**, 047301 (2012). arXiv:1111.0605 [astro-ph.CO]
90. C. Boehm, M. Dolan, C. McCabe, Increasing $N_{eff}$ with particles in thermal equilibrium with neutrinos. JCAP **12**, 027 (2012). arXiv:1207.0497 [astro-ph.CO]
91. S. Weinberg, Goldstone bosons as fractional cosmic neutrinos. Phys. Rev. Lett. **110**, 241301 (2013). arXiv:1305.1971 [astro-ph.CO]
92. F.-Y. Cyr-Racine, R. de Putter, A. Raccanelli, K. Sigurdson, Constraints on large-scale dark acoustic oscillations from cosmology. Phys. Rev. D **89**, 063517 (2014). arXiv:1310.3278 [astro-ph.CO]
93. H. Vogel, J. Redondo, Dark radiation constraints on minicharged particles in models with a hidden photon. JCAP **02**, 029 (2014). arXiv:1311.2600 [hep-ph]
94. M. Millea, L. Knox, B. Fields, New bounds for axions and axion-like particles with keV-GeV masses. Phys. Rev. D **92**, 023010 (2015). arXiv:1501.04097 [astro-ph.CO]
95. Z. Chacko, Y. Cui, S. Hong, T. Okui, Hidden dark matter sector, dark radiation and the CMB. Phys. Rev. D **92**, 055033 (2015). arXiv:1505.04192 [hep-ph]
96. D. Baumann, D. Green, J. Meyers, B. Wallisch, Phases of new physics in the CMB. JCAP **01**, 007 (2016). arXiv:1508.06342 [astro-ph.CO]
97. Adapted from [96] with permission of IOP Publishing. Copyright by Sissa Medialab srl. All rights reserved
98. W. Hu, Structure formation with generalized dark matter. Astrophys. J. **506**, 485 (1998). arXiv:astro-ph/9801234 [astro-ph]
99. M. Archidiacono, E. Giusarma, A. Melchiorri, O. Mena, Neutrino and dark radiation properties in light of recent CMB observations. Phys. Rev. D **87**, 103519 (2013). arXiv:1303.0143 [astro-ph.CO]
100. B. Audren et al., Robustness of cosmic neutrino background detection in the cosmic microwave background. JCAP **03**, 036 (2015). arXiv:1412.5948 [astro-ph.CO]
101. E. Sellentin, R. Durrer, Detecting the cosmological neutrino background in the CMB. Phys. Rev. D **92**, 063012 (2015). arXiv:1412.6427 [astro-ph.CO]
102. N. Bell, E. Pierpaoli, K. Sigurdson, Cosmological signatures of interacting neutrinos. Phys. Rev. D **73**, 063523 (2006). arXiv:astro-ph/0511410 [astro-ph]

103. A. Friedland, K. Zurek, S. Bashinsky, Constraining models of neutrino mass and neutrino interactions with the Planck satellite. arXiv:0704.3271 [astro-ph]
104. C. Brust, Y. Cui, K. Sigurdson, Cosmological constraints on interacting light particles. JCAP **08**, 020 (2017). arXiv:1703.10732 [astro-ph.CO]
105. D. Alves et al. (LHC New Physics Working Group), Simplified models for LHC new physics searches. J. Phys. G **39**, 105005 (2012). arXiv:1105.2838 [hep-ph]
106. E. Kolb, M. Turner, *The Early Universe* (Addison-Wesley, Reading, 1990)
107. S. Hannestad, What is the lowest possible reheating temperature? Phys. Rev. D **70**, 043506 (2004). arXiv:astro-ph/0403291 [astro-ph]
108. P. Graham, I. Irastorza, S. Lamoreaux, A. Lindner, K. van Bibber, Experimental searches for the axion and axion-like particles. Ann. Rev. Nucl. Part. Sci. **65**, 485 (2015). arXiv:1602.00039 [hep-ex]
109. G. Raffelt, Neutrinos and the stars. Proc. Int. Sch. Phys. Fermi **182**, 61 (2012). arXiv:1201.1637 [astro-ph.SR]
110. I. Irastorza, J. Redondo, New experimental approaches in the search for axion-like particles. Prog. Part. Nucl. Phys. **102**, 89 (2018). arXiv:1801.08127 [hep-ph]
111. A. Arvanitaki, S. Dubovsky, Exploring the string axiverse with precision black hole physics. Phys. Rev. D **83**, 044026 (2011). arXiv:1004.3558 [hep-th]
112. H. Yoshino, H. Kodama, Gravitational radiation from an axion cloud around a black hole: superradiant phase. Progr. Theor. Exp. Phys. **2014**, 043E02 (2014). arXiv:1312.2326 [gr-qc]
113. R. Brito, V. Cardoso, P. Pani, Black holes as particle detectors: evolution of superradiant instabilities. Class. Quant. Grav. **32**, 134001 (2015). arXiv:1411.0686 [gr-qc]
114. R. Brito, S. Ghosh, E. Barausse, E. Berti, V. Cardoso, I. Dvorkin, A. Klein, P. Pani, Gravitational wave searches for ultralight bosons with LIGO and LISA. Phys. Rev. D **96**, 064050 (2017). arXiv:1706.06311 [gr-qc]
115. D. Baumann, H.S. Chia, R. Porto, Probing ultralight bosons with binary black holes. Phys. Rev. D **99**, 044001 (2018). arXiv:1804.03208 [gr-qc]
116. A. Ayala, I. Domínguez, M. Giannotti, A. Mirizzi, O. Straniero, Revisiting the bound on axion-photon coupling from globular clusters. Phys. Rev. Lett. **113**, 191302 (2014). arXiv:1406.6053 [astro-ph.SR]
117. M. Buen-Abad, G. Marques-Tavares, M. Schmaltz, Non-Abelian dark matter and dark radiation. Phys. Rev. D **92**, 023531 (2015). arXiv:1505.03542 [hep-ph]
118. T. Smith, E. Pierpaoli, M. Kamionkowski, A new cosmic microwave background constraint to primordial gravitational waves. Phys. Rev. Lett. **97**, 021301 (2006). arXiv:astro-ph/0603144 [astro-ph]
119. L. Boyle, A. Buonanno, Relating gravitational wave constraints from primordial nucleosynthesis, pulsar timing, laser interferometers and the CMB: implications for the early universe. Phys. Rev. D **78**, 043531 (2008). arXiv:0708.2279 [astro-ph]
120. P. D. Meerburg, R. Hlo žek, B. Hadzhiyska, J. Meyers, Multiwavelength constraints on the inflationary consistency relation. Phys. Rev. D **91**, 103505 (2015). arXiv:1502.00302 [astro-ph.CO]
121. P. Lasky et al., Gravitational-wave cosmology across 29 decades in frequency. Phys. Rev. X **6**, 011035 (2016). arXiv:1511.05994 [astro-ph.CO]
122. W. Fischler, J. Meyers, Dark radiation emerging after big bang nucleosynthesis? Phys. Rev. D **83**, 063520 (2011). arXiv:1011.3501 [astro-ph.CO]
123. D. Hooper, F. Queiroz, N. Gnedin, Non-thermal dark matter mimicking an additional neutrino species in the early universe. Phys. Rev. D **85**, 063513 (2012). arXiv:1111.6599 [astro-ph.CO]
124. J. Hasenkamp, J. Kersten, Dark radiation from particle decay: cosmological constraints and opportunities. JCAP **08**, 024 (2013). arXiv:1212.4160 [hep-ph]
125. Z. Hou, R. Keisler, L. Knox, M. Millea, C. Reichardt, How massless neutrinos affect the cosmic microwave background damping tail. Phys. Rev. D **87**, 083008 (2013). arXiv:1104.2333 [astro-ph.CO]
126. R. Keisler et al., A measurement of the damping tail of the cosmic microwave background power spectrum with the South Pole Telescope. Astrophys. J. **743**, 28 (2011). arXiv:1105.3182 [astro-ph.CO]

127. S. Bashinsky, U. Seljak, Neutrino perturbations in CMB anisotropy and matter clustering. Phys. Rev. D **69**, 083002 (2004). arXiv:astro-ph/0310198 [astro-ph]
128. D. Baumann, D. Green, B. Wallisch, Searching for light relics with large-scale structure. JCAP **08**, 029 (2018). arXiv:1712.08067 [astro-ph.CO]
129. D. Baumann, D. Green, M. Zaldarriaga, Phases of new physics in the BAO spectrum. JCAP **11**, 007 (2017). arXiv:1703.00894 [astro-ph.CO]
130. S. Sarkar, Big bang nucleosynthesis and physics beyond the Standard Model. Rep. Prog. Phys. **59**, 1493 (1996). arXiv:hep-ph/9602260 [hep-ph]
131. M. Pospelov, J. Pradler, Big bang nucleosynthesis as a probe of new physics. Ann. Rev. Nucl. Part. Sci. **60**, 539 (2010). arXiv:1011.1054 [hep-ph]
132. G. Steigman, Neutrinos and big bang nucleosynthesis. Adv. High Energy Phys. **2012**, 268321 (2012). arXiv:1208.0032 [hep-ph]

# Chapter 4
# New Target for Cosmic Axion Searches

We have seen that many interesting extensions of the Standard Model of particle physics predict new light species. These particles have to be more weakly coupled than neutrinos, but may have been in thermal equilibrium in the early universe. In addition, we established that cosmological observations are, in principle, very sensitive to these types of particles. Given the Moore's law-like improvements in CMB detector sensitivity [1, 2], cosmology will push the sensitivity to new light particles beyond the strength of weak-scale interactions and has the potential to explore a fundamentally new territory of BSM physics. Specifically, observations may become sensitive to any light relics that have ever been in thermal equilibrium with the Standard Model. In this chapter, we demonstrate that even the absence of a detection would be informative since it would allow us to put constraints on the Standard Model interactions of light BSM particles, such as axions. In many cases, the constraints achievable from cosmology will surpass existing bounds from laboratory experiments and astrophysical observations by orders of magnitude.

The outline of this chapter, which is based on [3, 4], is as follows. In Sect. 4.1, we lay out the general philosophy underlying these new constraints on light relics. In principle, this is applicable to all effective field theories introduced in Sect. 3.2.3. As an example, we then focus on the EFT of pseudo-Nambu-Goldstone bosons and consider the couplings to gauge bosons, to fermions and to neutrinos in turn. In Sect. 4.2, we study the axion interactions focussing on the photon and gluon couplings, which are of particular phenomenological interest. In Sect. 4.3, we derive two types of bounds on familon interactions. Apart from constraints arising from the possibility that these pNGBs could have frozen out in the early universe, we also consider the case in which these particles might have frozen in, i.e. came back into thermal contact at later times. In Sect. 4.4, we finally discuss the same two classes of constraints for majorons. We conclude this chapter in Sect. 4.5 with a summary of the derived constraints. Some technical details underlying these bounds are relegated to Appendix A.

© Springer Nature Switzerland AG 2019
B. Wallisch, *Cosmological Probes of Light Relics*, Springer Theses,
https://doi.org/10.1007/978-3-030-31098-1_4

## 4.1  Novel Constraints on Light Relics

In Sect. 3.3.2, we derived the minimal contribution of any light thermal relic to the radiation density of the universe to be $\Delta N_{\text{eff}} = 0.027$. The fact that this contribution may be detectable in future cosmological observations has interesting consequences. First, the level $\Delta N_{\text{eff}} = 0.027$ provides a natural observational target (see e.g. [5–9] for related discussions). Second, even the absence of a detection would be very informative, because it would strongly constrain the EFT couplings between the extra light relics and the SM degrees of freedom, which can schematically be written as

$$\mathcal{L} \supset \sum \frac{\mathcal{O}_\phi \mathcal{O}_{\text{SM}}}{\Lambda^\Delta} . \tag{4.1}$$

These new bounds arise since a thermal abundance can be avoided[1] if the reheating temperature of the universe, $T_R$, is below the would-be freeze-out temperature, i.e. $T_R < T_F$. In that case, the extra particles have never been in thermal equilibrium and their densities therefore do not have to be detectable. In the absence of a detection, requiring $T_F(\Lambda) > T_R$ would place very strong bounds on the scale(s) in (4.1), i.e. $\Lambda > T_F^{-1}(T_R)$. As we will see, in many cases the cosmological bounds will be much stronger than existing bounds from laboratory experiments and astrophysical observations. We note that these constraints make no assumption about the nature of dark matter because the thermal population of axions arises independently of a possible cold population. On the other hand, we have to assume that the effective description of the pNGBs with interactions of the form of (4.1) holds up to $T_F \ll \Lambda$. This is equivalent to assuming that the ultraviolet completion of the effective theory is not too weakly coupled. Moreover, we also require the absence of any significant dilution of $\Delta N_{\text{eff}}$ after freeze-out. In practice, this means that we are restricting to scenarios in which the number of additional relativistic degrees of freedom at the freeze-out temperature is bounded by $\Delta g_*(T_F) \lesssim g_*^{\text{SM}}(T_F) \approx 10^2$ (cf. our discussion in Sect. 3.3.2).

The couplings of pNGBs to SM fermions $\psi$ can lead to a more complicated thermal evolution than the simple freeze-out scenario. Below the scale of electroweak symmetry breaking, the approximate chiral symmetry of the fermions makes the interactions with the pNGBs effectively marginal. The temperature dependence of the interaction rate is then weaker than that of the Hubble expansion rate, leading to a recoupling (i.e. freeze-in) of the pNGBs at low temperatures. To avoid a large density of pNGBs requires that the freeze-in temperature $T_{\tilde{F}}$ is smaller than the mass of the fermions participating in the interactions, $T_{\tilde{F}} < m_\psi$, so that the interaction rate becomes Boltzmann suppressed before freeze-in can occur. Again, this constraint can

---

[1] We remind ourselves that a thermal abundance may be diluted below the level of Fig. 3.2 if extra massive particles are added to the SM. As we have discussed in Sect. 3.3.2, however, a significant change to our conclusions would require a very large number of new particles or a significant amount of non-equilibrium photon production. In addition, the possibility that dark sectors never reach thermal equilibrium with the SM (see e.g. [10–14]) is strongly constrained by the physics of reheating [15].

be expressed as a bound on the scale(s) $\Lambda$ that couple the pNGBs to the SM fermions. Although the freeze-in constraints are usually weaker than the freeze-out constraints, they have the advantage that they do not make any assumptions about the reheating temperature (as long as reheating occurs above $T \sim m_\psi$). Furthermore, freeze-in produces larger contributions to $\Delta N_{\text{eff}}$ which are detectable with a less sensitive experiment.

In the rest of this chapter, we will show that cosmology is highly sensitive to axions, and other pNGBs, when $\Delta N_{\text{eff}} = 0.027$ is detectable. To simplify the narrative, we will assume that this sensitivity will be reached with CMB-S4, either on its own or in conjunction with other data (cf. Chaps. 5 and 6; see also [16, 17]). Alternatively, our arguments could be viewed as strong motivation for reaching this critical level of sensitivity in future experiments. In the following, we will derive bounds on the couplings of pNGBs to the SM arising from the absence of a detection. We will assume the mass range $0 \leq m_\phi < 1\,\text{MeV}$, so that the only possible decays of the pNGBs are to photons or neutrinos. This regime is probed by measurements of $N_{\text{eff}}$ for $m_\phi \leq T_{\text{rec}}$ and by warm dark matter constraints for $m_\phi > T_{\text{rec}}$ (see e.g. [18, 19]), where $T_{\text{rec}} \approx 0.26\,\text{eV}$ is the temperature at recombination.

## 4.2  Constraints on Axions

Axions arise naturally in many areas of high-energy physics, the QCD axion being a particularly well-motivated example. Besides providing a solution to the strong CP problem [20–22], the QCD axion also serves as a natural dark matter candidate [23–25]. Moreover, light axions appear prolifically in string theory [26–28] and have been proposed to explain the small mass of the inflation [29] as well as to solve the hierarchy problem [30]. Finally, axions are a compelling example of a new particle that is experimentally elusive [31, 32] because of its weak coupling rather than due to kinematic constraints.

What typically distinguishes axions from other pNGBs are their unique couplings to the SM gauge fields. Prior to EWSB, we consider the following effective theory with shift-symmetric couplings of the axion to the SM gauge sector:

$$\mathcal{L}_{\phi EW} = -\frac{1}{4}\frac{\phi}{\Lambda}\left(c_1\,B_{\mu\nu}\tilde{B}^{\mu\nu} + c_2\,W^a_{\mu\nu}\tilde{W}^{\mu\nu,a} + c_3\,G^a_{\mu\nu}\tilde{G}^{\mu\nu,a}\right), \qquad (4.2)$$

where $X_{\mu\nu} \equiv \{B_{\mu\nu}, W^a_{\mu\nu}, G^a_{\mu\nu}\}$ are the field strengths associated with the gauge groups $\{U(1)_Y, SU(2)_L, SU(3)_C\}$, and $\tilde{X}^{\mu\nu} \equiv \frac{1}{2}\epsilon^{\mu\nu\rho\sigma}X_{\rho\sigma}$ are their duals. Axion models will typically include couplings to all SM gauge fields, but only the coupling to gluons is strictly necessary to solve the strong CP problem.

At high energies, the rate of axion production through the gauge field interactions in (4.2) can be expressed as [6] (see also [33–36])

$$\Gamma(T, \Lambda_n) = \sum_{n=1}^{3} \gamma_n(T) \frac{T^3}{\Lambda_n^2}, \qquad (4.3)$$

where $\Lambda_n \equiv \Lambda/c_n$. The prefactors $\gamma_n(T)$ have their origin in the running of the couplings and are only weakly dependent on temperature. For simplicity of presentation, we will treat these functions as constants in the main text, but take them into account in Appendix A. We see that the production rate, $\Gamma \propto T^3$, decreases faster than the expansion rate during the radiation era, $H \propto T^2$. The axions therefore freeze out when the production rate drops below the expansion rate, with the freeze-out temperature $T_F$ determined by $\Gamma(T_F) = H(T_F)$. This thermal abundance can be avoided if the reheating temperature of the universe $T_R$ was below the would-be freeze-out temperature, i.e. $T_R < T_F$. In that case, the temperature of the universe was simply never high enough to bring the axions into thermal equilibrium. We can express this condition as

$$\Gamma(T_R, \Lambda_n) < H(T_R) = \frac{\pi}{\sqrt{90}} \sqrt{g_{*,R}} \frac{T_R^2}{M_{\mathrm{pl}}}, \qquad (4.4)$$

where $g_{*,R} \equiv g_*(T_R)$ denotes the effective number of relativistic species at $T_R$. For a given reheating temperature, this is a constraint on the couplings $\Lambda_n$ in (4.3). Treating the different axion couplings separately, we can write

$$\Lambda_n > \left(\frac{\pi^2}{90} g_{*,R}\right)^{-1/4} \sqrt{\gamma_{n,R} T_R M_{\mathrm{pl}}}, \qquad (4.5)$$

where $\gamma_{n,R} \equiv \gamma_n(T_R)$. In the following, we will evaluate these bounds for the couplings to photons (Sect. 4.2.1) and gluons (Sect. 4.2.2), and compare them to existing laboratory and astrophysical constraints.

### 4.2.1  Coupling to Photons

The operator that has been most actively investigated experimentally is the coupling to photons,

$$\mathcal{L}_{\phi EW} \supset \mathcal{L}_{\phi\gamma} = -\frac{1}{4} \frac{\phi}{\Lambda_\gamma} F_{\mu\nu} \tilde{F}^{\mu\nu}. \qquad (4.6)$$

The photon coupling $\Lambda_\gamma$ is related to the electroweak couplings $\Lambda_1$ and $\Lambda_2$ via $\Lambda_\gamma^{-1} = \cos^2\theta_w \Lambda_1^{-1} + \sin^2\theta_w \Lambda_2^{-1}$, where $\theta_w \approx 30°$ is Weinberg's mixing angle. Photons are easily produced in large numbers in both the laboratory and in many astrophysical settings which makes this coupling a particularly fruitful target for axion searches.

In Appendix A, we show in detail how the constraints (4.5) on the couplings to the electroweak gauge bosons map into a constraint on the coupling to photons. This constraint is a function of the relative size of the couplings to the $SU(2)_L$ and

$U(1)_Y$ sectors, as measured by the ratio $c_2/c_1$ in (4.2). To be conservative, we will here present the weakest constraint which arises for $c_2 = 0$ when the axion only couples to the $U(1)_Y$ gauge field. A specific axion model is likely to also couple to the $SU(2)_L$ sector, i.e. have $c_2 \neq 0$, and the constraint on $\Lambda_\gamma$ would then be stronger (as can be seen explicitly in Appendix A). Using $\gamma_{1,R} \approx \gamma_1(10^{10}\,\text{GeV}) = 0.017$ and $g_{*,R} = 106.75 + 1$, we find

$$\Lambda_\gamma > 1.4 \times 10^{13}\,\text{GeV} \left( \frac{T_R}{10^{10}\,\text{GeV}} \right)^{1/2} . \qquad (4.7)$$

For a reheating temperature of about $10^{10}\,\text{GeV}$, the bound in (4.7) is three orders of magnitude stronger than the best current constraints (cf. Fig. 4.1). Even for a reheating temperature as low as $10^4\,\text{GeV}$ the bound from the CMB would still marginally improve over existing constraints.

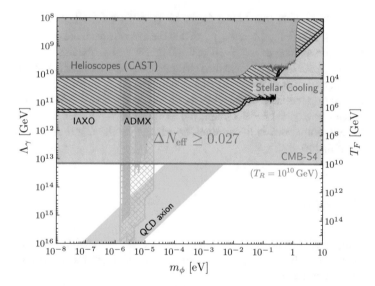

**Fig. 4.1** Comparison between current constraints on the axion-photon coupling and the sensitivity of a future cosmological survey (after [3, 4, 37]). Future laboratory constraints (IAXO and ADMX) are shown as shaded regions. The yellow band indicates a range of representative models for the QCD axion (not assuming that it provides all of the dark matter). The future CMB bound is a function of the reheating temperature $T_R$ and the displayed constraint conservatively assumes that the photon coupling derives only from the coupling $U(1)_Y$ above the electroweak scale. Specific axion models typically also involve a coupling to $SU(2)_L$ in which case the bound would strengthen by an order of magnitude or more (see Appendix A). We note that ADMX assumes that the axion is all of the dark matter, while all other constraints do not have this restriction

Massive axions are unstable to decay mediated by the operator $\phi F \tilde{F}$. However, for the range of parameters of interest, these decays occur after recombination and, hence, do not affect the CMB. To see this, we consider the decay rate for $m_\phi \gtrsim T$ [38],

$$\Gamma_{D,\gamma} = \frac{1}{64\pi} \frac{m_\phi^3}{\Lambda_\gamma^2}. \tag{4.8}$$

The decay time is $\tau_D = \Gamma_{D,\gamma}^{-1}$ and the temperature at decay is determined by $H(T_D) \approx \tau_D^{-1} = \Gamma_{D,\gamma}$. We will not consider the regime $m_\phi < T_D$ as it does not arise in the range of parameters of interest. Assuming that the universe is matter dominated at the time of the decay, we get

$$\frac{T_D}{T_{\text{rec}}} \approx 9.5 \times 10^{-10} \left(\frac{\Lambda_\gamma}{10^{10}\,\text{GeV}}\right)^{-4/3} \left(\frac{m_\phi}{T_{\text{rec}}}\right)^2. \tag{4.9}$$

Using the stellar cooling constraint, $\Lambda_\gamma > 1.3 \times 10^{10}\,\text{GeV}$ [39], we therefore infer that $T_D < 7.1 \times 10^{-10}\,T_{\text{rec}}\,(m_\phi/T_{\text{rec}})^2$, so that the axions are stable on the time-scale of recombination as long as $m_\phi \lesssim 10\,\text{keV}$. CMB-S4 will probe this regime through sensitivity to $N_{\text{eff}}$ for $m_\phi \lesssim T_{\text{rec}}$ and through sensitivity to warm dark matter for larger masses. Warm dark matter is already highly constrained by cosmology, with current CMB data limiting the mass of the QCD axion to $m_\phi < 0.53\,\text{eV}$ (95 % c.l.) [19]. The regime $10\,\text{keV} < m_\phi < 1\,\text{MeV}$ (where the axion decays between neutrino decoupling and recombination) is constrained by effects on the CMB and on big bang nucleosynthesis [40–42].

### 4.2.2   Coupling to Gluons

The coupling to gluons is especially interesting for the QCD axion since it has to be present in order to solve the strong CP problem. The axion production rate associated with the interaction $\phi G \tilde{G}$ is [6]

$$\Gamma_g \simeq 0.41 \frac{T^3}{\Lambda_g^2}, \tag{4.10}$$

where $\Lambda_g \equiv \Lambda/c_3$. As before, we have dropped a weakly temperature-dependent prefactor, but account for it in Appendix A. The bound (4.5) then implies

$$\Lambda_g > 5.4 \times 10^{13}\,\text{GeV} \left(\frac{T_R}{10^{10}\,\text{GeV}}\right)^{1/2}. \tag{4.11}$$

Laboratory constraints on the axion-gluon coupling are usually phrased in terms of the induced electric dipole moment (EDM) of nucleons: $d_n = g_d \phi_0$, where $\phi_0$ is the

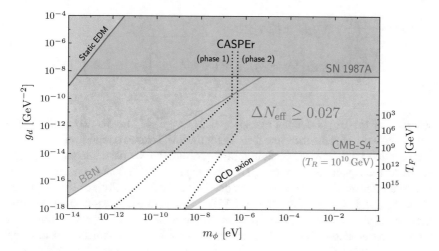

**Fig. 4.2** Comparison between current constraints on the axion-gluon coupling and the sensitivity of a future CMB-S4 experiment (after [3, 44–46]). The dotted lines are the projected sensitivities of the NMR experiment CASPEr [47]. We note that CASPEr, the static EDM [44] and BBN constraints [45] assume that the axion is all of the dark matter, while SN 1987A [48] and the future CMB constraint do not have this restriction

value of the local axion field. The coupling $g_d$ is given for the QCD axion with an uncertainty of about 40% by [43, 44]

$$ g_d \approx \frac{2\pi}{\alpha_s} \times \frac{3.8 \times 10^{-3} \text{GeV}^{-1}}{\Lambda_g} < 1.3 \times 10^{-14} \text{GeV}^{-2} \left( \frac{T_R}{10^{10} \text{GeV}} \right)^{-1/2} . \quad (4.12) $$

Constraints on $g_d$ (and hence $\Lambda_g$) are shown in Fig. 4.2. We see that future cosmological observations will improve over existing constraints on $\Lambda_g$ by up to six orders of magnitude if $T_R = \mathcal{O}(10^{10} \text{ GeV})$. Even if the reheating temperature is as low as $10^4$ GeV, the future CMB constraints will be tighter by three orders of magnitude. In Fig. 4.2, we also show the projected sensitivities of the proposed EDM experiment CASPEr [47]. We see that CASPEr and CMB-S4 probe complementary ranges of axion masses. It should be noted that CASPEr is only sensitive to axion dark matter, while the CMB constrains a separate thermal population of axions which does not require assumptions about the dark matter.

## 4.3 Constraints on Familons

Spontaneously broken global symmetries have also been envoked to explain the approximate $U(3)^5$ flavour symmetry of the Standard Model. As we discussed in Sect. 3.2.3, the associated pNGBs—called *familons* [49–51]—couple to the SM through Yukawa couplings,

$$\mathcal{L}_{\phi\psi} = -\frac{\partial_\mu \phi}{\Lambda_\psi}\, \bar{\psi}_i \gamma^\mu \big(g_V^{ij} + g_A^{ij}\gamma^5\big)\psi_j$$

$$\rightarrow \frac{\phi}{\Lambda_\psi}\bigg(\mathrm{i}H\,\bar{\psi}_{L,i}\Big[(\lambda_i - \lambda_j)g_V^{ij} + (\lambda_i + \lambda_j)g_A^{ij}\Big]\psi_{R,j} + \mathrm{h.c.}\bigg) + \mathcal{O}(\phi^2)\,,$$

(4.13)

where $H$ is the Higgs doublet and $\psi_{L,R} \equiv \frac{1}{2}(1 \mp \gamma^5)\psi$. The $SU(2)_L$ and $SU(3)_C$ structures in (4.13) take the same form as for the SM Yukawa couplings [38], but this has been left implicit to avoid clutter. In the second line we have integrated by parts and used the equations of motion. The subscripts $V$ and $A$ denote the couplings to the vector and axial-vector currents, respectively, and $\lambda_i \equiv \sqrt{2}m_i/v$ are the Yukawa couplings, with $v = 246\,\mathrm{GeV}$ being the Higgs vacuum expectation value. We note that the diagonal couplings, $i = j$, are only to the axial part, as expected from vector current conservation. Due to the chiral anomaly, a linear combination of the axial couplings is equivalent to the coupling of axions to gauge bosons. In this section, we only consider the effects of the couplings to matter with no contribution from anomalies.

In Table 4.1, we have collected accelerator and astrophysics constraints on the effective couplings $\Lambda_{ij}^I \equiv \Lambda_\psi/g_I^{ij}$ and $\Lambda_{ij} \equiv \Lambda_\psi/[(g_V^{ij})^2 + (g_A^{ij})^2]^{1/2}$. We see that current data typically constrain the couplings to the first generation fermions much more than those to the second and third generations. We wish to compare these constraints to the reach of future CMB observations. We will find distinct behaviour above and below the EWSB scale, due to the presence of the Higgs. The effective scaling of the operator (4.13) changes from irrelevant to marginal and we therefore have both freeze-out and freeze-in contributions.

### 4.3.1  Freeze-Out

At high energies, the flavour structure of (4.13) is unimportant since all SM particles are effectively massless. The role of the flavour is only to establish the strength of the interaction by the size of the Yukawa coupling. Above the EWSB scale, the production of the familon $\phi$ is determined by a four-point interaction. This allows the following processes: $\bar{\psi}_i + \psi_j \rightarrow H + \phi$ and $\psi_i + H \rightarrow \psi_j + \phi$. The total production rate is derived in Appendix A,

$$\Gamma_{ij}^I \simeq 0.37\,N_\psi\,\frac{(\lambda_i \mp \lambda_j)^2}{8\pi}\,\frac{T^3}{(\Lambda_{ij}^I)^2}\,,$$

(4.14)

where $N_\psi = 1$ for charged leptons and $N_\psi = 3$ for quarks. The '$-$' and '$+$' signs in (4.14) apply to $I = V$ and $I = A$, respectively. We see that the rate vanishes for the diagonal vector coupling, as it should by current conservation. Deriving the freeze-out temperature and imposing $T_F > T_R$, we find

**Table 4.1** Current experimental constraints on Goldstone-fermion couplings (taken from [5, 52, 53]) and future CMB constraints. In some cases, the current constraints are only on the coupling to right-handed particles (namely for $\Lambda_{\tau\tau}$, $\Lambda_{bb}$, $\Lambda_{tt}$) and to left-handed particles (namely for $\Lambda_{tu}$, $\Lambda_{tc}$). The quoted freeze-out bounds are for $T_R = 10^{10}$ GeV and require that a future CMB experiment excludes $\Delta N_{\text{eff}} = 0.027$. In contrast, the freeze-in bounds from avoiding recoupling of the familons to the SM at low temperatures do not depend on $T_R$ and assume weaker exclusions $\Delta \tilde{N}_{\text{eff}}$ (see the last column for estimates of the freeze-in contributions associated with the different couplings, $\Delta \tilde{N}_{\text{eff}} \simeq \Delta N_{\text{eff}}(\frac{1}{4} m_i)$). Hence, they may be detectable with a less sensitive experiment. Qualitatively, the bounds from the CMB are stronger for the second and third generations, while laboratory and stellar constraints are strongest for the first generation (with the exception of the constraint on $\Lambda_{tt}$)

| Coupling | Current constraints | | Future CMB constraints | | |
|---|---|---|---|---|---|
| | Bound (GeV) | Origin | Freeze-out (GeV) | Freeze-in (GeV) | $\Delta \tilde{N}_{\text{eff}}$ |
| $\Lambda_{ee}$ | $1.2 \times 10^{10}$ | White dwarfs | $6.0 \times 10^{7}$ | $2.7 \times 10^{6}$ | 1.3 |
| $\Lambda_{\mu\mu}$ | $2.0 \times 10^{6}$ | Stellar cooling | $1.2 \times 10^{10}$ | $3.4 \times 10^{7}$ | 0.5 |
| $\Lambda_{\tau\tau}$ | $2.5 \times 10^{4}$ | Stellar cooling | $2.1 \times 10^{11}$ | $9.5 \times 10^{7}$ | 0.05 |
| $\Lambda_{bb}$ | $6.1 \times 10^{5}$ | Stellar cooling | $9.5 \times 10^{11}$ | – | 0.04 |
| $\Lambda_{tt}$ | $1.2 \times 10^{9}$ | Stellar cooling | $3.5 \times 10^{13}$ | – | 0.03 |
| $\Lambda_{\mu e}^{V}$ | $5.5 \times 10^{9}$ | $\mu^{+} \to e^{+} \phi$ | $6.2 \times 10^{9}$ | $4.8 \times 10^{7}$ | 0.5 |
| $\Lambda_{\mu e}$ | $3.1 \times 10^{9}$ | $\mu^{+} \to e^{+} \phi\gamma$ | $6.2 \times 10^{9}$ | $4.8 \times 10^{7}$ | 0.5 |
| $\Lambda_{\tau e}$ | $4.4 \times 10^{6}$ | $\tau^{-} \to e^{-} \phi$ | $1.0 \times 10^{11}$ | $1.3 \times 10^{8}$ | 0.05 |
| $\Lambda_{\tau\mu}$ | $3.2 \times 10^{6}$ | $\tau^{-} \to \mu^{-} \phi$ | $1.0 \times 10^{11}$ | $1.3 \times 10^{8}$ | 0.05 |
| $\Lambda_{cu}^{A}$ | $6.9 \times 10^{5}$ | $D^{0}$-$\bar{D}^{0}$ | $1.3 \times 10^{11}$ | $2.0 \times 10^{8}$ | 0.05 |
| $\Lambda_{bd}^{A}$ | $6.4 \times 10^{5}$ | $B^{0}$-$\bar{B}^{0}$ | $4.8 \times 10^{11}$ | $3.7 \times 10^{8}$ | 0.04 |
| $\Lambda_{bs}$ | $6.1 \times 10^{7}$ | $b \to s\phi$ | $4.8 \times 10^{11}$ | $3.7 \times 10^{8}$ | 0.04 |
| $\Lambda_{tu}$ | $6.6 \times 10^{9}$ | Mixing | $1.8 \times 10^{13}$ | $2.1 \times 10^{9}$ | 0.03 |
| $\Lambda_{tc}$ | $2.2 \times 10^{9}$ | Mixing | $1.8 \times 10^{13}$ | $2.1 \times 10^{9}$ | 0.03 |

$$\Lambda_{ij}^{l} > \begin{cases} 1.0 \times 10^{11}\,\text{GeV}\, \dfrac{m_i \mp m_j}{m_\tau} \left( \dfrac{T_R}{10^{10}\,\text{GeV}} \right)^{1/2} & i, j = \text{leptons,} \\[3mm] 1.8 \times 10^{13}\,\text{GeV}\, \dfrac{m_i \mp m_j}{m_t} \left( \dfrac{T_R}{10^{10}\,\text{GeV}} \right)^{1/2} & i, j = \text{quarks,} \end{cases} \tag{4.15}$$

where $m_\tau \approx 1.8$ GeV and $m_t \approx 173$ GeV. In Table 4.1, we show how these bounds compare to current laboratory and astrophysics constraints for a fiducial reheating temperature of $10^{10}$ GeV. Except for the coupling to electrons, the constraints from future CMB experiments are orders of magnitude stronger than existing constraints. For lower reheating temperatures the constraints would weaken proportional to $\sqrt{T_R}$. We note that while laboratory and astrophysical constraints are considerably weaker for second and third generation particles because of kinematics, the cosmological constraints are strengthened for the higher mass fermions due to the larger effective strength of the interactions. The exception to this pattern is the top quark which is strongly constrained by stellar cooling due to a loop correction to the coupling of $W^\pm$ and $Z$ to $\phi$, with the loop factor suppression being offset by the large Yukawa coupling of the top quark.

### 4.3.2   Freeze-In

Below the EWSB scale, the leading coupling of the familon to fermions becomes marginal after replacing the Higgs in (4.13) with its vacuum expectation value. As the temperature decreases, the production rate will therefore grow relative to the expansion rate and we may get a thermal freeze-in abundance. By observationally excluding such a relic density, we can then put constraints on the familon interactions (cf. Sect. 4.1). The leading familon production mechanism will depend on whether the coupling is diagonal or off-diagonal in the mass eigenbasis.

*Diagonal couplings.*—For the diagonal couplings in (4.13), the production rate is dominated by a Compton-like process, $\{\gamma, g\} + \psi_i \to \psi_i + \phi$, and by fermion/anti-fermion annihilation, $\bar{\psi}_i + \psi_i \to \{\gamma, g\} + \phi$, where $\{\gamma, g\}$ is either a photon or gluon depending on whether the fermion is a lepton or quark. The full expression for the corresponding production rate is given in Appendix A. Since freeze-in occurs at low temperatures, the quark production becomes sensitive to strong coupling effects. Although qualitative bounds could still be derived for the quark couplings, we choose not to present them and instead focus on the quantitative bounds for the lepton couplings. Below the scale of EWSB, but above the lepton mass, the production rate is

$$\tilde{\Gamma}_{ii} \simeq 5.3\,\alpha\,\frac{|\tilde{\epsilon}_{ii}|^2}{8\pi}\,T\,, \tag{4.16}$$

where $\tilde{\epsilon}_{ii} \equiv 2m_i/\Lambda_{ii}$. The freeze-in temperature $T_{\tilde{F}}$ follows from $\tilde{\Gamma}_{ii}(T_{\tilde{F}}) = H(T_{\tilde{F}})$. To avoid producing a large familon abundance requires that the fermion abundance becomes Boltzmann suppressed before freeze-in could occur. This implies $T_{\tilde{F}} < m_i$, or

$$\Lambda_{ii} > 9.5 \times 10^7\,\mathrm{GeV} \left(\frac{g_{*,i}}{g_{*,\tau}}\right)^{-1/4} \left(\frac{\alpha_i}{\alpha_\tau}\right)^{1/2} \left(\frac{m_i}{m_\tau}\right)^{1/2}\,, \quad i = \mathrm{lepton}, \tag{4.17}$$

where $g_{*,i}$ and $\alpha_i$ are the effective number of relativistic species and the fine-structure constant at $T = m_i$. The scalings in (4.17) have been normalized with respect to $g_*$ and $\alpha$ at $T = m_\tau$, i.e. we use $g_{*,\tau} = 81.0$ and $\alpha_\tau = 134^{-1}$. In Table 4.1, these bounds are compared to the current astrophysical constraints. Except for the coupling to electrons, these new bounds are significantly stronger than the existing constraints.

*Off-diagonal couplings.*—For the off-diagonal couplings in (4.13), we have the possibility of a freeze-in population of the familon from the decay of the heavy fermion, $\psi_i \to \psi_j + \phi$. For $m_i \gg m_j$, the production rate associated with this process is

$$\tilde{\Gamma}_{ij} \simeq 0.31 \, N_\psi \, \frac{|\tilde{\epsilon}_{ij}|^2}{8\pi} \frac{m_i^2}{T} , \tag{4.18}$$

where $\tilde{\epsilon}_{ij} \approx m_i/\Lambda_{ij}$. Requiring the corresponding freeze-in temperature to be below the mass of the heavier fermion, $T_{\tilde{F}} < m_i$, we get

$$\Lambda_{ij} > \begin{cases} 1.3 \times 10^8 \, \text{GeV} \left( \dfrac{g_{*,i}}{g_{*,\tau}} \right)^{-1/4} \left( \dfrac{m_i}{m_\tau} \right)^{1/2} & i, j = \text{leptons}, \\[2ex] 2.1 \times 10^9 \, \text{GeV} \left( \dfrac{g_{*,i}}{g_{*,t}} \right)^{-1/4} \left( \dfrac{m_i}{m_t} \right)^{1/2} & i, j = \text{quarks}. \end{cases} \tag{4.19}$$

We see that this improves over existing constraints for the third generation leptons and for the second and third generation quarks (except the top).

The freeze-in abundance is created *after* the annihilation of the most massive fermion in the coupling. In the presence of a single massive fermion, the prediction for a freeze-in scenario is the same as that for a freeze-out scenario with $T_F \ll m_i$ since decoupling occurs after most of the fermions $\psi_i$ have annihilated and their abundance is exponentially suppressed. This then results in a relatively large contribution to $N_{\text{eff}}$. Of course, the SM contains fermions with different masses. To capture the energy injection from the relevant fermion annihilation without incorrectly including the effects from the annihilation of much lighter fermions, we take the decoupling temperature to be $\frac{1}{4}m_i$. This choice of decoupling temperature gives good agreement with numerical solutions to the Boltzmann equations and leads to the following estimate for the freeze-in contributions:

$$\Delta\tilde{N}_{\text{eff}} \simeq \Delta N_{\text{eff}}(\tfrac{1}{4}m_i) = \frac{4}{7} \left( \frac{43}{4 \, g_*(\tfrac{1}{4}m_i)} \right)^{4/3} . \tag{4.20}$$

When the heaviest fermion is a muon (electron), one finds $\Delta\tilde{N}_{\text{eff}} \simeq 0.5 \, (1.3)$ which is excluded by Planck at about $3\sigma \, (7\sigma)$. It is worth noting that the Planck constraint on the diagonal muon coupling, $\Lambda_{\mu\mu} > 3.4 \times 10^7 \, \text{GeV}$, improves on the current experimental bound by more than an order of magnitude. Couplings involving the tau and the charm or bottom quark produce $\Delta\tilde{N}_{\text{eff}} \sim 0.05$ which will become accessible when the sensitivity of CMB experiments reaches $\sigma(N_{\text{eff}}) \lesssim 0.025$.

## 4.4 Constraints on Majorons

In the Standard Model, the masses of Majorana neutrinos do not arise from renormalizable couplings to the Higgs, but instead must be written as irrelevant operators suppressed by a scale of about $10^{15} \, \text{GeV}$. Moreover, the existence of neutrino masses and mixings point to structure in the flavour physics of neutrinos. Much like in the case of familons, it is plausible that this structure could arise from the spontaneous breaking of the neutrino flavour symmetry. The associated Goldstone bosons are often referred to as *majorons* [54, 55].

Assuming that neutrinos are indeed Majorana fermions, the leading coupling of the majoron is

$$
\begin{aligned}
\mathcal{L}_{\phi\nu} &= -\frac{1}{2} \left( e^{i\phi T_{ik}/(2\Lambda_\nu)} m_{kl} e^{i\phi T_{lj}/(2\Lambda_\nu)} \nu_i \nu_j + \text{h.c.} \right) \\
&= -\frac{1}{2} \left[ \left( m_{ij}\nu_i\nu_j + i\tilde{\epsilon}_{ij}\phi\nu_i\nu_j - \frac{1}{2\Lambda_\nu}\epsilon_{ij}\phi^2\nu_i\nu_j + \cdots \right) + \text{h.c.} \right], \quad (4.21)
\end{aligned}
$$

where $\nu_i$ are the two-component Majorana neutrinos in the mass eigenbasis, $m_{ij}$ is the neutrino mass matrix and $T_{ij}$ are generators of the neutrino flavour symmetry. After expanding the exponentials, we have defined the dimensionless couplings $\tilde{\epsilon}_{ij} \equiv (T_{ik}m_{kj} + m_{ik}T_{kj})/(2\Lambda_\nu)$ and $\epsilon_{ij} \equiv (m_{ik}T_{kl}T_{lj} + 2T_{ik}m_{kl}T_{lj} + T_{ik}T_{kl}m_{lj})/(4\Lambda_\nu)$. For numerical estimates, we will use the cosmological upper limit on the sum of the neutrino masses [56], $\sum m_i < 0.23$ eV, and the mass splittings $m_2^2 - m_1^2 \approx 7.5 \times 10^{-5}\text{eV}^2$ and $|m_3^2 - m_1^2| \approx 2.4 \times 10^{-3}\text{eV}^2$ from neutrino oscillation measurements [57]. The couplings in $\mathcal{L}_{\phi\nu}$ are identical to the familon couplings after a chiral rotation, except that there is no analogue of the vector current in the case of Majorana neutrinos. The representation of the coupling in (4.21) is particularly useful as it makes manifest both the marginal and irrelevant couplings between $\phi$ and $\nu$. As a result, we will get both a freeze-out[2] and a freeze-in production of the majorons.

### 4.4.1  Freeze-Out

Thermalization at high energies is dominated by the dimension-five operator $\phi^2\nu_i\nu_j$ in (4.21). In Appendix A, we show that the corresponding production rate is

$$
\Gamma_{ij} \simeq 0.047\, s_{ij} \frac{|\epsilon_{ij}|^2}{8\pi} \frac{T^3}{\Lambda_\nu^2}, \quad (4.22)
$$

where $s_{ij} \equiv 1 - \frac{1}{2}\delta_{ij}$ is the symmetry factor for identical particles in the initial state. This leads to a freeze-out temperature of

$$
T_F \simeq 0.23\,\text{MeV}\, s_{ij}^{-1} \left( \frac{g_{*,F}}{10} \right)^{1/2} \left( \frac{\mu_{ij}}{0.1\,\text{eV}} \right)^{-2} \left( \frac{\Lambda_\nu}{10\,\text{MeV}} \right)^4, \quad (4.23)
$$

---

[2]Technically speaking the operator in (4.21) is only well-defined below the EWSB scale. However, in Sect. 4.4.1 we will find that in order for freeze-out to occur in the regime of a consistent effective field theory description ($T < \Lambda_\nu$), we require $T_F \lesssim 33$ MeV and, therefore, the operator as written will be sufficient for our purposes.

where $\mu_{ij} \equiv |\epsilon_{ij}|\Lambda_\nu$. Consistency of the effective field theory description requires $T_F$ to be below the cutoff $\Lambda_\nu$ associated with the interactions in (4.21). Using (4.23), this implies

$$T_F < \Lambda_\nu < 35\,\text{MeV}\, s_{ij}^{1/3} \left(\frac{g_{*,F}}{10}\right)^{-1/6} \left(\frac{\mu_{ij}}{0.1\,\text{eV}}\right)^{2/3} . \tag{4.24}$$

Taking $\mu_{ij} \lesssim m_3 < 0.1\,\text{eV}$ from both the mass splittings and the bound on the sum of neutrino masses, and $g_* \approx 14$, we obtain $T_F \lesssim 33\,\text{MeV}$. Such a low freeze-out temperature would lead to $\Delta N_{\text{eff}} \gtrsim 0.44$ (cf. Fig. 3.2) which is ruled out by *current* CMB measurements at more than $2\sigma$. To avoid this conclusion, we require $\Lambda_\nu > 33\,\text{MeV}$, so that the would-be freeze-out is pushed outside the regime of validity of the EFT. Moreover, we have to assume that the production of majorons is suppressed in this regime. This logic leads to the following constraint:

$$\Lambda_\nu > 33\,\text{MeV} \xrightarrow{\;\mu_{ij} \lesssim 0.1\,\text{eV}\;} |\epsilon_{ij}| < 3 \times 10^{-9} . \tag{4.25}$$

Somewhat stronger bounds can be derived for individual elements of $\epsilon_{ij}$. This simple bound is much stronger than existing constraints from neutrinoless double beta decay [58, 59] and supernova cooling [60], $\epsilon_{ij} \lesssim 10^{-7}$. Note also that the constraints on $\epsilon_{ij}$ are stronger for smaller values of $\mu_{ij}$.

### 4.4.2 Freeze-In

At low energies, the linear coupling $\phi\nu_i\nu_j$ in (4.21) will dominate. The corresponding two-to-one process is kinematically constrained and we therefore get qualitatively different behaviour depending on whether the majoron mass is larger or smaller than that of the neutrinos.

*Low-mass regime.*—For $m_\phi \ll m_i - m_j$, with $m_i > m_j$, the off-diagonal couplings allow the decay $\nu_i \to \nu_j + \phi$, while other decays are kinematically forbidden. As a result, only the off-diagonal couplings are constrained by freeze-in. Including the effect of time dilation at finite temperature, the rate is

$$\tilde{\Gamma}_{ij} \simeq 0.31\, \frac{|\tilde{\epsilon}_{ij}|^2}{8\pi} \frac{m_i^2}{T} , \tag{4.26}$$

where we have assumed $m_i \gg m_j$, which is guaranteed for the minimal mass normal hierarchy (for the general result see Appendix A). When the freeze-in occurs at $T_{\tilde{F}} > m_i$, then the majorons and neutrinos are brought into thermal equilibrium, while the comoving energy density is conserved. However, since the momentum exchange at each collision is only $\Delta p^2 \simeq m_i^2 \ll T^2$, the neutrino-majoron radiation is free-streaming at the onset of the freeze-in and is difficult[3] to distinguish from conventional neutrinos. As the temperature drops below $T_{\text{fluid}}$, with $\bar{\Gamma}_{ij}(T_{\text{fluid}}) = (T_{\text{fluid}}/m_i)^2 \, H(T_{\text{fluid}})$, enough momentum is exchanged between the neutrinos and they will behave as a relativistic fluid rather than free-streaming particles [61–63]. From the rate (4.26), we find

$$T_{\text{fluid}} \simeq 0.10 \, T_{\text{eq}} \times \left(\frac{\tilde{\epsilon}_{ij}}{10^{-13}}\right)^{2/5} \left(\frac{m_i}{0.05\,\text{eV}}\right)^{4/5}, \tag{4.27}$$

where we used $g_{*,\tilde{F}} \approx 3.4$ and $T_{\text{eq}} \approx 0.79\,\text{eV}$ for the temperature at matter-radiation equality. In this regime, the majoron scenario predicts both free-streaming and non-free-streaming radiation with $N_{\text{eff}} \leq 2$ and $N_{\text{fluid}} \geq 1$ (with equality when the majoron couples to one single neutrino species), which is inconsistent with the constraints from Planck data that we will find in the next chapter. To avoid this conclusion requires $T_{\text{fluid}} < T_{\text{eq}}$,[4] which puts a bound on the neutrino-majoron coupling[5]

$$\tilde{\epsilon}_{ij} < 3.2 \times 10^{-11} \times \left(\frac{m_i}{0.05\,\text{eV}}\right)^{-2}. \tag{4.28}$$

This constraint has been pointed out previously in [61–63, 65].

*High-mass regime.*—For $m_\phi \gg m_i \geq m_j$, the majoron decays into neutrinos, $\phi \to \nu_i + \nu_j$, and is produced by the inverse decay. For $T \gg m_\phi$, the production rate of the majoron is identical to the rate in (4.26) after making the replacement $m_i \to m_\phi/\sqrt{1 - 4/\pi^2}$ and the corresponding freeze-in temperature is

$$T_{\tilde{F}} \simeq 1.0 \, T_{\text{eq}} \times s_{ij}^{1/3} \left(\frac{\tilde{\epsilon}_{ij}}{10^{-13}} \frac{m_\phi}{T_{\text{eq}}}\right)^{2/3}. \tag{4.29}$$

If $T_{\tilde{F}} > m_\phi$, then freeze-in occurs while the majorons are relativistic, and the neutrinos and the majorons are brought into thermal equilibrium. How this affects the CMB will depend on whether $m_\phi$ is greater or smaller than $T_{\text{eq}}$. For $m_\phi > T_{\text{eq}}$, the majorons decay to neutrinos before matter-radiation equality. To compute the

---

[3] Since neutrinos have been converted to majorons with $m_\phi \ll m_i$, this scenario predicts that the cosmological measurement of the sum of the neutrino masses would be significantly lower than what would be inferred from laboratory measurements.

[4] The imprint of dark radiation is suppressed during matter domination since its contribution to the total energy density is sub-dominant. As a result, constraints on $N_{\text{eff}}$ are driven by the high-$\ell$ modes of the CMB which are primarily affected by the evolution of fluctuations during radiation domination (see Sect. 3.4).

[5] The effect of the linear coupling between a massless majoron and neutrinos on the CMB was also studied in [64] and a flavour-independent constraint of $\tilde{\epsilon}_{ij} < 8.2 \times 10^{-7}$ was obtained. This constraint is substantially weaker than our bound (4.28) because it only accounted for the scattering of neutrinos through the exchange of a *virtual* Goldstone boson. The neutrino cross section in that case is suppressed by a factor of $|\bar{\epsilon}_{ij}|^4$ which is much smaller than the rate for the production of real Goldstone bosons in (4.26).

effect on the CMB, we note that the initial (relativistic) freeze-in process conserves the comoving energy density and, once in equilibrium, the decay will conserve the comoving entropy density. This information allows us to derive the final neutrino temperature analytically (see Appendix A) and to determine the extra contribution to the radiation density,

$$\Delta N_{\text{eff}} \geq \left(1 + \frac{4}{7}\right)^{1/3} - 1 = 0.16. \tag{4.30}$$

This extra radiation density is easily falsifiable (or detectable) with future CMB experiments. If $m_\phi \ll T_{\text{eq}}$, on the other hand, the neutrinos and the majorons could effectively form a fluid at matter-radiation equality leading to a similar constraint as (4.28) with $m_i \rightarrow m_\phi$.

Assuming that future experiments do not detect the above effects would require either that the would-be freeze-in temperature is below the mass of the majoron, $T_{\tilde{F}} < m_\phi$, or that freeze-in occurs after matter-radiation equality, $T_{\tilde{F}} < T_{\text{eq}}$. Converting these constraints into a bound on the coupling, we find

$$\tilde{\epsilon}_{ij} < 9.9 \times 10^{-14} \, s_{ij}^{-1/2} \left(\frac{m_\phi}{T_{\text{eq}}}\right)^{1/2}, \quad \text{for} \quad m_\phi > T_{\text{eq}}. \tag{4.31}$$

A similar constraint, of the same order of magnitude, applies in the narrow range $m_i \ll m_\phi < T_{\text{eq}}$. This bound is stronger than the freeze-out constraint (4.25) over the full range of allowed masses up to the neutrino decoupling temperature $T_{F,\nu} \simeq 1\,\text{MeV}$ (note that although in general $\epsilon_{ij} \neq \tilde{\epsilon}_{ij}$, the two parameters are related by the symmetry under which the majoron transforms). For $m_\phi > T_{F,\nu}$, the decay of the majorons occurs while the neutrinos are still in equilibrium with the SM and, therefore, it has no impact on $N_{\text{eff}}$.

## 4.5 Summary

Light pseudo-Nambu-Goldstone bosons arise naturally in many proposals for physics beyond the Standard Model and are an exciting window into the early universe. In this chapter, we showed that future cosmological surveys will either detect these new particles, or place very strong constraints on their couplings to the SM. These constraints arise because the couplings to the SM can bring the Goldstone bosons into thermal equilibrium in the early universe. At the same time, cosmological experiments are becoming sensitive enough to detect thermal relics up to arbitrarily high freeze-out temperatures (see Fig. 3.2). To avoid producing this detectable relic abundance requires that the reheating temperature of the universe was below the would-be freeze-out temperature. In that case, the temperature in the universe simply was never high enough to bring the extra particles into thermal equilibrium with the SM. For a given reheating temperature $T_R$, this puts bounds on the scales $\Lambda_i$ in the effective interactions between the Goldstone boson $\phi$ and the SM fields,

$$\mathcal{L}_{\phi SM} = -\frac{1}{4} \frac{\phi}{\Lambda_\gamma} F \tilde{F} - \frac{1}{4} \frac{\phi}{\Lambda_g} \text{Tr}(G \tilde{G}) - \frac{\partial_\mu \phi}{\Lambda_\psi} \bar{\psi} \gamma^\mu \gamma^5 \psi + \cdots. \tag{4.32}$$

The bounds on the couplings to photons and gluons are

$$\Lambda_\gamma > 1.4 \times 10^{13} \text{ GeV} \sqrt{T_{R,10}}, \tag{4.33}$$

$$\Lambda_g > 5.4 \times 10^{13} \text{ GeV} \sqrt{T_{R,10}}, \tag{4.34}$$

where $T_{R,10} \equiv T_R/10^{10}$ GeV. When considering the interactions with fermions, we distinguish between the couplings to charged leptons and quarks. The resulting bounds are

$$\Lambda_\psi > \begin{cases} 2.1 \times 10^{11} \text{ GeV } m_{\psi,\tau} \sqrt{T_{R,10}} & \psi = \text{lepton}, \\ 3.5 \times 10^{13} \text{ GeV } m_{\psi,t} \sqrt{T_{R,10}} & \psi = \text{quark}, \end{cases} \tag{4.35}$$

where $m_{\psi,\tau} \equiv m_\psi/1.8 \text{ GeV}$ and $m_{\psi,t} \equiv m_\psi/173 \text{ GeV}$. For all reasonable reheating temperatures these bounds improve significantly over existing constraints, sometimes by many orders of magnitude. Moreover, while some of the current constraints only apply if the new particles are identified with the dark matter, our bounds do not have this restriction.

Below the scale of electroweak symmetry breaking, the couplings to the SM fermions become effectively marginal which can bring the decoupled Goldstone bosons back into thermal equilibrium leading to a detectable freeze-in abundance. Furthermore, the coupling to the light Goldstone boson can lead to a new force between the fermions which becomes relevant at low temperatures [61–63, 65]. As we will show in the next chapter, both of these effects are highly constrained, even with current data. These arguments are particularly relevant for the couplings to neutrinos,

$$\mathcal{L}_{\phi\nu} = -\frac{1}{2} \left( i \tilde{\epsilon}_{ij} \phi \nu_i \nu_j + \text{h.c.} \right) + \cdots. \tag{4.36}$$

For the off-diagonal couplings, the following constraints apply:

$$\tilde{\epsilon}_{ij} < \begin{cases} 3.2 \times 10^{-11} \times \left( \dfrac{m_i}{0.05 \text{ eV}} \right)^{-2} & m_\phi \ll m_i, \\ 9.9 \times 10^{-14} \times \left( \dfrac{m_\phi}{T_{\text{eq}}} \right)^{1/2} & m_\phi > T_{\text{eq}}, \end{cases} \tag{4.37}$$

where $m_i$ is the mass of the heavier neutrino in the off-diagonal interaction. A combination of freeze-in and freeze-out also constrain the diagonal couplings $\tilde{\epsilon}_{ii}$. These constraints are orders of magnitude stronger than existing laboratory and astrophysics constraints.

It is also interesting to consider a scenario in which one of the many ongoing searches directly detects axions, familons or majorons. This would determine the coupling strength to at least one of the SM fields (depending on the detection channel) and would predict the freeze-out temperature of these particles; cf. Figs. 4.1 and 4.2. Excitingly, the cosmological estimation of $\Delta N_{\mathrm{eff}}$ would then provide information about the reheating temperature of the universe: the absence of a detection of $N_{\mathrm{eff}} \neq 3.046$ would put an upper bound on $T_R$ [see e.g. (4.33)–(4.35)], while a measurement of $\Delta N_{\mathrm{eff}} \geq 0.027$ would imply a lower bound on $T_R$. The combination of a cosmological measurement of $N_{\mathrm{eff}}$ and a direct detection could therefore be used to probe the energy scale of the beginning of the hot big bang.

In closing, we would like to re-emphasize that $\Delta N_{\mathrm{eff}} = 0.027$ is an important theoretical threshold. Remarkably, we will show in the next two chapters that this target is within reach of future cosmological surveys. These observations therefore have the potential to probe for light thermal relics up to arbitrarily high decoupling temperatures. We consider this to be a unique opportunity to detect new particles, or place very strong constraints on their couplings to the Standard Model.

# References

1. K. Abazajian et al., Neutrino physics from the cosmic microwave background and large-scale structure. Astropart. Phys. **63**, 66 (2015). arXiv:1309.5383 [astro-ph.CO]
2. W. Wu, J. Errard, C. Dvorkin, C. Kuo, A. Lee, P. McDonald, A. Slosar, O. Zahn, A guide to designing future ground-based cosmic microwave background experiments. Astrophys. J. **788**, 138 (2014). arXiv:1402.4108 [astro-ph.CO]
3. D. Baumann, D. Green, B. Wallisch, New target for cosmic axion searches. Phys. Rev. Lett. **117**, 171301 (2016). arXiv:1604.08614 [astro-ph.CO]
4. Adapted from [3] with permission. Copyright by the American Physical Society
5. C. Brust, D.E. Kaplan, M. Walters, New light species and the CMB. JHEP **12**, 058 (2013). arXiv:1303.5379 [hep-ph]
6. A. Salvio, A. Strumia, W. Xue, Thermal axion production. JCAP **01**, 011 (2014). arXiv:1310.6982 [hep-ph]
7. M. Kawasaki, M. Yamada, T. Yanagida, Observable dark radiation from a cosmologically safe QCD axion. Phys. Rev. D **91**, 125018 (2015). arXiv:1504.04126 [hep-ph]
8. Z. Chacko, Y. Cui, S. Hong, T. Okui, Hidden dark matter sector, dark radiation and the CMB. Phys. Rev. D **92**, 055033 (2015). arXiv:1505.04192 [hep-ph]
9. R. Ferreira, A. Notari, Observable windows for the QCD axion through the number of relativistic species. Phys. Rev. Lett. **120**, 191301 (2018). arXiv:1801.06090 [hep-ph]
10. Z. Berezhiani, A. Dolgov, R. Mohapatra, Asymmetric inflationary reheating and the nature of mirror universe. Phys. Lett. B **375**, 26 (1996). arXiv:hep-ph/9511221 [hep-ph]
11. H. Hodges, Mirror baryons as the dark matter. Phys. Rev. D **47**, 456 (1993)
12. J. Feng, H. Tu, H.-B. Yu, Thermal relics in hidden sectors. JCAP **10**, 043 (2008). arXiv:0808.2318 [hep-ph]
13. R. Foot, S. Vagnozzi, Dissipative hidden sector dark matter. Phys. Rev. D **91**, 023512 (2015). arXiv:1409.7174 [hep-ph]
14. M. Reece, T. Roxlo, Non-thermal production of dark radiation and dark matter. JHEP **09**, 096 (2016). arXiv:1511.06768 [hep-ph]
15. P. Adshead, Y. Cui, J. Shelton, Chilly dark sectors and asymmetric reheating. JHEP **06**, 016 (2016). arXiv:1604.02458 [hep-ph]
16. A. Font-Ribera, P. McDonald, N. Mostek, B. Reid, H.-J. Seo, A. Slosar, DESI and other dark energy experiments in the era of neutrino mass measurements. JCAP **05**, 023 (2014). arXiv:1308.4164 [astro-ph.CO]

17. A. Manzotti, S. Dodelson, Y. Park, External priors for the next generation of CMB experiments. Phys. Rev. D **93**, 063009 (2016). arXiv:1512.02654 [astro-ph.CO]
18. M. Archidiacono, T. Basse, J. Hamann, S. Hannestad, G. Raffelt, Y.Y.Y. Wong, Future cosmological sensitivity for hot dark matter axions. JCAP **05**, 050 (2015). arXiv:1502.03325 [astro-ph.CO]
19. E. Di Valentino, E. Giusarma, M. Lattanzi, O. Mena, A. Melchiorri, J. Silk, Cosmological axion and neutrino mass constraints from Planck 2015 temperature and polarization data. Phys. Lett. B **752**, 182 (2016). arXiv:1507.08665 [astro-ph.CO]
20. R. Peccei, H. Quinn, CP conservation in the presence of pseudoparticles. Phys. Rev. Lett. **38**, 1440 (1977)
21. S. Weinberg, A new light boson? Phys. Rev. Lett. **40**, 223 (1978)
22. F. Wilczek, Problem of strong P and T invariance in the presence of instantons. Phys. Rev. Lett. **40**, 279 (1978)
23. J. Preskill, M. Wise, F. Wilczek, Cosmology of the invisible axion. Phys. Lett. B **120**, 127 (1983)
24. L. Abbott, P. Sikivie, A cosmological bound on the invisible axion. Phys. Lett. B **120**, 133 (1983)
25. M. Dine, W. Fischler, The not so harmless axion. Phys. Lett. B **120**, 137 (1983)
26. P. Svrcek, E. Witten, Axions in string theory. JHEP **06**, 051 (2006). arXiv:hep-th/0605206 [hep-th]
27. A. Arvanitaki, S. Dimopoulos, S. Dubovsky, N. Kaloper, J. March-Russell, String axiverse. Phys. Rev. D **81**, 123530 (2010). arXiv:0905.4720 [hep-th]
28. D. Baumann, L. McAllister, *Inflation and String Theory* (Cambridge University Press, Cambridge, UK, 2015). arXiv:1404.2601 [hep-th]
29. K. Freese, J. Frieman, A. Olinto, Natural inflation with pseudo-Nambu-Goldstone bosons. Phys. Rev. Lett. **65**, 3233 (1990)
30. P. Graham, D.E. Kaplan, S. Rajendran, Cosmological relaxation of the electroweak scale. Phys. Rev. Lett. **115**, 221801 (2015). arXiv:1504.07551 [hep-ph]
31. R. Essig et al., Working group report: new light weakly coupled particles. arXiv:1311.0029 [hep-ph]
32. P. Graham, I. Irastorza, S. Lamoreaux, A. Lindner, K. van Bibber, Experimental searches for the axion and axion-like particles. Ann. Rev. Nucl. Part. Sci. **65**, 485 (2015). arXiv:1602.00039 [hep-ex]
33. E. Braaten, T. Yuan, Calculation of screening in a hot plasma. Phys. Rev. Lett. **66**, 2183 (1991)
34. M. Bolz, A. Brandenburg, W. Buchmüller, Thermal production of gravitinos, Nucl. Phys. B **606**, 518 (2001) [Erratum: Nucl. Phys. B **790**, 336 (2008)]. arXiv:hep-ph/0012052 [hep-ph]
35. E. Masso, F. Rota, G. Zsembinszki, On axion thermalization in the early universe. Phys. Rev. D **66**, 023004 (2002). arXiv:hep-ph/0203221 [hep-ph]
36. P. Graf, F. Steffen, Thermal axion production in the primordial Quark-Gluon plasma. Phys. Rev. D **83**, 075011 (2011). arXiv:1008.4528 [hep-ph]
37. G. Carosi, A. Friedland, M. Giannotti, M. Pivovaroff, J. Ruz, J. Vogel, Probing the axion-photon coupling: phenomenological and experimental perspectives. arXiv:1309.7035 [hep-ph]
38. M. Peskin, D. Schroeder, *An Introduction to Quantum Field Theory* (Addison-Wesley, Reading, 1995)
39. A. Friedland, M. Giannotti, M. Wise, Constraining the axion-photon coupling with massive stars. Phys. Rev. Lett. **110**, 061101 (2013). arXiv:1210.1271 [hep-ph]
40. D. Cadamuro, S. Hannestad, G. Raffelt, J. Redondo, Cosmological bounds on sub-MeV mass axions. JCAP **02**, 003 (2011). arXiv:1011.3694 [hep-ph]
41. D. Cadamuro, J. Redondo, Cosmological bounds on pseudo-Nambu-Goldstone bosons. JCAP **02**, 032 (2012). arXiv:1110.2895 [hep-ph]
42. M. Millea, L. Knox, B. Fields, New bounds for axions and axion-like particles with keV-GeV masses. Phys. Rev. D **92**, 023010 (2015). arXiv:1501.04097 [astro-ph.CO]
43. M. Pospelov, A. Ritz, Theta-induced electric dipole moment of the neutron via QCD sum rules. Phys. Rev. Lett. **83**, 2526 (1999). arXiv:hep-ph/9904483 [hep-ph]

44. P. Graham, S. Rajendran, New observables for direct detection of axion dark matter. Phys. Rev. D **88**, 035023 (2013). arXiv:1306.6088 [hep-ph]
45. K. Blum, R. D'Agnolo, M. Lisanti, B. Safdi, Constraining axion dark matter with big bang nucleosynthesis. Phys. Lett. B **737**, 30 (2014). arXiv:1401.6460 [hep-ph]
46. Adapted from [3, 44] with permission. Copyright by the American Physical Society
47. D. Budker, P. Graham, M. Ledbetter, S. Rajendran, A. Sushkov, Proposal for a cosmic axion spin precession experiment (CASPEr). Phys. Rev. X **4**, 021030 (2014). arXiv:1306.6089 [hep-ph]
48. G. Raffelt, *Stars as Laboratories for Fundamental Physics* (University of Chicago Press, Chicago, 1996)
49. F. Wilczek, Axions and family symmetry breaking. Phys. Rev. Lett. **49**, 1549 (1982)
50. D. Reiss, Can the family group be a global symmetry? Phys. Lett. B **115**, 217 (1982)
51. J. Kim, Light pseudoscalars, particle physics and cosmology. Phys. Rep. **150**, 1 (1987)
52. J. Feng, T. Moroi, H. Murayama, E. Schnapka, Third-generation familons, B factories and neutrino cosmology. Phys. Rev. D **57**, 5875 (1998). arXiv:hep-ph/9709411 [hep-ph]
53. B. Hansen, H. Richer, J. Kalirai, R. Goldsbury, S. Frewen, J. Heyl, Constraining neutrino cooling using the hot white dwarf luminosity function in the globular cluster 47 Tucanae. Astrophys. J. **809**, 141 (2015). arXiv:1507.05665 [astro-ph.SR]
54. Y. Chikashige, R. Mohapatra, R. Peccei, Are there real Goldstone bosons associated with broken lepton number? Phys. Lett. B **98**, 265 (1981)
55. Y. Chikashige, R. Mohapatra, R. Peccei, Spontaneously broken lepton number and cosmological constraints on the neutrino mass spectrum. Phys. Rev. Lett. **45**, 1926 (1980)
56. P.A.R. Ade et al. (Planck Collaboration), Planck 2015 results. XIII. Cosmological parameters. Astron. Astrophys. **594**, A13 (2016). arXiv:1502.01589 [astro-ph.CO]
57. C. Patrignani et al. (Particle Data Group), Review of particle physics. Chin. Phys. C **40**, 100001 (2016)
58. A. Gando et al. (KamLAND-Zen Collaboration), Limits on majoron-emitting double-beta decays of 136Xe in the KamLAND-Zen experiment. Phys. Rev. C **86**, 021601 (2012). arXiv:1205.6372 [hep-ex]
59. J. Albert et al. (EXO-200 Collaboration), Search for majoron-emitting modes of double-beta decay of 136Xe with EXO-200. Phys. Rev. D **90**, 092004 (2014). arXiv:1409.6829 [hep-ex]
60. Y. Farzan, Bounds on the coupling of the majoron to light neutrinos from supernova cooling. Phys. Rev. D **67**, 073015 (2003). arXiv:hep-ph/0211375 [hep-ph]
61. Z. Chacko, L. Hall, T. Okui, S. Oliver, CMB signals of neutrino mass generation. Phys. Rev. D **70**, 085008 (2004). arXiv:hep-ph/0312267 [hep-ph]
62. S. Hannestad, G. Raffelt, Constraining invisible neutrino decays with the cosmic microwave background. Phys. Rev. D **72**, 103514 (2005). arXiv:hep-ph/0509278 [hep-ph]
63. A. Friedland, K. Zurek, S. Bashinsky, Constraining models of neutrino mass and neutrino interactions with the Planck satellite. arXiv:0704.3271 [astro-ph]
64. F. Forastieri, M. Lattanzi, P. Natoli, Constraints on secret neutrino interactions after Planck. JCAP **07**, 014 (2015). arXiv:1504.04999 [astro-ph.CO]
65. M. Archidiacono, S. Hannestad, Updated constraints on non-standard neutrino interactions from Planck. JCAP **07**, 046 (2014). arXiv:1311.3873 [astro-ph.CO]

# Chapter 5
# Searching for Light Relics with the CMB

Fluctuations in the cosmic neutrino background are known to produce a phase shift in the acoustic peaks of the CMB. It is through the sensitivity to this effect that the recent CMB data has provided a robust detection of free-streaming neutrinos [1]. In this chapter, we revisit the phase shift of the CMB anisotropy spectrum as a probe of new physics. The phase shift is particularly interesting because its physical origin is strongly constrained by the analytic properties of the Green's function of the gravitational potential. For adiabatic fluctuations, a phase shift requires modes that propagate faster than the speed of fluctuations in the photon-baryon plasma. This possibility is realized by free-streaming relativistic particles, such as neutrinos or other forms of dark radiation. Alternatively, a phase shift can arise from isocurvature fluctuations. We present simple models to illustrate each of these effects and provide observational constraints from the Planck temperature and polarization data on additional forms of radiation. We also estimate the capabilities of future CMB Stage-4 experiments. Whenever possible, we give analytic interpretations of our results.

The outline of this chapter, which is based on the research first presented in [2, 3], is as follows. In Sect. 5.1, we briefly review the phase shift in the acoustic oscillations and preview some of this chapter's results. In Sect. 5.2, we analytically determine the effects of neutrinos and other relativistic species on the perturbations of the photon density. We derive the precise physical conditions that give rise to a phase shift in the CMB anisotropy spectrum and illustrate these effects through a toy model which is exactly solvable. In addition, we calculate the phase shift for a simple model with isocurvature fluctuations and for free-streaming relativistic particles, such as neutrinos. In Sect. 5.3, we perform a numerical analysis to confirm some of these pen-and-paper results. We infer new constraints on (both free-streaming and non-free-streaming) dark radiation from the Planck 2015 data [4] and present forecasts of the sensitivity of a future CMB-S4 experiment [5]. In Sect. 5.4, we consider possible implications of this future sensitivity of the CMB on different scenarios beyond the Standard Model. Section 5.5 contains a summary. In Appendix B, we complement our analytic treatment with comments on the impact of matter and the inclusion of polarization.

© Springer Nature Switzerland AG 2019
B. Wallisch, *Cosmological Probes of Light Relics*, Springer Theses,
https://doi.org/10.1007/978-3-030-31098-1_5

## 5.1  Phases of New Physics in the CMB Spectrum

After accounting for the difference in the sound horizon, the main effect of increasing the radiation density of the early universe on the CMB anisotropy spectrum is to enhance the damping of the high-$\ell$ multipoles. As we discussed in Sect. 3.4.2, neutrinos and other free-streaming particles also induce a distinct shift in the temporal phase of sound waves in the primordial plasma [6]. This subtle effect manifests itself in the CMB spectrum as a coherent shift in the locations of the acoustic peaks (see Fig. 5.1).[1]

In this chapter, we will revisit the analytic treatment of the CMB anisotropies with an eye towards BSM applications. While numerical codes are ultimately necessary in order to make precise predictions for any particular model, analytic results still play a vital role. It is through the physical understanding of the data that we can devise new tests and motivate new models. For example, the use of B-modes in the search for primordial gravitational waves arose from a clear analytic understanding of CMB polarization [11–14]. Similarly, we wish to identify CMB observables that are sensitive to well-motivated forms of BSM physics, but are not strongly degenerate

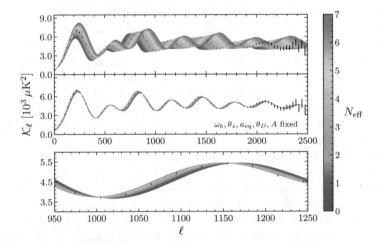

**Fig. 5.1** Variation of the CMB power spectrum as a function of $N_{\mathrm{eff}}$. The spectra have been undamped, i.e. the exponential diffusion damping was removed. Following [1], the physical baryon density $\omega_b$, the scale factor at matter-radiation equality $a_{\mathrm{eq}} \equiv \omega_m/\omega_r$, the angular size of the sound horizon $\theta_s$ and the angular size of the damping scale $\theta_D$ are held fixed in the second panel. In addition, the spectra are normalized at the fourth peak. The remaining variation is the phase shift $\phi$ with a zoom-in shown in the bottom panel. To illustrate the sensitivity of the Planck 2015 high-$\ell$ temperature data, we also display their $1\sigma$ error bars [4]

---

[1] This phase shift refers to a coherent shift in the locations of the high-$\ell$ acoustic peaks. We emphasize that this is a distinct effect from the locations of the first few acoustic peaks which are sensitive to many cosmological parameters, as studied e.g. in [7–9]. A detailed study of the CMB peak locations recently appeared in [10].

with other cosmological parameters. We will advocate the phase shift of the acoustic peaks of the CMB spectrum as an observable with the desired characteristics. As we will see below, the physical conditions that lead to a phase shift are rather restrictive and determined by the analytic properties of the Green's function of the gravitational potential. For adiabatic fluctuations, a phase shift requires fluctuations that travel faster than the sound speed of the photon-baryon fluid, which arises naturally for free-streaming relativistic particles. Alternatively, a shift in the temporal phase of the cosmic sound waves can also arise from isocurvature fluctuations. The phase shift therefore probes an interesting regime in the parameter space of BSM models.

We have previously seen that it is useful to characterise the effects of extra light species in terms of free-streaming species ($X$) and non-free-streaming species ($Y$). We parametrize their respective energy density in terms of the parameters $N_{\text{eff}}$ and $N_{\text{fluid}}$, i.e. with respect to the energy density of a single SM neutrino species (cf. Sect. 3.3). Only recently, CMB observations have become sensitive enough to discriminate between these different types of relativistic particles. The underlying reason for this is that they affect the CMB damping tail in the same way since they contribute equally to the background density of the universe [15]. The ability to separate $N_{\text{eff}}$ and $N_{\text{fluid}}$ therefore demands that subtle differences in the evolution of perturbations can be measured. Free-streaming particles, such as neutrinos, create significant anisotropic stress which induces a characteristic phase shift in the CMB anisotropies [6] that has been detected for the first time not long ago [1]. On the other hand, we will show that non-free-streaming particles (e.g. [16–20]) generally do not generate a phase shift (at least as long as the perturbations are adiabatic and their sound speed is not larger than that of the photons).

Based on these analytic insights, we will subsequently analyse the capabilities of the Planck satellite and a future CMB-S4 experiment in this respect, focussing on the ability to discriminate the parameters $N_{\text{eff}}$ and $N_{\text{fluid}}$. By analysing the Planck 2015 temperature and polarization data, we find that the current data is already sensitive to the free-streaming nature of the fluctuations[2]:

$$N_{\text{eff}} = 2.80^{+0.24}_{-0.23} \ (1\sigma), \qquad N_{\text{fluid}} < 0.67 \ (2\sigma). \qquad (5.1)$$

Moreover, we will describe the significant role that is played by the polarization data in breaking not only the degeneracy between $N_{\text{eff}}$ and $N_{\text{fluid}}$, but also that with the helium fraction $Y_p$. Interestingly, we will also observe that a CMB-S4 experiment has the potential to improve these constraints by up to an order of magnitude for a number of observational set-ups. In addition, we will inspect various ways to optimize these measurements which includes the application of delensing to sharpen the acoustic peaks. Finally, we will discuss to what degree present and future constraints are driven by measurements of the phase shift.

---

[2]While these constraints assume that the primordial helium fraction $Y_p$ is given by requiring consistency with BBN, we present the results that marginalize over $Y_p$ in Sect. 5.3.2.

## 5.2  Physical Origin of the Phase Shift

We will now study under which circumstances a phase shift is imprinted in the CMB anisotropies. The structure of the acoustic peaks in the CMB is determined to a large extent by the propagation of perturbations in the photon-baryon plasma. In turn, the physics of the cosmic sound waves is that of a harmonic oscillator with a time-dependent gravitational force term,

$$\ddot{d}_\gamma - c_\gamma^2 \, \nabla^2 d_\gamma = \nabla^2 \Phi_+ \,, \tag{5.2}$$

where $c_\gamma^2 \approx \frac{1}{3}$. Following [6], we wrote this equation in terms of the overdensity in the particle number with respect to the coordinate volume, $d_a \equiv \delta\rho_a/(\bar{\rho}_a + \bar{P}_a) - 3\Psi = \delta_a/(1 + w_a) - 3\Psi$. Moreover, it will be convenient to work with the sum and the difference of the two metric potentials,

$$\Phi_\pm \equiv \Phi \pm \Psi \,. \tag{5.3}$$

A non-trivial evolution of $\Phi_+$ is sourced either by anisotropic stress $\sigma$ or by pressure perturbations $\delta P$ (see Fig. 5.2). Under certain conditions, which we will identify, this induces a contribution to the overdensity $d_\gamma$ which is out of phase with its freely oscillating part.

In this section, we will give an analytic description of these effects, building on the pioneering work of Bashinsky and Seljak [6]. We first extract the two possible origins of a phase in the acoustic oscillations (Sect. 5.2.1). We then study a simple toy model to illustrate these abstract conditions (Sect. 5.2.2). Finally, we explicitly derive the phase shift in a curvaton-like model with isocurvature fluctuations and for light free-streaming particles, such as neutrinos (Sect. 5.2.3).

**Fig. 5.2** Illustration of the coupled perturbations in the primordial plasma (after [2, 3])

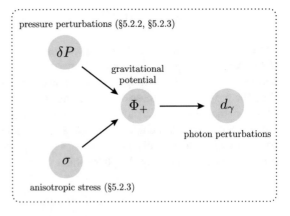

## 5.2.1 Conditions for a Phase Shift

In the following, we will analyse the evolution of fluctuations in the photon-baryon plasma. We have seen in Sect. 2.3 that the CMB couples gravitationally to perturbations in the matter densities. In terms of the variable $d_\gamma$, the evolution equation (2.27) for the photon fluctuations is given by

$$\ddot{d}_\gamma + \chi_\gamma \dot{d}_\gamma - c_\gamma^2 \nabla^2 d_\gamma = \nabla^2(\Phi + 3c_\gamma^2 \Psi), \tag{5.4}$$

where we omitted the term $\nabla^4 \sigma_\gamma$, since the anisotropic stress is absent before decoupling, $\sigma_\gamma \approx 0$. For simplicity, we will ignore the small effect due to the baryons,[3] so that the speed of the fluctuations in the photon density is $c_\gamma^2 \approx \frac{1}{3}$. This implies that the Hubble drag rate in (5.4) vanishes, $\chi_\gamma \approx 0$, which results in the evolution equation (5.2). The solution for $d_\gamma$ can then be expressed as

$$d_\gamma(y) = d_{\gamma,\text{in}} \cos y - c_\gamma^{-2} \int_0^y dy'\, \Phi_+(y') \sin(y - y'), \tag{5.5}$$

where $y \equiv c_\gamma k \tau$. The first term is the homogeneous solution with constant superhorizon adiabatic initial condition $d_{\gamma,\text{in}} \equiv d_\gamma(y_{\text{in}} \ll 1)$. (Primordial isocurvature modes would be straightforward to include here and in the following by an additional sine contribution.) The second term is the inhomogeneous correction generated by the evolution of metric perturbations. By employing the trigonometric addition identity $\sin(y - y') = \sin y \cos y' - \cos y \sin y'$, we can rewrite (5.5) as

$$d_\gamma(y) = \left[ d_{\gamma,\text{in}} + c_\gamma^{-2} A(y) \right] \cos y - c_\gamma^{-2} B(y) \sin y, \tag{5.6}$$

where we defined

$$A(y) \equiv \int_0^y dy'\, \Phi_+(y') \sin y', \tag{5.7}$$

$$B(y) \equiv \int_0^y dy'\, \Phi_+(y') \cos y'. \tag{5.8}$$

In general, we wish to evaluate (5.6) at recombination, $y \to y_{\text{rec}}$. It is however a good approximation for the high-$\ell$ CMB modes to take the limit $y \to \infty$ and assume that the background is radiation dominated (see Appendix B for additional details).

---

[3] While we neglect the contributions of baryons and dark matter to the energy density, we implicitly include the baryons by assuming that the photons are not free-streaming particles.

If the integral in (5.8) converges in this limit, a constant phase shift $\phi$ relative to the homogeneous solution,

$$\sin \phi = \frac{B}{\sqrt{\left(A + c_\gamma^2 d_{\gamma,\text{in}}\right)^2 + B^2}}, \tag{5.9}$$

is generated by a non-zero value of $B \equiv \lim_{y\to\infty} B(y)$. This phase shift will be reflected in a shift of the acoustic peaks of the CMB anisotropy spectrum.

In order to identify the precise physical conditions for which such a non-zero phase shift is generated, it is convenient to combine $A$ and $B$ into a complex field

$$B + iA = \int_0^\infty dy\, e^{iy}\, \Phi_+(y) = \frac{1}{2}\int_{-\infty}^\infty dy\, e^{iy}\left[\Phi_+^{(s)}(y) + \Phi_+^{(a)}(y)\right], \tag{5.10}$$

where $\Phi_+^{(s)}(y)$ and $\Phi_+^{(a)}(y)$ are even and odd functions of $y$, respectively. It is straightforward to realize that the even part of $\Phi_+$ determines $B$, while the odd part controls $A$:

$$B = \frac{1}{2}\int_{-\infty}^\infty dy\, e^{iy}\, \Phi_+^{(s)}(y), \qquad iA = \frac{1}{2}\int_{-\infty}^\infty dy\, e^{iy}\, \Phi_+^{(a)}(y). \tag{5.11}$$

As long as $e^{iy}\Phi_+^{(s)}(y)$ vanishes faster than $y^{-1}$ for $|y| \to \infty$ and $\Phi_+^{(s)}(y)$ is an analytic function, we will therefore get $B = 0.$[4] This suggests two ways of generating a non-zero $B$ and, hence, a phase shift in the solution for the photon density:

  i. rapid growth of $\Phi_+^{(s)}(\pm iy)$        $\longrightarrow$   mode travelling faster than $c_\gamma$,

  ii. non-analytic behaviour of $\Phi_+^{(s)}(y)$   $\longrightarrow$   non-adiabatic fluctuations.

The mathematical requirements listed on the left are mapped directly into physical conditions, shown on the right.

- The first condition is easy to understand physically: in (5.10), the Green's function of $d_\gamma$, i.e. $\sin(y - y')$, leads to exponential suppression for $y \to i\infty$. To have a growing solution at $y = i\infty$, we therefore need a term in $\Phi_+$ of the form $e^{-ic_s k\tau} = e^{-i(c_s/c_\gamma)y}$ with $c_s > c_\gamma.$[5]
- The second possibility, non-analyticity, is easy to understand mathematically, but the physical requirements are less transparent. First of all, the equations of motion for any mode should be analytic around any finite value of $k\tau$ in the radiation-dominated era since there is no preferred time. Hence, the only moment at which

----

[4]The odd part $\Phi_+^{(a)}(y)$ is not analytic around $y = 0$ since the equations are symmetric in $y \to -y$, which is the reason why we always find contributions to $A$.

[5]Note that $c_s$ is just a parameter of the wave-like solution and is not necessarily the sound speed of a fluid. Indeed, in the case of free-streaming radiation, it corresponds to the propagation speed of the individual particles.

non-analytic behaviour is possible is around $k\tau = 0$, i.e. where the initial conditions are defined. Let us first show that adiabatic initial conditions are analytic at $k\tau = 0$. By definition, for adiabatic initial conditions, any long-wavelength mode is locally generated by a diffeomorphism [21]. In the limit $k\tau \to 0$, we then have $\Phi_+ = \Phi_{+,\text{in}} + \mathcal{O}(k^2\tau^2)$. This expansion is necessarily analytic in $k^2$ (by locality and rotational invariance), but also in $k^2\tau^2$, because the scaling $k \to \lambda k$ and $\tau \to \lambda^{-1}\tau$ can be absorbed into the overall normalization of the scale factor $a$ which has no physical effect.[6] Hence, $\Phi_+^{(s)}(y)$ must be analytic around $y = c_\gamma k\tau = 0$, as long as the modes are adiabatic. Conversely, any violation of analyticity requires a source of non-adiabaticity.

In the following sections, we will illustrate the different physical origins of the CMB phase shift through a number of simple examples.

### 5.2.2 Intuition from a Toy Model

We can exactly solve the just described system of equations in a simple toy model which will also help to build some intuition for them. To be specific, we consider a scenario in which the photons propagate at a different speed than the metric fluctuations $\Phi_+$, $c_s \ne c_\gamma$. Our goal is to understand under which conditions this mismatch results into a phase shift in the photon oscillations.

The Einstein equations for the metric potentials, Eq. (2.29), can be rewritten in terms of the fields $\Phi_\pm$ as

$$\ddot{\Phi}_+ + 3\mathcal{H}\dot{\Phi}_+ + (2\dot{\mathcal{H}} + \mathcal{H}^2)\Phi_+ = 8\pi Ga^2\,\delta P + \mathcal{S}[\Phi_-], \qquad (5.12)$$

where the source term on the right-hand side is defined as

$$\mathcal{S}[\Phi_-] \equiv \ddot{\Phi}_- + \mathcal{H}\dot{\Phi}_- - \left(2\dot{\mathcal{H}} + \mathcal{H}^2 + \frac{2}{3}\nabla^2\right)\Phi_-. \qquad (5.13)$$

The field $\Phi_-$ is related to the total anisotropic stress $\sigma$ via the constraint equation (2.30). In the standard model, only neutrinos provide a source for $\sigma$ (and

---

[6]In a universe with a preferred time, this rescaling would also require a shift in this preferred time to keep the density fluctuations fixed. For adiabatic modes, the curvature perturbation $\zeta$ is conserved outside the horizon even in the presence of such a preferred time and so this is unlikely to have an impact on gauge-invariant observables.

hence $\Phi_-$), while both photons and neutrinos contribute to $\delta P$. BSM particles may lead to additional anisotropic stress and/or pressure. In general, we can express the pressure perturbation as [22]

$$\delta P = c_s^2 \delta\rho + \delta P_{\text{en}}, \tag{5.14}$$

where $c_s$ is the speed controlling the propagation of the total density perturbation $\delta\rho$ and $\delta P_{\text{en}}$ designates the non-adiabatic entropy perturbation. In the case of adiabatic perturbations, one has $\delta P_{\text{en}} = 0$ and $c_s^2 = w - [3\mathcal{H}(1+w)]^{-1}\dot{w}$, where we emphasize that we are *not* assuming that $P = w\rho$, but $\bar{P} = w\bar{\rho}$. By employing the relativistic generalization of the Poisson equation (2.28), we can substitute the density perturbation $\delta\rho$ and rewrite (5.12) as

$$\ddot{\Phi}_+ + 3\mathcal{H}(1+c_s^2)\dot{\Phi}_+ - 3\mathcal{H}^2(w-c_s^2)\Phi_+ + c_s^2 k^2\Phi_+ = 8\pi G a^2 \delta P_{\text{en}}. \tag{5.15}$$

Here, we additionally assumed vanishing anisotropic stress which implies that the source term $\mathcal{S}[\Phi_-]$ is absent.

Until now, this discussion has been fairly general and we have only assumed the absence of anisotropic stress. At this point, in particular the parameters $w$ and $c_s$ are still general and possibly time-dependent. Having said this, we will now make a few simplifying assumptions so that we can derive an analytic solution for the evolution of $\Phi_+(\tau)$. First, we take the equation of state $w$ to be nearly constant. This allows us to integrate the combined Friedmann equations (2.5) to arrive at

$$\mathcal{H} = \frac{2}{1+3w}\frac{1}{\tau}. \tag{5.16}$$

Second, we assume $c_s^2 \approx \text{const}$ and $\delta P_{\text{en}} \approx 0$ so that we can solve (5.15) analytically. We remark that these assumptions are not guaranteed to be easily realizable in a physical model for arbitrary $w$ and $c_s$. Our analysis only serves as a simple illustration of some of the effects that give rise to phase shifts in the cosmic sound waves. We will discuss more concrete examples of these effects in Sect. 5.2.3.

For the following discussion, it will be convenient to define $z \equiv c_s k\tau$ and express (5.15) as

$$\frac{d^2}{dz^2}\Phi_+ + \frac{1-2\alpha}{z}\frac{d}{dz}\Phi_+ + \left[1 - \frac{\beta}{z^2}\right]\Phi_+ = 0, \tag{5.17}$$

where we introduced

$$\alpha \equiv \frac{1}{2} - \frac{3(1+c_s^2)}{1+3w}, \qquad \beta \equiv \frac{12(w-c_s^2)}{(1+3w)^2}. \tag{5.18}$$

In the physically interesting parameter regime, $0 \leq (c_s^2, w) \leq 1$, we have $-\frac{11}{2} \leq \alpha \leq -\frac{1}{4}$, with equality for $(c_s^2, w) = (1, 0)$ and $(0, 1)$, respectively. The general solution of (5.17) is

$$\Phi_+(z) = z^\alpha \left( c_1 J_\kappa(z) + c_2 Y_\kappa(z) \right), \qquad \kappa \equiv \sqrt{\alpha^2 + \beta}, \qquad (5.19)$$

where $J_\kappa(z)$ and $Y_\kappa(z)$ are Bessel functions of the first and second kind, respectively. Note that $\kappa$ is strictly positive and real-valued for the physically relevant parameter range, $\frac{1}{2}\sqrt{3} \leq \kappa \leq \frac{1}{2}\sqrt{73}$, with the minimum at $(c_s^2, w) = (\frac{1}{3}, 1)$ and the maximum at $(c_s^2, w) = (1, 0)$.

The next step is to impose the initial conditions on this solution. To this end, we consider the superhorizon limit, $z \ll 1$,

$$\Phi_+(z) \simeq \frac{2^{-\kappa} \left( c_1 + c_2 \cot(\pi\kappa) \right)}{\Gamma(1 + \kappa)} z^{\alpha+\kappa} - \frac{2^\kappa c_2 \Gamma(\kappa)}{\pi} z^{\alpha-\kappa} + \cdots . \qquad (5.20)$$

The "growing mode" solution corresponds to $c_2 \equiv 0$ because $\alpha + \kappa > \alpha - \kappa$ which implies that

$$\Phi_+(z) = c_1 z^\alpha J_\kappa(z) . \qquad (5.21)$$

The nature of the initial conditions (adiabatic or entropic) will dictate the overall normalization in (5.21). Since the superhorizon limit of $\Phi_+$ is a constant for $c_s^2 = w$, which we can match to the superhorizon value of the primordial curvature perturbation $\zeta$, the normalization is given by $c_1 = 2\sqrt{2\pi}\,\zeta$. Although the normalization of non-adiabatic modes is in principle model-dependent, we maintain this normalization even for $c_s^2 \neq w$. In this way, the $y \to \infty$ limit of (5.8) is given by

$$B = 2\sqrt{2\pi}\,\zeta \int_0^\infty dy \left( \frac{c_s}{c_\gamma} y \right)^\alpha J_\kappa\left( \frac{c_s}{c_\gamma} y \right) \cos y , \qquad (5.22)$$

and similarly for $A$. Together with $d_{\gamma,\text{in}} = -3\zeta$, this allows us to compute the phase shift via (5.9). Figure 5.3 illustrates the dependence of the phase shift $\phi$ on the parameters $c_s^2$ and $w$. Let us reiterate that we do not imagine that all of the displayed combinations of $c_s^2$ and $w$ can be covered in physically realistic models. To conclude the discussion of this toy model and show that the most important features of the figure can be understood analytically, we now take slices through the parameter space.

First, we look at the special case of adiabatic fluctuations which are characterised by $c_s^2 = w$. The parameters in (5.18) and (5.19) then reduce to

$$\alpha = -\frac{5 + 3c_s^2}{2(1 + 3c_s^2)}, \qquad \beta = 0, \qquad \kappa = |\alpha| . \qquad (5.23)$$

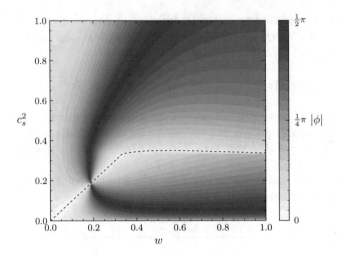

**Fig. 5.3** Phase shift $\phi$ for varying speed of sound $c_s$ and equation of state $w$ (after [2, 3]). The dashed line indicates $\phi = 0$. The phase shift is positive above this line, while below it is negative

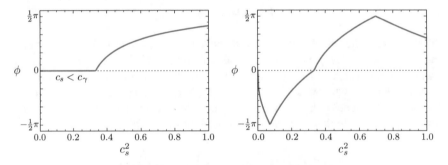

**Fig. 5.4** Phase shift $\phi$ for varying $c_s^2 = w$ (*left*), and for varying $c_s^2$ at fixed $w = \frac{1}{3}$ (*right*; after [2, 3])

For this case, the left panel of Fig. 5.4 shows that the phase shift is positive for $c_s > c_\gamma$ and, in particular, vanishes for $c_s \leq c_\gamma = \frac{1}{\sqrt{3}}$. This is consistent with our abstract reasoning of the previous section. At large $z = c_s k \tau$, the solution (5.21) behaves as $z^{\alpha - \frac{1}{2}} \cos(z) \propto \cos(c_s/c_\gamma \, y)$, with $y = c_\gamma k \tau$. We cannot conclude that $\phi = 0$ because the contour at infinity in (5.11) will not vanish when $c_s > c_\gamma$ (see condition *i.* in Sect. 5.2.1). It was still not guaranteed to find $\phi \neq 0$, but there was no reason to expect otherwise. By contrast, the phase shift $\phi$ vanishes for $c_s \leq c_\gamma$ for exactly the reasons discussed above. To be specific, the solution (5.21) is now of the form $z^\alpha J_{|\alpha|}(z)$ with $\alpha < 0$. Since the non-analytic behaviour of $z^\alpha$ near $z = 0$ cancels that of the Bessel function, the solution is analytic around that point (cf. condition *ii.* in Sect. 5.2.1). For adiabatic modes, this is of course precisely what we expected.

Second, we consider non-adiabatic fluctuations, i.e. fluctuations which contain an isocurvature component, which corresponds to $c_s^2 \neq w$. In this case, we expect a non-zero phase shift to occur for any values of $c_s^2$ and $w$. For illustrative purposes, we take $w = \frac{1}{3}$ and only allow $c_s^2$ to vary which results in

$$\alpha = -1 - \frac{3c_s^2}{2}, \qquad \beta = 1 - 3c_s^2, \qquad \kappa = \frac{1}{2}\sqrt{8 + 9c_s^4}. \qquad (5.24)$$

In the right panel of Fig. 5.4, we show the corresponding phase shift and observe that it now only vanishes at the special point $c_s^2 = w = \frac{1}{3}$. The reason for this is of course that the fluctuations are adiabatic for this combination of parameters. At the same time, this is also where the parameter $\beta$ changes sign, which is the reason for the change in the sign of the phase shift, $\phi \lessgtr 0$ for $c_s^2 \lessgtr w$. Now, the phase shift is linked to the non-analytic behaviour of $\Phi_+(z)$ near the origin. By considering the $z \to 0$ limit of (5.21), we are able to see this explicitly:

$$\Phi_+(z) = \frac{c_1}{2^\kappa \Gamma(1 + \kappa)} z^{\alpha + \kappa} \left[1 + \mathcal{O}(z^2)\right]. \qquad (5.25)$$

Since $\alpha + \kappa < 2$ for $0 \leq c_s^2 \leq 1$, the function $\Phi_+(z)$ has a non-analyticity at $z = 0$ (cf. condition *ii.* in Sect. 5.2.1), unless we take the adiabatic limit for which $\alpha + \kappa = 0$. Consequently, this corresponds to the non-trivial superhorizon evolution of $\Phi_+$ in the presence of isocurvature modes.

### 5.2.3 Simple Examples

The discussion of our toy model in the previous section indicates that isocurvature perturbations and free-streaming particles produce a phase shift. To study this further, it is useful to consider more realistic models. In the following, we will therefore compute the phase shift in a curvaton-like model, and for neutrinos and neutrino-like species.

**Isocurvature Perturbations**

Our minimal curvaton-like model only includes three species to simplify the calculations as much as possible: photons $(\gamma)$, a dark radiation fluid $(Y)$ and a matter component $(m)$ which decays into the dark radiation and will eventually carry the isocurvature fluctuations. Since we are only interested in the gravitational effects on the photons, i.e. we want to avoid the direct heating of the photons due to the decaying matter, we include the dark radiation. We are particularly interested in small deviations from the adiabatic limit $c_s^2 = w$, which is what this model allows us to explore.

By introducing a constant decay rate $\Gamma$, the background densities of $m$ and $Y$ are governed by the coupled equations

$$\frac{1}{a^3} \frac{d}{d\tau} (a^3 \bar{\rho}_m) = -\Gamma a \bar{\rho}_m , \qquad (5.26)$$

$$\frac{1}{a^4} \frac{d}{d\tau} (a^4 \bar{\rho}_Y) = +\Gamma a \bar{\rho}_m . \qquad (5.27)$$

We will work perturbatively in $\epsilon_m \equiv \bar{\rho}_m / \bar{\rho}$ to further simplify the calculations. At zeroth order in $\epsilon_m$, the universe is radiation dominated which implies that $a = \tau / \tau_{\mathrm{in}}$. Integrating (5.26) and setting the initial value $\bar{\rho}_{m,\mathrm{in}} \equiv \bar{\rho}_m (\tau_{\mathrm{in}})$, we obtain

$$\bar{\rho}_m (a) = \frac{\bar{\rho}_{m,\mathrm{in}}}{a^3} e^{-\frac{1}{2} \Gamma \tau_{\mathrm{in}} (a^2 - 1)} . \qquad (5.28)$$

We could get the solution for $\bar{\rho}_Y (a)$ by substituting (5.28) into (5.27). This will however not be required for our purposes.

We are interested in investigating the implications of the decaying matter on the evolution of the metric perturbation $\Phi_+$ as given by (5.12). Since the pressure perturbations only receive contributions from $\gamma$ and $Y$, they are given by

$$\delta P = c_\gamma^2 (\delta \rho_\gamma + \delta \rho_Y) = \frac{1}{3} \delta \rho - \frac{1}{3} \delta \rho_m , \qquad (5.29)$$

where we used $c_Y^2 = c_\gamma^2 = \frac{1}{3}$. We can then write (5.12) in the absence of anisotropic stress as

$$\ddot{\Phi}_+ + 4\mathcal{H} \dot{\Phi}_+ - \frac{1}{3} \nabla^2 \Phi_+ = (3w - 1) \mathcal{H}^2 \Phi_+ - \mathcal{H}^2 \epsilon_m \delta_m \qquad (5.30)$$

by employing the Poisson equation (2.28). To infer the desired insights, we have to solve this at linear order in $\epsilon_m$.

By decomposing $\Phi_+ = \Phi_+^{\mathrm{ad}} + \Phi_+^{\mathrm{iso}}$ (similarly for $\delta_m$), we can isolate the isocurvature contribution and shortcut this computation. Equation (5.30) then implies

$$\Phi_+^{\mathrm{iso}\,\prime\prime} + \frac{4}{y} \Phi_+^{\mathrm{iso}\,\prime} + \Phi_+^{\mathrm{iso}} = -\frac{\epsilon_m}{y^2} \delta_m^{\mathrm{iso}} + \mathcal{O}(\epsilon_m^2) , \qquad (5.31)$$

where the primes denote derivatives with respect to $y \equiv c_\gamma k \tau$. In addition, we used that $\Phi_+^{\mathrm{iso}} \sim \mathcal{O}(\epsilon_m)$, which implies that all terms multiplying $\Phi_+^{\mathrm{iso}}$ can be evaluated at zeroth order in $\epsilon_m$. Since the evolution of $\delta_m^{\mathrm{iso}}$ is governed by $\delta_m^{\mathrm{iso}\,\prime\prime} + \frac{1}{y} \delta_m^{\mathrm{iso}\,\prime} = 0 + \mathcal{O}(\epsilon_m)$ according to (2.27), we only need the homogeneous solution to continue, which is

$$\delta_m^{\mathrm{iso}} (y) = c_1 + c_2 \ln y , \qquad (5.32)$$

where $c_{1,2}$ are constants that may depend on $k$, because the right-hand side of (5.31) is proportional to $\epsilon_m$. By means of the Green's function $G_{\Phi_+} (y, y')$, we can then solve (5.31). Substituting (5.28) and (5.32), we get

$$\Phi_+^{\text{iso}}(y) = \frac{1}{c_\gamma k \tau_{\text{in}}} \, \tilde{\epsilon}_{m,\text{in}} \underbrace{\int_{y_{\text{in}}}^{y} \mathrm{d}y' \, G_{\Phi_+}(y, y') \left( -\exp\left[ -\frac{1}{2} \frac{(y')^2}{(c_\gamma k \tau_{\text{dec}})^2} \right] \frac{c_1 + c_2 \ln(y')}{y'} \right)}_{\equiv \mathcal{I}(y)} ,$$

$$(5.33)$$

where we introduced the "decay time scale" $\tau_{\text{dec}}^2 \equiv \tau_{\text{in}} / \Gamma$ and defined $\tilde{\epsilon}_{m,\text{in}} \equiv \epsilon_{m,\text{in}} e^{\frac{1}{2}(\tau_{\text{in}}/\tau_{\text{dec}})^2}$.

A couple of comments on the features of this solution are in order. First, we notice that the solution has an overall factor of $(c_\gamma k \tau_{\text{in}})^{-1}$ which reflects the growth of $\epsilon_m$ from the initial time, $\tau_{\text{in}}$, to the time of horizon crossing, $(c_\gamma k)^{-1}$. Second, we observe that the integral is highly suppressed when $k\tau_{\text{dec}} \ll 1$. The reason for this can be easily understood: the integral would have been dominated by contributions around the time of horizon crossing, $y \sim \mathcal{O}(1)$, but this is long after $\rho_m$ has decayed for $k\tau_{\text{dec}} \ll 1$.

In order to interpret (5.33), it is useful to introduce $\tau_{\text{eq}}$ as the time at which the matter and radiation background densities $\bar{\rho}_m$ and $\bar{\rho}_\gamma + \bar{\rho}_Y$ would be equal if there was no decay. Since this is given by $\tau_{\text{eq}} \simeq \tau_{\text{in}} / \tilde{\epsilon}_{m,\text{in}}$, we can rewrite (5.33) as

$$\Phi_+^{\text{iso}}(y) = \frac{1}{c_\gamma k \tau_{\text{eq}}} \mathcal{I}(y) .$$

$$(5.34)$$

By substituting (5.34) into (5.7) and (5.8), and taking the limits $y_{\text{in}} \to 0$ and $y \to \infty$,

$$A^{\text{iso}} \equiv \frac{1}{c_\gamma k \tau_{\text{eq}}} \int_0^\infty \mathrm{d}y' \, \mathcal{I}(y') \sin y' ,$$

$$(5.35)$$

$$B^{\text{iso}} \equiv \frac{1}{c_\gamma k \tau_{\text{eq}}} \int_0^\infty \mathrm{d}y' \, \mathcal{I}(y') \cos y' .$$

$$(5.36)$$

we could compute the phase shift. Furthermore, in the additional limit $y_{\text{dec}} \gg 1$, we could drop the exponential in (5.33) which simplifies the result to

$$A^{\text{iso}} \xrightarrow{y_{\text{dec}} \gg 1} \frac{\pi}{4} c_2 \frac{1}{c_\gamma k \tau_{\text{eq}}} ,$$

$$(5.37)$$

$$B^{\text{iso}} \xrightarrow{y_{\text{dec}} \gg 1} \frac{1}{2} (c_1 - c_2 \gamma_{\text{E}}) \frac{1}{c_\gamma k \tau_{\text{eq}}} ,$$

$$(5.38)$$

where $\gamma_{\text{E}} \approx 0.5772$ is the Euler–Mascheroni constant. Figure 5.5 shows the numerical result for $B^{\text{iso}}$ as a function of $y_{\text{dec}} \equiv c_\gamma k \tau_{\text{dec}}$. Moreover, for $y_{\text{dec}} \gtrsim 5$, we observe that the analytic result (5.38) becomes a good approximation.

Summing up this calculation, we confirmed within a simple model that isocurvature perturbations may lead to a phase shift, as we expected from condition *ii.* of Sect. 5.2.1. Interestingly, the phase shift has a nontrivial scale dependence according to Fig. 5.5. This is likely a general feature of isocurvature models and probably allows it to be distinguished from other sources for a phase shift.

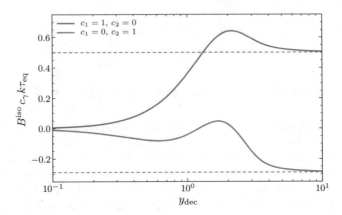

**Fig. 5.5** Numerical value of $B^{\text{iso}} c_\gamma k \tau_{\text{eq}}$ as a function of $y_{\text{dec}}$ (after [2, 3]). The blue and red solid lines illustrate the effect from non-zero values of $c_1$ and $c_2$, respectively. The asymptotic values of (5.38) are denoted by the dashed lines

### Free-Streaming Particles

In the toy model of Sect. 5.2.2, we also observed that a phase shift arises if perturbations in the gravitational potential travel faster than the speed of sound in the photon-baryon fluid. Free-streaming relativistic particles, such as neutrinos, provide a simple way to physically realize this [6].[7] In the rest of this section, we present a new derivation of this result. We will then show in Sect. 5.3 that CMB observations have become accurate enough so that we are able to detect this effect in the data (see also [1]).

Our discussion in this section will ignore both the matter and baryon content of the universe since most of the modes relevant to both current and future measurements of the CMB entered the horizon during the radiation-dominated era. We however show in Appendix B that this is a good approximation for high-$\ell$ modes, which are relevant here, and also discuss some of the implications of a finite matter density.

---

[7] From the physical point of view, it should be clear that free-streaming radiation travels at the speed of light. However, in the equations for the density perturbations of this radiation, this property is not necessarily transparent. Usually, the fact that free-streaming particles can induce a significant anisotropic stress is more apparent. Since this then serves as a source for $\Phi_+$ and, consequently, the evolution of $d_\gamma$ through (5.2), the origin of the phase shift is often identified with the presence of anisotropic stress. In principle, however, it is possible to imagine scenarios with no supersonic propagation modes, but significant anisotropic stress (e.g. non-relativistic, free-streaming particles), in which case we would not expect a phase shift. This therefore illustrates that the phase shift is generated by the propagation speed exceeding $c_s$ and not by the anisotropic stress itself.

In the absence of pressure perturbations $\delta P_a = c_a^2 \delta \rho_a$, with $c_a \neq c_\gamma$, the evolution equation (5.12) is given by

$$\Phi_+'' + \frac{4}{y}\Phi_+' + \Phi_+ = \tilde{\mathcal{S}}[\Phi_-]$$

$$\equiv \Phi_-'' + \frac{2}{y}\Phi_-' + 3\Phi_- . \tag{5.39}$$

The solution for $\Phi_+$ can be written as

$$\Phi_+(y) = 3\Phi_{+,\text{in}} \frac{\sin y - y \cos y}{y^3} + \int_{y_{\text{in}}}^y \mathrm{d}y'\, \tilde{\mathcal{S}}[\Phi_-(y')]\, G_{\Phi_+}(y, y') , \tag{5.40}$$

where we introduced the Green's function

$$G_{\Phi_+}(y, y') = \Theta(y - y')\frac{y'}{y^3}\left[(y' - y)\cos(y' - y) - (1 + yy')\sin(y' - y)\right], \tag{5.41}$$

with the Heaviside step function $\Theta$. We proceed by following [6] and writing the solution for $\Phi_+$ as an expansion in powers of the fractional energy density that is contributed by the species of free-streaming particles, $X$, as measured by the dimensionless ratio

$$\epsilon_X \equiv \frac{\rho_X}{\rho_\gamma + \rho_X} = \frac{N_{\text{eff}}}{a_\nu + N_{\text{eff}}} . \tag{5.42}$$

We have $\epsilon_\nu \approx 0.41$ for the SM neutrinos. By matching to the constant superhorizon solution for adiabatic initial conditions [6], we obtain the superhorizon initial condition of the homogeneous solution,

$$\Phi_{+,\text{in}} = \frac{20 + 4\epsilon_X}{15 + 4\epsilon_X}\zeta = \frac{4}{3}\zeta\left(1 - \frac{1}{15}\epsilon_X + \mathcal{O}(\epsilon_X^2)\right) , \tag{5.43}$$

where $\zeta$ is the conserved curvature perturbation.

We require an expression for $\Phi_-(y)$ to find the inhomogeneous part of the solution (5.40). This is governed by the anisotropic stress $\sigma_X$ induced by the free-streaming particles,

$$\Phi_-(y) = -\frac{2k^2\epsilon_X}{y^2}\sigma_X(y) , \tag{5.44}$$

which follows from the Einstein constraint equation (2.30) with $\sigma = \epsilon_X \sigma_X$. An evolution equation for the anisotropic stress $\sigma_X$ can be derived fromthe Boltzmann equation for the distribution function $f_X(\tau, x, q, \hat{n})$ with comoving momenta $q = q\hat{n}$. We decompose $f_X$ into a background component $\bar{f}_X$ and a perturbation $\delta f_X \equiv f_X - \bar{f}_X$. It is useful to integrate out the momentum dependence of the distribution function in the case of massless particles and define

$$\int dq\, q^3 \left(\delta f_X + q\, \partial_q \bar{f}_X \Psi\right) = \frac{4}{3} D_X(\tau, x, \hat{n}) \times \int dq\, q^3 \bar{f}_X(q)\,. \tag{5.45}$$

The linearised, collisionless Boltzmann equation then takes the form

$$\dot{D}_X + ik\mu D_X = -3ik\mu\, \Phi_+\,, \tag{5.46}$$

where we introduced $\mu = \hat{n} \cdot \hat{k}$. We note that $D_X$ only depends on $\Phi_+$, but not on $\Phi_-$. It is useful to expand the momentum-integrated distribution function $D_X$ into multipole moments,

$$D_X = \sum_{\ell=0}^{\infty} (-i)^\ell (2\ell + 1) D_{X,\ell}\, P_\ell(\mu)\,, \tag{5.47}$$

with the Legendre polynomials $P_\ell(\mu)$. The monopole moment $D_{X,0}$ determines the overdensity $d_X$, while the quadrupole moment $D_{X,2}$ is associated with the anisotropic stress $\sigma_X$. This can be seen by writing the perturbed stress-energy tensor in terms of the perturbed distribution function,

$$\delta T^\mu{}_{\nu,X} = a^{-4} \int d\Omega_{\hat{n}}\, \hat{n}^\mu \hat{n}_\nu \int dq\, q^3\, \delta f_X\,. \tag{5.48}$$

Comparing this expression to (2.23), we find

$$D_{X,0} = d_X\,, \quad D_{X,1} = ku_X\,, \quad D_{X,2} = \frac{3}{2} k^2 \sigma_X\,. \tag{5.49}$$

The quadrupole moment of (5.46) then provides the missing evolution equation for the anisotropic stress and we can rewrite (5.44) as

$$\Phi_-(y) = -\frac{4}{3}\frac{\epsilon_X}{y^2} D_{X,2}(y)\,. \tag{5.50}$$

Defining $D_{X,\text{in}} \equiv D_X(\tau_\text{in})$ at some time $\tau_\text{in}$, the solution to (5.46) is

$$D_X(\tau) = e^{-ik\mu(\tau - \tau_\text{in})} D_{X,\text{in}} - 3ik\mu \int_{\tau_\text{in}}^{\tau} d\tau'\, e^{-ik\mu(\tau - \tau')} \Phi_+(\tau')\,. \tag{5.51}$$

Our goal is to extract the quadruple moment $D_{X,2}$ of the equation. We will only keep the monopole term $D_{X,0}(\tau_\text{in})$ in the homogeneous part of the solution since

$D_{X,\ell}(\tau_{in}) \propto \tau_{in}^{\ell}$. This is made possible by the fact that we can set the initial conditions at a sufficiently early time $\tau_{in} \ll k^{-1}$ so that the modes with $\ell > 0$ become subdominant. From now on, we in fact take $k\tau_{in} \to 0$. By assuming adiabatic initial conditions, i.e. $D_{X,0}(\tau_{in}) = d_{X,in} = -3\zeta$, we obtain

$$D_{X,2}(y) = -3\zeta\, j_2\left[c_\gamma^{-1}y\right] + \frac{3}{c_\gamma}\int_0^y dy'\,\Phi_+(y')\left\{\frac{2}{5}j_1\left[c_\gamma^{-1}(y-y')\right] - \frac{3}{5}j_3\left[c_\gamma^{-1}(y-y')\right]\right\},$$

(5.52)

where the Bessel functions $j_\ell$ arise from the Rayleigh expansion of the exponentials. By substituting this into (5.50), we are able to provide the desired direct link between the two gravitational potentials $\Phi_+$ and $\Phi_-$. The most important feature of the solution (5.52) is that it contains modes that travel at the speed of light. Specifically, recall that $c_\gamma^{-1}y = k\tau$ and, therefore, the Bessel functions describe oscillatory solutions with a speed of propagation of $c_s = 1$. This is exactly the property of the free-streaming radiation that makes a phase shift possible, as we have emphasized before.

Since we have now obtained a closed set of equations, we can solve them perturbatively in $\epsilon_X$:

$$d_\gamma \equiv \sum_n d_\gamma^{(n)}, \quad \Phi_\pm \equiv \sum_n \Phi_\pm^{(n)},$$

(5.53)

where the superscripts on $d_\gamma^{(n)}$ and $\Phi_\pm^{(n)}$ count the order in $\epsilon_X$. In the following, we provide the solution up to first order:

• Zeroth order in $\epsilon_X$: because $\Phi_-^{(0)}(y) = 0$, the function $\Phi_+^{(0)}(y)$ is given by the homogeneous solution,

$$\Phi_+^{(0)}(y) = 4\zeta\,\frac{\sin y - y\cos y}{y^3}.$$

(5.54)

Plugging this equation into (5.7) and (5.8), we obtain

$$A^{(0)}(y) = 2\zeta - 2\zeta\,\frac{\sin^2(y)}{y^2} \xrightarrow{y\to\infty} 2\zeta,$$

(5.55)

$$B^{(0)}(y) = 2\zeta\,\frac{y - \cos y \sin y}{y^2} \xrightarrow{y\to\infty} 0.$$

(5.56)

The photon density perturbation at zeroth order is therefore given by

$$d_\gamma^{(0)}(y) \approx 3\zeta\cos y.$$

(5.57)

In the absence of anisotropic stress, we conclude that $B$ vanishes as expected and the correction due to $\Phi_+$ is in phase with the homogeneous solution.

• First order in $\epsilon_X$: since the source in (5.50) already contains an overall factor of $\epsilon_X$, we only have to compute the zeroth-order solution of the anisotropic stress, $\sigma_X^{(0)}$.

This means that we can rewrite (5.50) and (5.52) as

$$\Phi_-^{(1)}(y) = 4\zeta \frac{\epsilon_X}{y^2} j_2[c_\gamma^{-1} y] \tag{5.58}$$

$$- \frac{4}{c_\gamma} \frac{\epsilon_X}{y^2} \int_0^y dy' \, \Phi_+^{(0)}(y') \left\{ \frac{2}{5} j_1[c_\gamma^{-1}(y-y')] - \frac{3}{5} j_3[c_\gamma^{-1}(y-y')] \right\},$$

where $\Phi_+^{(0)}$ is given by (5.54). Inserting (5.58) into (5.40), we find

$$\Phi_+^{(1)}(y) = -\frac{4}{15} \zeta \, \epsilon_X \, \frac{\sin y - y \cos y}{y^3} + \int_0^y dy' \, \tilde{\mathcal{S}}[\Phi_-^{(1)}(y')] \, G_{\Phi_+}(y, y'). \tag{5.59}$$

Substituting this into (5.7) and (5.8), we finally arrive at expressions for $A^{(1)}$ and $B^{(1)}$. Evaluating these numerically, we get

$$A^{(1)} \approx -0.268 \, \zeta \, \epsilon_X, \qquad B^{(1)} \approx 0.600 \, \zeta \, \epsilon_X. \tag{5.60}$$

The expected phase shift manifests itself in the non-zero value of $B^{(1)}$.

Employing (5.9) together with the initial condition $d_{\gamma,\text{in}} = -3\zeta$, we obtain

$$\phi \approx 0.191\pi \, \epsilon_X + \mathcal{O}(\epsilon_X^2), \tag{5.61}$$

which is consistent with the result of Bashinsky and Seljak [6].

The phase shift will naturally play an important role in the observational constraints discussed in the following section since it is a clean signature of free-streaming particles. Let us employ the analytic result of this section to put these constraints into context by relating variations in $N_{\text{eff}}$ to shifts $\Delta\ell$ in the peaks of the CMB spectra. As discussed in Appendix B, the E-mode spectrum exhibits precisely the same phase shift as the temperature spectrum and, therefore, our analytic estimates are applicable in either case. We can relate a shift in angle $\phi$ in the small-angle approximation to a multipole shift by $\Delta\ell \simeq (\phi/\pi) \, \Delta\ell_{\text{peak}}$, where $\Delta\ell_{\text{peak}} \sim 330$ [4] is the distance between peaks in the temperature anisotropy spectrum for modes entering the horizon during radiation domination. Compared to a neutrinoless universe, we find that the shift of the peaks arising from ordinary neutrinos ($N_{\text{eff}} = N_\nu = 3.046$) is $\Delta\ell_\nu \approx 26$ based on (5.61). In a similar fashion, small variations $\Delta N_{\text{eff}}$ around the standard value $N_\nu = 3.046$ will give rise to a multipole shift of order

$$\Delta\ell_{\Delta N_{\text{eff}}} \approx 5.0 \times \Delta N_{\text{eff}}, \tag{5.62}$$

where we have expanded to linear order in $\Delta N_{\text{eff}}$. Although this result might be subject to an error of 20–30%, it is reliable enough to see that a sensitivity of $\sigma(N_{\text{eff}}) \sim 0.1$ will constrain a phase shift of order $\Delta\ell \lesssim 1$. Current constraints on $N_{\text{eff}}$ therefore imply $\Delta\ell \sim \mathcal{O}(1)$. We will observe in Sect. 5.3 that future CMB measurements are expected to constrain (or measure) shifts of order $\Delta\ell \sim \mathcal{O}(0.1)$. This is consistent with the rough expectation from measuring $\mathcal{O}(10)$ peaks and troughs in the E-mode power spectrum.


## 5.3   Current and Future Constraints

Many distinct sources of BSM physics can possibly be distinguished by observations of the CMB: new free-streaming or non-free-streaming particles, isocurvature perturbations and/or non-standard thermal histories. Having said that, both the size of the effect and whether it is degenerate with other cosmological parameters will ultimately determine the observability of any new physics. In this section, we infer new constraints on the density of free-streaming and non-free-streaming radiation from the data obtained by the Planck satellite [4]. Moreover, we also consider the capabilities of a proposed CMB-S4 experiment [5]. We will carry out a full likelihood analysis in order to obtain precise quantitative results.[8] In addition, we will complement these results, whenever possible, by some approximate analytic understanding of their qualitative origin.

   As we have discussed in Sect. 3.4.1, a change in the radiation density is degenerate with a shift in the primordial helium fraction $Y_p$ because they are anti-correlated at fixed $\theta_s$. This is why we expect that the CMB temperature constraints on $N_{\mathrm{eff}} + N_{\mathrm{fluid}}$ and $Y_p$ weaken considerably if we vary both of these parameters in the analysis. Since $N_{\mathrm{eff}}$ and $N_{\mathrm{fluid}}$ are not distinguished by their effects on the damping tail, we also have to break the degeneracy between these dark radiation parameters. We are however in the fortunate position that future datasets will be much less sensitive to these degeneracies especially due to the following two reasons:

- First, the amplitude of the CMB polarization signal, $\Theta_{P,\ell}$, is proportional to $n_e^{-1}$, but not $H$, as we demonstrate in Appendix B. This implies that it is sensitive to $Y_p$ alone. Physically speaking, this can be understood as follows: while the damping tail of the temperature spectrum is the integrated effect of the quadrupole on the monopole, polarization is a direct measurement of the quadrupole at the surface of last-scattering. This difference allows us to break the degeneracy between $Y_p$ and $N_{\mathrm{eff}} + N_{\mathrm{fluid}}$.
- Second, the CMB is sensitive to the perturbations in the free-streaming particles, as we showed in Sect. 5.2, and not only their contribution to the background evolution. Figure 5.6 illustrates this key feature which we will explore in more detail in the next subsection. The important point here is that the phase shift associated with free-streaming particles is not expected to be degenerate with other effects and,

---

[8]In this section, we prefer the use of Markov chain Monte Carlo (MCMC) techniques over forecasts based on the Fisher information matrix since the latter can underestimate the impact of degeneracies on the posterior distributions [23]. We attribute the (small) differences between our results and those of [24] to this.

in principle, it is measurable out to very high multipoles. Furthermore, the same phase shift appears in both temperature and polarization (see Appendix B for the derivation) which means that there is significant room for improvement in the sensitivity to this effect. As a consequence, the issues of degeneracies for $N_{\text{eff}}$ will largely be eliminated.

In the future, the perturbations in the radiation density will play an increasingly important role (and, consequently, the second point) by moving towards more sensitive experiments. In the rest of this section, we will demonstrate the potential of current and future experiments to detect the free-streaming nature of relativistic species.

### 5.3.1  MCMC Methodology

Our analysis employed the following observational and mock datasets:

- Planck data: A detailed description of the Planck 2015 likelihoods is provided in [4]. We include both the temperature and polarization data (even in the cases labelled "TT-only") in the low-$\ell$ likelihood ($2 \leq \ell \leq 29$). For the remaining multipoles ($\ell \geq 30$), we use the `plik` joint TT+TE+EE likelihood.[9] The TT spectrum is available up to $\ell_{\text{max}} = 2508$, while the TE and EE spectra are included up to $\ell_{\text{max}} = 1996$. The lensing potential is reconstructed in the multipole range $40 \leq \ell \leq 400$ using `SMICA` temperature and polarization maps. Finally, we use the `plik` TT-only likelihood with range $30 \leq \ell \leq 2508$ and the lensing reconstruction which only involves the temperature map for the TT-only constraints.
- CMB-S4 mock data: Our forecasts for a CMB-S4 experiment consider $10^6$ polarization-sensitive detectors with a sky coverage $f_{\text{sky}} = 0.75$ and a 1 arcmin beam. We choose a default observing time of five years which results in a sensitivity of $\sigma_T = \sigma_P/\sqrt{2} = 0.558 \, \mu\text{K}$ arcmin matching the most optimistic experimental setup studied in [24]. Our analysis included multipoles up to $\ell_{\text{max}} = 5000$ for both temperature and polarization.[10] We also study the implications of variations of the beam size and the maximum multipole on our results. Lastly, the fiducial cosmology used for all forecasts in this section is described by the following parameters: $\Omega_b h^2 = 0.022$, $\Omega_c h^2 = 0.120$, $h = 0.67$, $A_s = 2.42 \times 10^{-9}$, $n_s = 0.965$, $\tau = 0.078$, $N_{\text{eff}} = 3.046$ and $N_{\text{fluid}} = 0.0$.

We included an additional relativistic fluid, whose energy density is measured by the parameter $N_{\text{fluid}}$ defined in (3.22), in the analysis by modifyingthe Boltzmann

---

[9]Note that the high-$\ell$ polarization data was publicly released by the Planck collaboration, but labelled preliminary due to possible unresolved systematics. It should therefore be used with caution.

[10]These experimental specifications are slightly more optimistic (in particular the maximum temperature multipole of $\ell_{\text{max}}^T = 5000$ which will be hard to achieve because of astrophysical foregrounds) than those currently considered by the CMB-S4 collaboration [5]. We present forecasts on $N_{\text{eff}}$ using these specifications in Appendix C.

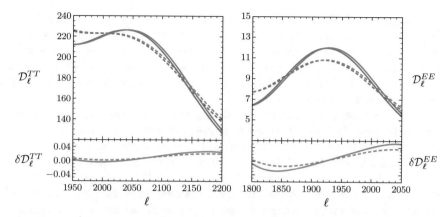

**Fig. 5.6** Illustration of the phase shift and the impact of lensing in the CMB power spectra (after [2, 3]). *Top:* Temperature spectrum $\mathcal{D}_\ell^{TT}$ (*left*) and polarization spectrum $\mathcal{D}_\ell^{EE}$ (*right*) for ($N_{\text{eff}} = 3.046$, $N_{\text{fluid}} = 0$) (blue) and ($N_{\text{eff}} = 2.046$, $N_{\text{fluid}} = 1.0$) (red) with $\mathcal{D}_\ell^X \equiv \ell(\ell+1)C_\ell^X/(2\pi)$ in units of $\mu\text{K}^2$. We note that we rescaled the TT and EE spectra represented by the red curves by the same constant factor which was chosen such that the height of the seventh peak of the TT spectrum matches for the red and blue curves. The unlensed and lensed data are shown with solid and dashed lines, respectively. In both the TT and EE spectra, the phase shift from $N_{\text{eff}}$ and the peak smearing from lensing can be seen. *Bottom:* Comparison of the relative difference $\delta\mathcal{D}_\ell^X \equiv \Delta\mathcal{D}_\ell^X/\mathcal{D}_\ell^X$ between the ($N_{\text{eff}} = 3.046$, $N_{\text{fluid}} = 0$) and ($N_{\text{eff}} = 2.046$, $N_{\text{fluid}} = 1.0$) spectra of the upper panels. The green solid and dashed lines are the differences in the unlensed and lensed data, respectively. It is apparent that the largest change occurs in the unlensed EE spectrum

code CLASS [25]. We fixed the equation of state and the sound speed of the fluid to $w_Y = c_Y^2 = \frac{1}{3}$, with initial conditions that were taken to be adiabatic. With this choice, our analytic results of Sect. 5.2 imply that this fluid does not contribute to the phase shift in the acoustic peaks. To derive constraints on the parameters $N_{\text{eff}}$ and $N_{\text{fluid}}$, we then run MontePython [26] and set the primordial helium abundance $Y_p$ to be consistent with the predictions of BBN (using the total relativistic energy density including both $N_{\text{eff}}$ and $N_{\text{fluid}}$ in determining the expansion rate), whenever it was not varied independently, which we denote as "$Y_p$ fixed". We run all Monte Charlo Markov chains until the variation in their means was small relative to the standard deviation ($R - 1 \lesssim 0.01$ in the Gelman–Rubin criterion [27]).

The described analysis employs the effects of gravitational lensing of the CMB in two distinct ways. First, we use the reconstruction of the power spectrum of the lensing potential from the measurements of the temperature and polarization four-point functions which is referred to as "lensing reconstruction". In the case of the CMB-S4 forecasts, we calculated the power spectrum of the lensing potential with CLASS. Second, as is illustrated in Fig. 5.6, CMB lensing also modifies the observed CMB power spectra (TT, TE and EE), primarily in the form of smearing the peaks [28]. A method called "delensing" removes the effect of lensing on these power spectra by means of the reconstructed lensing potential. We can trivially implement this in our forecasts in the limit of perfect delensing (we will simply compute spectra without including the lensing), but it is an involved procedure to implement on real data. While lensing moves information from the power spectra to higher-point functions, the utility of this delensing procedure is that it moves this information back to the power spectra. Rather than working with some more elaborate multi-point function likelihood, the information can easily be accounted for in our likelihood analysis, in this way.

While our Planck results do not include any delensing of the power spectra, we employ the publicly available lensing reconstruction likelihood. The lensing reconstruction noise for our CMB-S4 forecasts was calculated using the iterated delensing method described in [29] (based on [30–32]). Since the assumption of perfect delensing is good across a wide range of multipoles for a CMB-S4 experiment, this is what we assume for those forecasts using delensed spectra. A more careful analysis would be required to take lensing reconstruction noise into account when computing delensed spectra (see [5, 33] for subsequent studies).

### 5.3.2  Planck 2015 Results

We have reached an important threshold with the Planck 2015 dataset: the level of sensitivity is now sufficient to detect the free-streaming nature of the neutrinos (or any additional dark radiation). We display the marginalized constraints on $N_{eff}$ and $N_{fluid}$ in Table 5.1, while we show their posterior distributions in Fig. 5.7 and the two-dimensional joint constraints in Fig. 5.8. We present results both for the combined TT, TE and EE likelihoods, and for TT alone. In each case, we compare the results for fixed $Y_p$ with those when $Y_p$ is allowed to vary. When employing both temperature and polarization data from Planck, these results robustly demonstrate that there is very little degeneracy between $N_{eff}$, $N_{fluid}$ and $Y_p$.

The left panels in Figs. 5.7 and 5.8 clearly show that the constraints on $N_{eff}$ are mostly unaffected by the marginalization over $N_{fluid}$ and only mildly sensitive to the marginalization over $Y_p$ (even when the polarization data is removed). This robustness of the constraints suggests that they receive considerable constraining power from the phase shift because $N_{eff}$ is degenerate with $Y_p$ and $N_{fluid}$ in the damping tail. The measurement of the latter breaks the degeneracy between $N_{eff}$

**Table 5.1** Best-fit values and $1\sigma$ errors for $N_{eff}$, and $2\sigma$ upper limits for $N_{fluid}$ for the Planck 2015 data. In all cases, the lensing reconstruction and low-P likelihoods were used, and both $N_{fluid}$ and $N_{eff}$ are allowed to vary

|  | TT, TE, EE | | TT-only | |
|---|---|---|---|---|
|  | Varying $Y_p$ | Fixed $Y_p$ | Varying $Y_p$ | Fixed $Y_p$ |
| $N_{eff}$ | $2.68^{+0.29}_{-0.33}$ | $2.80^{+0.24}_{-0.23}$ | $2.89^{+0.49}_{-0.62}$ | $2.87^{+0.45}_{-0.37}$ |
| $N_{fluid}$ | <0.64 | <0.67 | <1.08 | <0.94 |

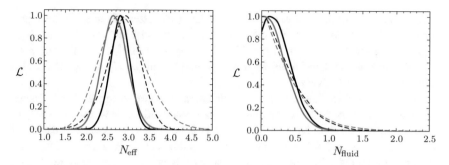

**Fig. 5.7** One-dimensional posterior distributions for the dark radiation parameters in our Planck analysis (after [2]). *Left:* Posteriors for $N_{\text{eff}}$ from Planck TT, TE and EE marginalized over $N_{\text{fluid}}$. While the blue curve involves the marginalization over $Y_p$, the black curve keeps $Y_p$ fixed to the value consistent with BBN. The hypothesis of no free-streaming radiation, $N_{\text{eff}} = 0$, is ruled out at high significance by both likelihoods. *Right:* Posterior for $N_{\text{fluid}}$ from Planck TT, TE and EE marginalized over $N_{\text{eff}}$. While the blue curve involves the marginalization over $Y_p$, the black curve keeps $Y_p$ fixed again. In both panels, the posterior distributions for Planck TT-only with the same marginalizations are shown as dashed lines

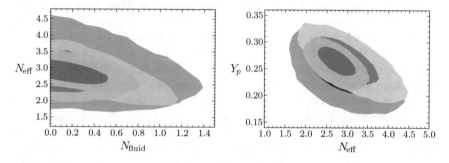

**Fig. 5.8** Two-dimensional posterior distributions for the dark radiation parameters in our Planck analysis (after [2]). *Left:* Constraints on $N_{\text{eff}}$ and $N_{\text{fluid}}$ using the Planck TT, TE and EE likelihoods for varying $Y_p$ (red) and when $Y_p$ is fixed (green). For comparison, we also show the Planck TT-only results for varying $Y_p$ (indigo) and when $Y_p$ is fixed (olive). *Right:* Constraints on $N_{\text{eff}}$ and $Y_p$ using the Planck TT, TE and EE likelihoods for varying $N_{\text{fluid}}$ (red) and when $N_{\text{fluid}}$ is fixed (cyan). In addition, we display the Planck TT-only results for varying $N_{\text{fluid}}$ (indigo) and when $N_{\text{fluid}}$ is fixed (gray). The lensing reconstruction and low-P likelihoods were used in all of the presented analyses

and both $N_{\text{fluid}}$ and $Y_p$ because $N_{\text{eff}}$ is the unique parameter capable of producing the phase shift. Moreover, we observe a large improvement in the constraint on $N_{\text{eff}}$ when adding polarization data. The reason for this is most likely related to the fact that the peaks of the E-mode spectrum are sharper, which makes the phase shift easier to measure [6]. We illustrate this point in Fig. 5.6, which displays the relative differences in the TT and EE spectra when varying $N_{\text{eff}}$ and $N_{\text{fluid}}$. Although the phase shift is visible in both cases, the E-mode data has a larger impact since the size of the effect is more pronounced in the polarization spectrum.

We note that similar analyses of WMAP data (and external datasets) were performed in [34, 35]. While their results indicate weaker constraints on $N_{\rm eff}$ and $N_{\rm fluid}$, they are qualitatively similar to our TT-only analysis with $Y_p$ fixed. By comparison, adding E-mode data further improves constraints in the $N_{\rm eff}$-$N_{\rm fluid}$ plane, also when $Y_p$ is allowed to vary.

Figure 5.8 exhibits a significant difference in the constraints in the $N_{\rm eff}$-$Y_p$ plane, with and without the polarization data. Without polarization, allowing $N_{\rm fluid}$ to vary weakens the constraints on $N_{\rm eff}$ and $Y_p$. This is consistent with our discussion of degeneracies in Sect. 3.4.1 and at the beginning of this section. When only employing the temperature data, $Y_p$ and $N_{\rm fluid}$ are measured mostly from the damping tail, on which their effects are degenerate. This is why the constraints on $Y_p$ and $N_{\rm eff}$ weaken if we also vary $N_{\rm fluid}$. When we add polarization data, the situation changes: as we vary the marginalization over additional parameters, there is very little difference in the constraints. This is most noticeable in the right panel of Fig. 5.8, where the constraints on $Y_p$ and $N_{\rm eff}$ become nearly independent of the treatment of $N_{\rm fluid}$. Given that we observed above that the degeneracy between $Y_p$ and $N_{\rm fluid}$ is broken by polarization, we of course anticipated this feature.

Recently, a constraint was placed on the effective number of free-streaming species by isolating the phase shift in the Planck 2013 temperature data: $N_{\rm eff} = 2.3^{+1.1}_{-0.4}$ (68 % c.l.) when keeping the damping tail fixed and $N_{\rm eff} = 3.5 \pm 0.65$ (68 % c.l.) while marginalizing over the effect on the damping tail [1, 36]. When removing the polarization data to compare to this analysis, we obtain $N_{\rm eff} = 2.89^{+0.49}_{-0.62}$, which is quite similar to the direct measurement of the phase shift.[11] Upon adding the TE and EE data, our constraint tightens by about a factor of two to $N_{\rm eff} = 2.68^{+0.29}_{-0.33}$. Based on the estimate [12] in (5.62), we deduce that these results correspond to a sensitivity to the phase shift at the level of $\delta\ell \approx 1$. This is therefore compatible with both expectations and forecasts which only include the isolated phase shift resulting in $\sigma(N_{\rm eff}) = 0.41$ for Planck with polarization [1]. Nonetheless, it is impressive that the data is sensitive to these small and subtle effects.

---

[11] Our analysis is most comparable to the marginalized result since we marginalize over $N_{\rm fluid}$ and $Y_p$: $N_{\rm eff} = 3.5 \pm 0.65$ [36]. We note that our approach includes information in the amplitude shift produced by the free-streaming species that is not included in the other analysis. However, due to the degeneracy with the amplitude of the primordial power spectrum, this effect is likely sub-dominant to the phase shift. As we will discuss in the next subsection, future CMB experiments get a considerable fraction of the sensitivity to $N_{\rm eff}$ from the phase shift when we allow $Y_p$ to vary. This is why we expect that our methods result in increasingly similar constraints.

[12] A more schematic estimate is based on the knowledge that $N_{\rm eff} = 3.046$ produces $\delta\ell \approx 10$ for $\ell \lesssim 3000$ relative to $N_{\rm eff} = 1$ [1]. Current constraints allow for roughly a 10 percent variation in $N_{\rm eff}$, which would imply $\delta\ell \approx 1$. In fact, the level of precision associated with direct measurements of the individual peak locations in [4] is similar.

### 5.3.3 CMB Stage-4 Forecasts

We have seen that current data is already sensitive to the free-streaming nature of neutrinos. Now, we turn to future experiments which are expected to improve these constraints by at least an order of magnitude. This is particularly interesting since an increase in the sensitivity to the $\sigma(N_{\text{eff}}) \sim 10^{-2}$ level probes a number of plausible BSM scenarios that are currently unconstrained, as we have been emphasizing. In the following, we forecast not only the constraints on $N_{\text{eff}}$, but also on $Y_p$ and $N_{\text{fluid}}$. This will allow us to more clearly identify the types of BSM physics that we might be sensitive to. To ensure that degeneracies are treated correctly, we perform a full likelihood analysis (rather than a Fisher forecast; see [23] for a discussion).

We summarize the results of our forecasts in Table 5.2. It is not surprising that $N_{\text{eff}}$ is easily distinguished from $N_{\text{fluid}}$ with a CMB-S4 experiment, given the constraints that we inferred from Planck data. As in that case, the constraints on $N_{\text{eff}}$ are significantly stronger than those on $N_{\text{fluid}}$, which further supports the interpretation that these parameters are being distinguished by differences in the perturbations for the two types of radiation. When both radiation components are included, the detailed matching of the acoustic peaks is very sensitive to the phase shifts due to $N_{\text{eff}}$, but is much less affected by $N_{\text{fluid}}$. This is illustrated by the left panel in Fig. 5.9, which shows the two-dimensional constraints on $N_{\text{eff}}$ and $N_{\text{fluid}}$ (both for varying and fixed $Y_p$). While there is a strong degeneracy between $N_{\text{eff}}$ and $N_{\text{fluid}}$ when $Y_p$ is fixed, it is absent when $Y_p$ is allowed to vary. Nevertheless, the contours close at roughly the same value of $N_{\text{eff}}$ in both cases, which points to the fact that the degeneracy is broken in a way that is insensitive to the damping tail.

We have been reiterating that better measurements of the phase shift are partly enabled by the sharpness of the peaks of the E-mode spectrum, resulting in the reported strong constraint on $N_{\text{eff}}$. We can apply the same reasoning to explain the strengthening of the constraints on $N_{\text{eff}}$ when the delensed power spectrum is used. One of the well-known effects of lensing is a smearing of the acoustic peaks [28]

**Table 5.2** Marginalized constraints for Planck and CMB-S4 when varying $N_{\text{eff}}$, $N_{\text{fluid}}$ and $Y_p$, with $1\sigma$ error bars for $N_{\text{eff}}$ and $Y_p$, and $2\sigma$ upper limits for $N_{\text{fluid}}$. While, we employ the TT, TE and EE power spectra as well as the lensing reconstruction likelihood for Planck, the CMB-S4 forecasts also study the influence of delensing and lensing reconstruction on the sensitivity. Since combining lensed spectra with lensing reconstruction is known to produce overly optimistic error forecasts due to a double counting of lensing information [37], we excluded this possibility in our forecast. For Planck, the double counting can be safely ignored, but it will become more important for future experiments [38]

| Experiment | Delensing | Reconstruction | $\sigma(N_{\text{eff}})$ | $\sigma(Y_p)$ | $N_{\text{fluid}}$ |
|---|---|---|---|---|---|
| Planck 2015 | No | Yes | 0.31 | 0.019 | <0.64 |
| CMB-S4 | No | No | 0.062 | 0.0053 | <0.18 |
| | Yes | No | 0.054 | 0.0044 | <0.17 |
| | Yes | Yes | 0.050 | 0.0043 | <0.16 |

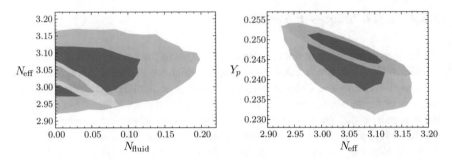

**Fig. 5.9** Forecasted CMB-S4 sensitivity in the considered three-parameter extension of $\Lambda$CDM (after [2, 3]). *Left*: Joint constraints on $N_{\text{eff}}$ and $N_{\text{fluid}}$ for $Y_p$ fixed (orange) and varying (blue). *Right*: Joint constraints on $N_{\text{eff}}$ and $Y_p$ for $N_{\text{fluid}} = 0$ fixed (purple) and varying (blue)

(see Fig. 5.6). By delensing the power spectra, we therefore sharpen the peaks which allows for a more precise measurement of the phase shift. Figure 5.6 illustrates the impact that delensing can have: a much larger relative change on the unlensed data is produced by changes in $N_{\text{eff}}$. The procedure of delensing is analogous to reconstructing the BAO peak to sharpen distance measurements [39]. In both cases, nonlinearities transfer information from the power spectrum to higher-point correlation functions. BAO reconstruction and delensing move that information back to the power spectrum, so that it is more easily accounted for in these analyses. One of the implications is a reduction of the error in the measurement of any quantity that is sensitive to the sharpness of the peak, such as the BAO scale and the phase shift. For these reasons, delensing provides an effective way to tighten the constraints on cosmological parameters in these future experiments.

So far, we have considered the three-parameter extension to $\Lambda$CDM comprised of $N_{\text{eff}}$, $N_{\text{fluid}}$ and $Y_p$, and especially focussed on understanding their mutual degeneracies. However, the actual physical models that we are interested to constrain may actually not exhibit these degeneracies. First of all, many models with additional light fields still only contain free-streaming radiation (i.e. $N_{\text{fluid}} = 0$) [40], which would in principle allow us to combine information from the damping tail and the phase shift to constrain $N_{\text{eff}}$. Moreover, while $Y_p$ is often affected by BSM physics at the time of BBN, the precise degeneracy needed to keep the damping tail fixed is unlikely to occur naturally. This implies that the constraints on such models would be significantly stronger than in the generic three-parameter case. In fact, an improvement by a factor of 3 to 4 in the constraints is possible when this degeneracy with changes in $Y_p$ is not present, as reported in Table 5.3.

Ultimately, the possibly most important feature of these forecasts is the dramatic improvement that can be expected from the studied experimental configuration of CMB-S4 in comparison to Planck 2015. Our projections suggest that improvements of a factor of 5 to 10 are possible. Having said this, we should also investigate the robustness of this conclusion to changes of the experimental specifications. To this end, we investigate variations of the maximal available multipole ($\ell_{\text{max}}$) and the beam size ($\theta_b$). These are important for two reasons: (*i*) the presence of foregrounds

**Table 5.3** Marginalized $1\sigma$ constraints for Planck (TT, TE, EE and lensing reconstruction) and forecasts for CMB-S4 for $N_{\text{eff}}$ and $Y_p$, when $N_{\text{fluid}} = 0$ is kept fixed. Both delensing and lensing reconstruction were assumed in the CMB-S4 forecasts. A dash in the $\sigma(Y_p)$ entry indicates that $Y_p$ was fixed by requiring consistency with BBN

| Experiment | $\sigma(N_{\text{eff}})$ | $\sigma(Y_p)$ |
|---|---|---|
| Planck 2015 | 0.30 | 0.018 |
| | 0.19 | – |
| CMB-S4 | 0.048 | 0.0027 |
| | 0.013 | – |

**Table 5.4** Comparison of forecasted constraints for a CMB-S4 experiment with varying beam size and maximum multipole assuming $10^6$ detectors. When varying the beam size $\theta_b$ (maximum multipole $\ell_{\text{max}}$), we take $\ell_{\text{max}} = 5000$ ($\theta_b = 1$ arcmin). For $N_{\text{eff}}$, we show $1\sigma$ error bars, while we display $2\sigma$ upper limits for $N_{\text{fluid}}$

| Parameter | $1'$ | $2'$ | $3'$ | $\ell_{\text{max}} = 3000$ | $\ell_{\text{max}} = 4000$ |
|---|---|---|---|---|---|
| $\sigma(N_{\text{eff}})$ ($Y_p$ fixed, $N_{\text{fluid}} = 0$) | 0.013 | 0.015 | 0.016 | 0.023 | 0.015 |
| $\sigma(N_{\text{eff}})$ ($Y_p$ fixed, $N_{\text{fluid}} \neq 0$) | 0.026 | 0.027 | 0.029 | 0.034 | 0.028 |
| $\sigma(N_{\text{eff}})$ ($Y_p$ varying, $N_{\text{fluid}} = 0$) | 0.048 | 0.051 | 0.055 | 0.058 | 0.052 |
| $\sigma(N_{\text{eff}})$ ($Y_p$ varying, $N_{\text{fluid}} \neq 0$) | 0.050 | 0.052 | 0.055 | 0.061 | 0.051 |
| $N_{\text{fluid}}$ ($Y_p$ varying) | <0.16 | <0.17 | <0.18 | <0.20 | <0.17 |
| $N_{\text{fluid}}$ ($Y_p$ fixed) | <0.068 | <0.072 | <0.076 | <0.090 | <0.072 |

or systematics makes it difficult to predict $\ell_{\text{max}}$ reliably beforehand and (*ii*) the beam size is ultimately a choice made within the context of limited resources. Assuming $10^6$ detectors, Table 5.4 compares the forecasted constraints for various values of $\theta_b$ and $\ell_{\text{max}}$.

From this table, we can draw two simple lessons. First, only a moderate improvement on our constraints is seen when measuring modes with $\ell > 3000$. Similarly, the benefit to reducing the beam size significantly is also limited. This shows that most of the improvement in the measurement of the phase shift is coming from high-precision measurements of E-modes with $\ell < 3000$. In contrast, the sensitivity to the beam size and maximum multipole is stronger when $Y_p$ is fixed and $N_{\text{fluid}} = 0$, which suggests[13] that we are gaining useful information from the damping tail at $\ell > 3000$.

In the presented forecasts, we employed a fixed number of $10^6$ detectors. The precise value will however play a very important role in the ultimate reach of CMB-S4 for most physics targets, not just for neutrinos and other light relics. For $N_{\text{eff}}$ specifically, we may still improve the constraints until we reach the limit set by cosmic variance[14] in the E-mode power spectrum since modes with $\ell \gtrsim 3700$ ($\ell \gtrsim 2300$)

---

[13]We note that our forecasts have not taken the presence of foregrounds in temperature or polarization into account. In a real experiment, the information in our forecasts that is encoded in the high-$\ell$ temperature data will likely not be available.

[14]The cosmic variance-limited constraint is roughly given by $\sigma(N_{\text{eff}}) \sim 0.008$ (cf. Appendix C).

are still dominated by noise for $10^6$ ($10^4$) detectors. We therefore improve constraints significantly by further reducing the detector noise and, hence, increasing the number of detectors within the considered range of $10^4$ to $10^4$. This intuition is confirmed by forecasts for $N_{\text{eff}}$ which varied the numbers of detectors [5, 24].

## 5.4  Consequences for BBN and BSM

With the substantially improved constraints on $N_{\text{eff}}$, $N_{\text{fluid}}$ and $Y_p$ expected from future CMB experiments, it is interesting to examine the possible impacts that these measurements might have on our understanding of the laws of Nature. In the following, we consider some consequences of combining these CMB observations with those of big bang nucleosynthesis (Sect. 5.4.1) and discuss a few opportunities to probe physics beyond the Standard Model (Sect. 5.4.2).

### 5.4.1  Interplay Between CMB and BBN

We have seen that measuring the fluctuations in the dark radiation eliminates the degeneracy between $Y_p$ and $N_{\text{eff}}$. Since $Y_p$ is sensitive to the total radiation density at the time of BBN (3 min after the big bang), while $N_{\text{eff}}$ and $N_{\text{fluid}}$ are related to these radiation densities around the time of recombination (373000 years after the big bang), this has important consequences for constraints on BSM physics. This means that CMB measurements of $Y_p$ and $N_{\text{eff}}$ allow us to probe scenarios where these densities change between those two times, for example due to the decay of a heavy particle or some other production mechanism [41].

For the following discussion, it will be convenient to translate constraints on $Y_p$ into bounds on the radiation density at the time of BBN [42],

$$Y_p \approx 0.247 + 0.014 \times \Delta N_{\text{eff+fluid}}^{\text{BBN}}, \qquad (5.63)$$

with $\Delta N_{\text{eff+fluid}}^{\text{BBN}} \equiv N_{\text{eff}}{}^{\text{BBN}} + N_{\text{fluid}}{}^{\text{BBN}} - 3.046$. For CMB-S4, we obtained a constraint on the helium fraction of $\sigma(Y_p) = 0.0043$ when marginalizing over $N_{\text{fluid}}$ (see Table 5.2), while we found $\sigma(Y_p) = 0.0027$ for fixed $N_{\text{fluid}} = 0$ (cf. Table 5.3). Employing (5.63), these standard deviations imply

$$\sigma(N_{\text{eff+fluid}}^{\text{BBN}}) = \begin{cases} 0.31 & N_{\text{fluid}} \neq 0, \\ 0.19 & N_{\text{fluid}} = 0. \end{cases} \qquad (5.64)$$

Notably, the constraint for $N_{\text{fluid}} = 0$ is stronger than the current limit from BBN alone, $\sigma(N_{\text{eff+fluid}}^{\text{BBN}}) = 0.28$ [43]. As a result, the CMB will yield independent measurements of $N_{\text{eff}}$ at two different times in a single experiment, with each one of them surpassing our current level of sensitivities from combining multiple probes.

Although a CMB-S4 experiment will constrain $N_{\text{eff+fluid}}^{\text{BBN}}$ only with a modest improvement over current measurements from primordial abundances, its unique advantage is that it can be combined with measurements of other cosmological parameters (e.g. $\Omega_b h^2$) without jointly analysing different datasets and it provides a clean measurement (i.e. it is not affected by astrophysical processes at later times). A common approach with current data to improve the overall sensitivity to $N_{\text{eff}}$ in the case where it is time independent is to combine the constraints from the CMB and primordial abundances. Table 5.3 indicates that we get very strong constraints on $N_{\text{eff}}$ due to the lack of degeneracies in the damping tail if we do not allow a variation in $N_{\text{eff}}$ between BBN and the CMB. Given that these results are sufficiently strong, it is unlikely that the constraints will improve much when additionally including information from primordial abundances.

In our discussion of the possible constraints on the time variation of $N_{\text{eff}}$, we have exclusively focussed on the model-independent measurement implied by independently varying $N_{\text{eff}}$ and $Y_p$. As a consequence, it is primarily the degeneracy between $N_{\text{eff}}$ and $Y_p$ in the damping tail that controls the inferred sensitivity. In the absence of this degeneracy, the constraints on $N_{\text{eff}}$ are much stronger. It could be the case in realistic models that $N_{\text{eff}}$ and $Y_p$ do not produce this degeneracy in the relevant range of parameters although they are changed independently. A dedicated analysis of CMB data would therefore likely offer a much larger gain over the current limits from primordial abundances for these specific models.

### 5.4.2  Implications for BSM Physics

Until now, we have concentrated on simple descriptions of BSM physics in terms of the effective parameters $N_{\text{eff}}$, $N_{\text{fluid}}$ and $Y_p$. Given that these parameters capture important aspects of our cosmological history and of CMB physics, we can however directly use them to test a number of scenarios for BSM physics (see Table 5.5):

**Table 5.5** Cosmological probes of BSM physics and their sensitivity to free-streaming and non-free-streaming radiation ($N_{\text{eff}}$ and $N_{\text{fluid}}$), the number of active neutrinos ($N_\nu$), the baryon-to-photon ratio ($\eta$) and the amount of energy injection ($E_{\text{inj}}$). We use the superscripts BBN, CMB, or post-BBN to indicate the time at which a quantity is being probed, with the latter referring to redshifts of $z \lesssim 10^6$ when spectral distortions become possible. While $\mathcal{GR}$ denotes modified gravity, the parameter $N_{\text{fluid}}^{\text{iso}}$ abstractly stands for contributions from isocurvature fluctuations

| Signature | Influenced by | Degeneracies broken by |
|---|---|---|
| CMB damping tail | $N_{\text{eff}}^{\text{CMB}} + N_{\text{fluid}}^{\text{CMB}}$, $Y_p$, $E_{\text{inj}}^{\text{CMB}}$ | Phase shift, polarization |
| Phase shift | $N_{\text{eff}}$, $N_{\text{fluid}}^{\text{iso}}$, $\mathcal{GR}$ | Scale dependence |
| Spectral distortions | $E_{\text{inj}}^{\text{post-BBN}}$ | |
| Primordial abundances | $N_{\text{eff}}^{\text{BBN}} + N_{\text{fluid}}^{\text{BBN}}$, $N_\nu$, $\eta^{\text{BBN}}$, $E_{\text{inj}}^{\text{BBN}}$ | CMB |

- As we discussed in Sect. 3.3.2, the parameter $N_{\text{eff}}$ is sensitive to the freeze-out of a particle for minimal extensions of the Standard Model with a light field. It is interesting to note that we can rule out some scenarios where particles freeze-out after the QCD phase transition at current levels of sensitivity, $\sigma(N_{\text{eff}}) \gtrsim 0.1$ [40]. Since the contribution to $N_{\text{eff}}$ is typically diluted by about an order of magnitude through freeze-out before the QCD phase transition, such models can easily evade current constraints. Having said that, some of these scenarios will likely be accessible with a CMB-S4 experiment [5]. For these cases, we are sensitive to sufficiently early times so that BSM physics above the TeV scale may be of importance and can be probed along the lines of Chap. 4.
- Any energy which is injected into the Standard Model particles after the time of neutrino decoupling may also lead to a contribution to the effective number of free-streaming particles at recombination, $N_{\text{eff}}^{\text{CMB}}$. This energy injection could, depending on its time and nature, alter the primordial abundances or introduce spectral distortions which would distinguish it from a new light field. For example, while the radiation density at BBN, $N_{\text{eff}}^{\text{BBN}}$, would be unchanged, a decay to photons after BBN would lower both $N_{\text{eff}}^{\text{CMB}}$ and the baryon-to-photon ratio at BBN, $\eta^{\text{BBN}}$ [44].
- While energy injection of many different kinds is a typical byproduct of changing $N_{\text{eff}}$, it may also be the dominant signature of BSM physics. The formation of nuclei may, for example, be disrupted by decays during BBN without substantially changing the total energy in radiation. Alternatively, through changes in the form of the visibility function, recombination itself is also very sensitive to energy injection [45].[15]
- Isocurvature perturbations may also produce phase shifts of the acoustic peaks (denoted by $N_{\text{fluid}}^{\text{iso}}$ in Table 5.5), as we explained in Sect. 5.2. There are many good reasons to imagine why isocurvature perturbations might arise in the dark sector. We discussed a simple curvaton-like example of this effect, but we anticipate that not purely adiabatic dark sectors may result in similar signatures so that we are also sensitive to such physics. This therefore motivates a future exploration of the observability of these effects.
- Finally, throughout this chapter, we assumed that the Einstein equations are valid. In the absence of anisotropic stress, this enforced that $\Phi_- = 0$. However, modified theories of gravity are often parametrized in terms of their change to the Einstein constraint equation and the corresponding effect on $\Phi_-$ (see e.g. [46]). It would be interesting to investigate how the result changes for specific modifications of general relativity (GR) since the field $\Phi_-$ played an important role in our analysis of the phase shift. At the same time, the phase shift of the CMB spectrum may constitute an interesting new probe of modified gravity.

---

[15]Both of these examples have in common that the tail of the Boltzmann distribution plays a critical role due to the large value of $\eta^{-1}$. This means that the change to the small number of high-energy photons is more important than the total energy density.

## 5.5 Summary

CMB observations have become precise enough to probe the gravitational imprints of BSM physics on the perturbations of the primordial plasma. In the upcoming era of CMB polarization experiments, our sensitivity to these subtle effects will increase significantly and will offer new opportunities in the search for new physics. It is therefore timely to re-evaluate how CMB data can inform our view of the laws of physics.

In this chapter, we have explored how the phase shift of the acoustic peaks might be used as such a probe. This phase shift is particularly interesting because analytic properties of the Green's function of the gravitational potential strongly limit the possible origins of such a shift to

    *i*. waves propagating faster than the sound speed of the photon-baryon fluid,

    *ii*. isocurvature fluctuations.

For adiabatic initial conditions, the phase shift is most easily generated by free-streaming radiation and becomes an excellent measure of the effective number of free-streaming relativistic species, $N_{\text{eff}}$, at the time of recombination. Realistic models of isocurvature fluctuations typically produce a scale-dependent phase shift, which allows them to be distinguished from changes to the energy density of the radiation.

What makes these results particularly compelling is that current and future CMB experiments are sensitive enough to detect these phase shifts at high significance [1]. We have demonstrated this with an analysis of the 2015 data from the Planck satellite and forecasts for a CMB Stage-4 experiment (see Fig. 5.10). Our results emphasize the crucial role played by the polarization data in breaking the degeneracy between the contributions from free-streaming and non-free-streaming

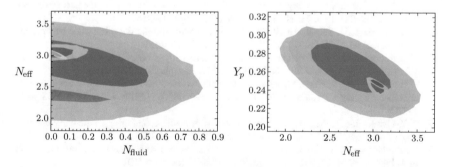

**Fig. 5.10** Comparison of the two-dimensional posterior distributions of Planck and CMB-S4 for $N_{\text{eff}}$, $N_{\text{fluid}}$ and $Y_p$ (after [2, 3]). *Left:* Planck constraints and CMB-S4 forecasts on the effective number of free-streaming and non-free-streaming relativistic species, $N_{\text{eff}}$ and $N_{\text{fluid}}$, when the helium fraction $Y_p$ is allowed to vary (red) and is kept fixed (green). *Right:* Planck constraints and CMB-S4 forecasts on $Y_p$ and $N_{\text{eff}}$ for varying $N_{\text{fluid}}$ (red) and fixed $N_{\text{fluid}} = 0$ (cyan). In both plots, the contours from the CMB-S4 forecasts were already presented in Fig. 5.9, but have been included to show the expected improvements in the constraining power

particles, $N_{\text{eff}}$ and $N_{\text{fluid}}$, as well as that with the helium fraction $Y_p$. We also provide a clear detection of the free-streaming nature of neutrinos which constitutes further evidence for the cosmic neutrino background.

# References

1. B. Follin, L. Knox, M. Millea, Z. Pan, First detection of the acoustic oscillation phase shift expected from the cosmic neutrino background. Phys. Rev. Lett. **115**, 091301 (2015). arXiv:1503.07863 [astro-ph.CO]
2. D. Baumann, D. Green, J. Meyers, B. Wallisch, Phases of new physics in the CMB. JCAP **01**, 007 (2016). arXiv:1508.06342 [astro-ph.CO]
3. Adapted from [2] with permission of IOP Publishing. Copyright by Sissa Medialab srl. All rights reserved
4. N. Aghanim et al. (Planck Collaboration), Planck 2015 results. XI. CMB power spectra, likelihoods and robustness of parameters. Astron. Astrophys. **594**, A11 (2016). arXiv:1507.02704 [astro-ph.CO]
5. K. Abazajian et al. (CMB-S4 Collaboration), CMB-S4 Science Book, 1st edn. arXiv:1610.02743 [astro-ph.CO]
6. S. Bashinsky, U. Seljak, Neutrino perturbations in CMB anisotropy and matter clustering. Phys. Rev. D **69**, 083002 (2004). arXiv:astro-ph/0310198 [astro-ph]
7. W. Hu, M. Fukugita, M. Zaldarriaga, M. Tegmark, CMB observables and their cosmological implications. Astrophys. J. **549**, 669 (2001). arXiv:astro-ph/0006436 [astro-ph]
8. M. Doran, M. Lilley, The location of CMB peaks in a universe with dark energy. Mon. Not. Roy. Astron. Soc. **330**, 965 (2002). arXiv:astro-ph/0104486 [astro-ph]
9. P. Corasaniti, A. Melchiorri, Testing cosmology with cosmic sound waves. Phys. Rev. D **77**, 103507 (2008). arXiv:0711.4119 [astro-ph]
10. Z. Pan, L. Knox, B. Mulroe, A. Narimani, Cosmic microwave background acoustic peak locations. Mon. Not. Roy. Astron. Soc. **459**, 2513 (2016). arXiv:1603.03091 [astro-ph.CO]
11. U. Seljak, Measuring polarization in the cosmic microwave background. Astrophys. J. **482**, 6 (1997). arXiv:astro-ph/9608131 [astro-ph]
12. U. Seljak, M. Zaldarriaga, Signature of gravity waves in polarization of the microwave background. Phys. Rev. Lett. **78**, 2054 (1997). arXiv:astro-ph/9609169 [astro-ph]
13. M. Zaldarriaga, U. Seljak, An all-sky analysis of polarization in the microwave background. Phys. Rev. D **55**, 1830 (1997). arXiv:astro-ph/9609170 [astro-ph]
14. M. Kamionkowski, A. Kosowsky, A. Stebbins, Statistics of cosmic microwave background polarization. Phys. Rev. D **55**, 7368 (1997). arXiv:astro-ph/9611125 [astro-ph]
15. Z. Hou, R. Keisler, L. Knox, M. Millea, C. Reichardt, How massless neutrinos affect the cosmic microwave background damping tail. Phys. Rev. D **87**, 083008 (2013). arXiv:1104.2333 [astro-ph.CO]
16. F.-Y. Cyr-Racine, K. Sigurdson, Limits on neutrino-neutrino scattering in the early universe. Phys. Rev. D **90**, 123533 (2014). arXiv:1306.1536 [astro-ph.CO]
17. M. Archidiacono, S. Hannestad, Updated constraints on non-standard neutrino interactions from Planck. JCAP **07**, 046 (2014). arXiv:1311.3873 [astro-ph.CO]
18. I. Oldengott, C. Rampf, Y.Y.Y. Wong, Boltzmann hierarchy for interacting neutrinos I: formalism. JCAP **04**, 016 (2015). arXiv:1409.1577 [astro-ph.CO]
19. M. Buen-Abad, G. Marques-Tavares, M. Schmaltz, Non-Abelian dark matter and dark radiation. Phys. Rev. D **92**, 023531 (2015). arXiv:1505.03542 [hep-ph]
20. Z. Chacko, Y. Cui, S. Hong, T. Okui, Hidden dark matter sector, dark radiation and the CMB. Phys. Rev. D **92**, 055033 (2015). arXiv:1505.04192 [hep-ph]
21. S. Weinberg, Adiabatic modes in cosmology. Phys. Rev. D **67**, 123504 (2003). arXiv:astro-ph/0302326 [astro-ph]

22. P. Peter, J.-P. Uzan, *Primordial Cosmology* (Oxford University Press, Oxford, 2013)
23. L. Perotto, J. Lesgourgues, S. Hannestad, H. Tu, Y.Y.Y. Wong, Probing cosmological parameters with the CMB: forecasts from full Monte Carlo simulations. JCAP **10**, 013 (2006). arXiv:astro-ph/0606227 [astro-ph]
24. W. Wu, J. Errard, C. Dvorkin, C. Kuo, A. Lee, P. McDonald, A. Slosar, O. Zahn, A guide to designing future ground-based cosmic microwave background experiments. Astrophys. J. **788**, 138 (2014). arXiv:1402.4108 [astro-ph.CO]
25. D. Blas, J. Lesgourgues, T. Tram, The Cosmic Linear Anisotropy Solving System (CLASS) II: approximation schemes. JCAP **07**, 034 (2011). arXiv:1104.2933 [astro-ph.CO]
26. B. Audren, J. Lesgourgues, K. Benabed, S. Prunet, Conservative constraints on early cosmology: an illustration of the MontePython cosmological parameter inference code. JCAP **02**, 001 (2013). arXiv:1210.7183 [astro-ph.CO]
27. A. Gelman, D. Rubin, Inference from iterative simulation using multiple sequences. Stat. Sci. **7**, 457 (1992)
28. U. Seljak, Gravitational lensing effect on cosmic microwave background anisotropies: a power spectrum approach. Astrophys. J. **463**, 1 (1996). arXiv:astro-ph/9505109 [astro-ph]
29. K. Smith, D. Hanson, M. LoVerde, C. Hirata, O. Zahn, Delensing CMB polarization with external datasets. JCAP **06**, 014 (2012). arXiv:1010.0048 [astro-ph.CO]
30. W. Hu, T. Okamoto, Mass reconstruction with cosmic microwave background polarization. Astrophys. J. **574**, 566 (2002). arXiv:astro-ph/0111606 [astro-ph]
31. T. Okamoto, W. Hu, Cosmic microwave background lensing reconstruction on the full sky. Phys. Rev. D **67**, 083002 (2003). arXiv:astro-ph/0301031 [astro-ph]
32. C. Hirata, U. Seljak, Reconstruction of lensing from the cosmic microwave background polarization. Phys. Rev. D **68**, 083002 (2003). arXiv:astro-ph/0306354 [astro-ph]
33. D. Green, J. Meyers, A. van Engelen, CMB delensing beyond the B-modes. JCAP **12**, 005 (2017). arXiv:1609.08143 [astro-ph.CO]
34. N. Bell, E. Pierpaoli, K. Sigurdson, Cosmological signatures of interacting neutrinos. Phys. Rev. D **73**, 063523 (2006). arXiv:astro-ph/0511410 [astro-ph]
35. A. Friedland, K. Zurek, S. Bashinsky, Constraining models of neutrino mass and neutrino interactions with the Planck satellite. arXiv:0704.3271 [astro-ph]
36. B. Follin, L. Knox, M. Millea, Z. Pan, Private communication (2015)
37. W. Hu, Dark synergy: gravitational lensing and the CMB. Phys. Rev. D **65**, 023003 (2002). arXiv:astro-ph/0108090 [astro-ph]
38. M. Schmittfull, A. Challinor, D. Hanson, A. Lewis, Joint analysis of CMB temperature and lensing-reconstruction power spectra. Phys. Rev. D **88**, 063012 (2013). arXiv:1308.0286 [astro-ph.CO]
39. D. Eisenstein, H.-J. Seo, E. Sirko, D. Spergel, Improving cosmological distance measurements by reconstruction of the baryon acoustic peak. Astrophys. J. **664**, 675 (2007). arXiv:astro-ph/0604362 [astro-ph]
40. C. Brust, D.E. Kaplan, M. Walters, New light species and the CMB. JHEP **12**, 058 (2013). arXiv:1303.5379 [hep-ph]
41. W. Fischler, J. Meyers, Dark radiation emerging after big bang nucleosynthesis? Phys. Rev. D **83**, 063520 (2011). arXiv:1011.3501 [astro-ph.CO]
42. J. Bernstein, L. Brown, G. Feinberg, Cosmological helium production simplified. Rev. Mod. Phys. **61**, 25 (1989)
43. R. Cyburt, B. Fields, K. Olive, T.-H. Yeh, Big bang nucleosynthesis: 2015. Rev. Mod. Phys. **88**, 015004 (2016). arXiv:1505.01076 [astro-ph.CO]
44. D. Cadamuro, S. Hannestad, G. Raffelt, J. Redondo, Cosmological bounds on sub-MeV mass axions. JCAP **02**, 003 (2011). arXiv:1011.3694 [hep-ph]
45. N. Padmanabhan, D. Finkbeiner, Detecting dark matter annihilation with CMB polarization: signatures and experimental prospects. Phys. Rev. D **72**, 023508 (2005). arXiv:astro-ph/0503486 [astro-ph]
46. P. Zhang, M. Liguori, R. Bean, S. Dodelson, Probing gravity at cosmological scales by measurements which test the relationship between gravitational lensing and matter overdensity. Phys. Rev. Lett. **99**, 141302 (2007). arXiv:0704.1932 [astro-ph]

# Chapter 6
# Searching for Light Relics with LSS

We have established in the previous chapter that the cosmic microwave background is a sensitive probe of light thermal relics. The forecasts for a CMB Stage-4 experiment indicate that the possible bounds on $N_{\mathrm{eff}}$ are tantalizingly close to well-motivated theoretical targets, in particular the threshold value $\Delta N_{\mathrm{eff}} = 0.027$ associated with the minimal abundance of any light thermal relic. Since the initial conditions for the clustering of matter are set by the same physics that leads to the acoustic oscillations in the CMB, we expect to find imprints of relativistic species also in the large-scale structure of the universe. In this chapter, we therefore explore to what degree the CMB observations can be enhanced by future LSS surveys. We carefully isolate the information encoded in the shape of the galaxy power spectrum and in the spectrum of baryon acoustic oscillations. We find that measurements of the shape of the power spectrum can significantly improve on current and near-term CMB experiments. We also propose a modified analysis of BAO data and show that the phase shift induced by relic neutrinos in the BAO spectrum can be detected at high significance in future experiments.

The outline of this chapter, which is based on [1], is as follows. In Sect. 6.1, we recapitulate the effects of free-streaming radiation on the matter power spectrum as well as the baryon acoustic oscillation signal, focussing especially on the phase shift in the BAO spectrum. In Sect. 6.2, we forecast combined CMB and LSS constraints on the number of relativistic species, $N_{\mathrm{eff}}$, for a range of future observations. In Sect. 6.3, we isolate the information encoded in the phase shift of the BAO spectrum and study the prospects for extracting this effect in upcoming surveys. A summary is presented in Sect. 6.4. We refer to Appendices C and D for technical details of our analysis, such as the experimental specifications of the surveys, and the methods used to extract the matter broadband spectrum and the phase shift.

© Springer Nature Switzerland AG 2019
B. Wallisch, *Cosmological Probes of Light Relics*, Springer Theses,
https://doi.org/10.1007/978-3-030-31098-1_6

## 6.1 Phases of New Physics in the BAO Spectrum

Having seen that the next generation of CMB experiments will be very sensitive to the radiation density, we will now explore the additional constraining power provided by current and future LSS experiments, such as (e)BOSS [2, 3], DES [4], DESI [5], LSST [6] and Euclid [7]. It was established in [8–10] that these surveys carry information about relativistic species. We will examine how this information is stored in both the power spectrum shape and the BAO spectrum.

In Sect. 3.4.3, we discussed the main effects of the radiation density in the early universe on the growth of structure. In brief, increasing the radiation density moves the maximum of the power spectrum to larger scales and the amplitude on small scales gets suppressed (cf. Fig. 6.1). On the other hand, a change in the size of the sound horizon (and therefore in the BAO frequency) has the dominant impact on the BAO spectrum. The phase shift is clearly sub-dominant, but a distinct feature of free-streaming particles (see Sect. 3.4.2). Currently, of all these effects, only the change in the BAO frequency is taken into account when analysing LSS data. In principle, however, there is much more additional information available that can be used to improve the constraints and learn about the physics of the early universe.

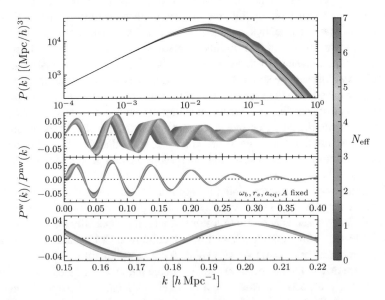

**Fig. 6.1** Variation of the matter power spectrum $P(k)$ (*top*) and the BAO spectrum $P^{\mathrm{w}}(k)/P^{\mathrm{nw}}(k)$ (*bottom*) as a function of $N_{\mathrm{eff}}$. The physical baryon density, $\omega_b$, the physical sound horizon at the drag epoch, $r_s$, the scale factor at matter-radiation equality, $a_{\mathrm{eq}} \equiv \omega_m/\omega_r$, and the BAO amplitude $A$ at the fourth peak are held fixed in the second BAO panel (as in Fig. 3.5). This panel and the bottom zoom-in show the remaining phase shift induced by free-streaming relativistic species

Although its accessibility is hampered by non-linear effects, we will demonstrate that some of these clues can still be robustly extracted from LSS observables.

We will pay particular attention to the information about $N_{\mathrm{eff}}$ contained in the BAO spectrum and propose a new analysis of this observable. To isolate the BAO signal, we split the power spectrum into a smooth ('no-wiggle') part and an oscillatory ('wiggle') part,

$$P(k) \equiv P^{\mathrm{nw}}(k) + P^{\mathrm{w}}(k)\,, \qquad (6.1)$$

with our method for performing this separation being described in Appendix D. We will demonstrate that the most robust information about $N_{\mathrm{eff}}$ lives in $P^{\mathrm{w}}(k)$. The peak locations of the BAO spectrum carry additional information about light relics that is immune to corrections to the overall shape of the power spectrum, such as those arising from non-linear gravitational evolution [11–13]. The reason underlying this is that the phase of the BAO spectrum is unaffected by the effects of non-linear gravitational evolution [14]. We claim that this information is preserved after non-linear corrections are taken into account and explore in detail how it can be isolated in the BAO spectrum. This protected information may play a useful role in elucidating apparent discrepancies between CMB and low-redshift measurements, and be a valuable tool in the search for exotic physics in the dark sector.

## 6.2 Current and Future Constraints

We have been arguing throughout this thesis that measuring the radiation density at the percent level provides an interesting window into early universe cosmology and BSM particle physics. We have already seen that upcoming CMB experiments will considerably improve the current constraints. In the following, we will further quantify the constraining power of future cosmological observations by including LSS surveys. We will consider two types of forecasts based on $P(k)$ and $P^{\mathrm{w}}(k)$, which we refer to as '$P(k)$-forecasts' and 'BAO-forecasts', respectively.

### 6.2.1 Fisher Methodology

In this section, we will use standard Fisher information theory to forecast the constraints of future observations. While Fisher forecasts have to be used with care, they provide useful guidance for the sensitivities and design of future experiments. In the following, we recall the basic elements of the Fisher methodology and its application to galaxy surveys [15, 16]. The relatively standard Fisher forecasting of CMB observations is summarized in Appendix C, where further details on the LSS forecasting can be found as well.

Given a likelihood function $\mathcal{L}(\vec{\theta})$ for the model parameters $\vec{\theta} \equiv \{\omega_b, \omega_c, \theta_s, \tau, A_s, n_s, N_{\mathrm{eff}}, Y_p\}$, we define the Fisher matrix as the average curvature of the log-likelihood around the fiducial point in parameter space,

$$F_{ij} = -\left\langle \frac{\partial^2 \ln \mathcal{L}}{\partial \theta_i\, \partial \theta_j} \right\rangle, \qquad (6.2)$$

where the expectation value denotes an average over all possible realizations of the data. If the likelihood is Gaussian, then the inverse Fisher matrix gives the covariance matrix. This means that $F_{ii}^{-1/2}$ is the error on the parameter $\theta_i$, when all other parameters $\theta_{j \neq i}$ are known, while $\sigma(\theta_i) = (F^{-1})_{ii}^{1/2}$ is the error on $\theta_i$ after marginalizing over the other parameters. More generally, the Cramér–Rao bound,

$$\sigma(\theta_i) \geq \sqrt{(F^{-1})_{ii}}, \qquad (6.3)$$

gives a lower limit on the marginalized constraints.

The Fisher matrix for a galaxy survey is [17]

$$F_{ij} = \int_{-1}^{1} \frac{d\mu}{2} \int_{k_{\min}}^{k_{\max}} \frac{dk\, k^2}{(2\pi)^2} \frac{\partial \ln P_g(k, \mu)}{\partial \theta_i} \frac{\partial \ln P_g(k, \mu)}{\partial \theta_j} V_{\mathrm{eff}}(k, \mu), \qquad (6.4)$$

where $P_g(k, \mu)$ is the anisotropic galaxy power spectrum, $\mu$ is the cosine between the wavevector $\boldsymbol{k}$ and the line-of-sight, and $V_{\mathrm{eff}}$ is the effective survey volume,

$$V_{\mathrm{eff}}(k, \mu) \equiv \int d^3 x \left[ \frac{n_g(\boldsymbol{x}) P_g(k, \mu)}{n_g(\boldsymbol{x}) P_g(k, \mu) + 1} \right]^2 \approx \left[ \frac{\bar{n}_g P_g(k, \mu)}{\bar{n}_g P_g(k, \mu) + 1} \right]^2 V. \qquad (6.5)$$

In the second equality, we have assumed that the comoving number density of galaxies is independent of position, $n_g(\boldsymbol{x}) \approx \bar{n}_g = \mathrm{const}$, and introduced the actual survey volume $V$. To derive the constraints from independent redshift bins, we take $V$ to be the volume within each bin and add the corresponding Fisher matrices. The minimum wavenumber accessible in a survey is given by the volume of the survey [1] as $k_{\min} = 2\pi\, [3V/(4\pi)]^{-1/3}$.

## Modelling the Power Spectrum

In Sect. 2.4.2, we introduced the linear matter power spectrum $P_{\mathrm{lin}}(k)$ as the main LSS observable and discussed some of the complications that arise because we cannot observe it directly. In order to obtain semi-realistic constraints on most parameters of the cosmological model, it is often sufficient to model the observed galaxy power spectrum as $P_g(k) \approx b^2 P_{\mathrm{lin}}(k)$, where $b$ is the linear biasing parameter. However, the constraints on extra relativistic species are particularly sensitive to the way degenera-

---

[1] We assume that the survey volume has a spherical geometry. The geometry of a given redshift bin (or the full survey volume) is neither spherical nor cubic, but we have checked that all of our results are essentially unaffected by this choice.

cies are broken and to the non-linear damping of the oscillatory feature, so we need to be more careful in the modelling of the signal [18–20]. Moreover, since observations only determine the angular positions and redshifts of objects, we need to take into account the corresponding redshift space distortions (RSD) and geometric projection effects.

Separating the spectrum into its smooth and oscillatory parts according to (6.1), our model for the observed galaxy power spectrum is the following remapping of the linear matter power spectrum:

$$P_g(k, \mu) = b^2 F^2(k, \mu) P^{nw}(k, \mu) \Big[ 1 + O(k, \mu) D(k, \mu) \Big] Z(k, \mu). \tag{6.6}$$

All functions in this expression have an implicit redshift dependence. We now define the different elements of (6.6):

- $O(k, \mu)$: This function encodes the BAO signal and can be written as

$$O(k, \mu) \equiv B(k) O_{\text{lin}}(k'(k, \mu)) + A(k), \tag{6.7}$$

where $O_{\text{lin}}(k') \equiv P^w_{\text{lin}}(k')/P^{nw}_{\text{lin}}(k')$ is the normalized wiggle spectrum evaluated at the rescaled wavenumbers [21]

$$k' = k \sqrt{(1 - \mu^2)/q_\perp^2 + \mu^2/q_\parallel^2}, \quad \text{with} \quad q_\perp \equiv \frac{D_A(z)}{D_A^{\text{fid}}(z)}, \quad q_\parallel \equiv \frac{H^{\text{fid}}(z)}{H(z)}. \tag{6.8}$$

This rescaling reflects the fact that the wavenumbers $k$ cannot be measured directly, but instead have to be derived from the measured angles and redshifts using the angular diameter distance $D_A^{\text{fid}}(z)$ and Hubble rate $H^{\text{fid}}(z)$ of a fiducial cosmology. This is often referred to as anisotropic geometric effects. In the limit of spherically-averaged clustering measurements, these become isotropic and $k' = k/q$, where $q = q_\perp^{2/3} q_\parallel^{1/3} = D_V(z)/D_V^{\text{fid}}(z)$, with the radial BAO dilation given by $D_V \propto (D_A^2/H)^{1/3}$.

To model uncertainties in the BAO extraction, we have introduced two free functions $B(k)$ and $A(k)$ in (6.7), which we take to be smooth polynomials in $k$ (see below). Ultimately, we will marginalize over these polynomials to remove any information that is not robust to the BAO signal itself.

- $b(z)$: The bias of the target galaxies (e.g. luminous red galaxies, emission line galaxies or quasars) sets the overall amplitude of the signal in each redshift bin. We will make the common assumption that $b(z) \propto 1/D_1(z)$, where $D_1(z)$ is the linear growth function. This means that the bias is larger at high redshifts, which implies that the galaxy power spectrum may get significant corrections from non-linear biasing even at high redshifts.
- $F(k, \mu)$: This function characterises the effect of redshift space distortions. Following [22], we write

$$F(k, \mu) = \frac{1}{(q_\perp^2 q_\parallel)^{1/2}} \left[1 + \beta \mu'(k, \mu)^2 R(k)\right], \tag{6.9}$$

where $\beta \equiv f/b$, with the linear growth rate $f \equiv d \ln D_1 / d \ln a$. The factors of $q_i$ account for differences in the cosmic volume in different cosmologies. Projection effects on the angle to the line-of-sight are included as [21]

$$\mu'(k, \mu) = \mu / \sqrt{\mu^2 + (1 - \mu^2) Q^2}, \tag{6.10}$$

where $Q \equiv q_\parallel / q_\perp$, which becomes unity in the isotropic case. BAO reconstruction removes redshift space distortions on large scales, which we have modelled by adding the factor $R(k) = 1 - \exp[-(k \Sigma_s)^2/2]$ in (6.9), where the value of $\Sigma_s$ depends on the experimental specifications, in particular the noise levels. In our baseline forecasts, we take $\Sigma_s \to \infty$, i.e. $R \equiv 1$, but we comment on finite values of $\Sigma_s$ in Sect. 6.2.2.

- $D(k, \mu)$: This function models the non-linear damping of the BAO signal [11, 23]

$$D(k, \mu) \equiv \exp\left[-\frac{1}{2}\left(k^2 \mu^2 \Sigma_\parallel^2 + k^2 (1 - \mu^2) \Sigma_\perp^2\right)\right], \tag{6.11}$$

where the damping scales perpendicular and parallel to the line-of-sight are given by

$$\Sigma_\perp(z) = 9.4 \, (\sigma_8(z)/0.9) \, h^{-1} \, \text{Mpc}, \tag{6.12}$$

$$\Sigma_\parallel(z) = (1 + f(z)) \, \Sigma_\perp(z), \tag{6.13}$$

with $\sigma_8$ being the amplitude of (linear) matter fluctuations at a scale of $8 \, h^{-1}$ Mpc. We account for BAO reconstruction by decreasing these damping scales by an appropriate factor, e.g. 0.5 for 50% reconstruction. Following [8, 24], we include the degradation in the reconstruction due to shot noise using a reconstruction multiplier $r(x)$, i.e. $\Sigma_i \to r(x) \Sigma_i$. We obtain $r(x)$ by interpolating over the table

$$
\begin{aligned}
r &= (1.0, \ 0.9, \ 0.8, \ 0.7, \ 0.6, \ 0.55, \ 0.52, \ 0.5), \\
x &= (0.2, \ 0.3, \ 0.5, \ 1.0, \ 2.0, \ 3.0, \ 6.0, \ 10.0),
\end{aligned} \tag{6.14}
$$

with $r(x < 0.2) = 1.0$ and $r(x > 10.0) = 0.5$, which depends on the number density $\bar{n}_g$ via $x \equiv \bar{n}_g P_g(k_0, \mu_0)/0.1734$ evaluated at $k_0 = 0.14 \, h \, \text{Mpc}^{-1}$ and $\mu_0 = 0.6$. This means that we assume 50% reconstruction at high number densities and no reconstruction for low densities.

- $P^{\text{nw}}(k, \mu)$: The linear no-wiggle spectrum $P_{\text{lin}}^{\text{nw}}(k, \mu)$ is determined from the linear power spectrum using the method described in Appendix D. Non-linear corrections to this spectrum can be parametrized as

$$P^{\text{nw}}(k, \mu) = \tilde{B}(k) P_{\text{lin}}^{\text{nw}}(k'(k, \mu)) + \tilde{A}(k), \tag{6.15}$$

where $\tilde{B}(k)$ and $\tilde{A}(k)$ are smooth functions (see below). For the purpose of our BAO-forecasts, $\tilde{A}(k)$ and $\tilde{B}(k)$ are degenerate with $A(k)$ and $B(k)$ in (6.7) and it is therefore consistent to use the linear spectrum.

- $Z(k, \mu)$: For photometric surveys, we take the uncertainty in the redshift determination of the targets into account through the following function:

$$Z(k, \mu) = \exp\left[-k^2\mu^2\Sigma_z^2\right], \tag{6.16}$$

where $\Sigma_z = c\,(1 + z)\,\sigma_{z0}/H(z)$ is given in terms of the root-mean-square redshift error $\sigma_{z0}$ [25, 26]. The redshift error, which depends on the experimental specifications, reduces the effective resolution for modes along the line-of-sight. We neglect this effect for spectroscopic surveys.

When evaluating the derivatives in the Fisher matrix (6.4), the parameters $b(z)$, $\beta$, $R(k)$, $D(k, \mu)$ and $Z(k, \mu)$ are always computed using the fiducial cosmology. We are assuming that, after accounting for modelling uncertainties, no relevant cosmological information can be recovered from these functions.

**Accounting for Broadband Effects**

Non-linear evolution and biasing can change the shape of the power spectrum at high wavenumbers in a way that cannot be modelled from first principles. We account for this uncertainty by marginalizing over polynomials in $k$ in both the $P(k)$- and BAO-forecasts. In particular, the functions introduced in (6.15) are defined as

$$\tilde{A}(k, z_i) = \sum_{n=0}^{N_a} \tilde{a}_{n,i}\, k^n\,, \qquad \tilde{B}(k, z_i) = \sum_{m=0}^{N_b} \tilde{b}_{m,i}\, k^{2m}\,. \tag{6.17}$$

As indicated, we allow independent polynomials in each redshift bin centred around $z_i$. The coefficients $\tilde{a}_{n,i}$ and $\tilde{b}_{m,i}$ are included in the list of parameters $\theta_i$. Derivatives with respect to these parameters are determined analytically, using the fiducial values $\tilde{b}_{0,i} = 1$ and $\tilde{a}_{n,i} = \tilde{b}_{m\neq0,i} = 0$. A more careful treatment would replace this polynomial model with a perturbative model for the dark matter and biasing, and would marginalize over the bias parameters. In practice, this has been shown to give qualitatively similar forecasts [27]. Our marginalization procedure is therefore sufficient to illustrate the sensitivity of our forecasts to broadband information.

Our BAO-forecasts will marginalize over the 'broadband corrections' in (6.7), with $A(k)$ and $B(k)$ defined as in (6.17).[2] At the level of the Fisher matrix, marginalizing over a polynomial and an exponential are equivalent. As a result, the func-

---

[2]To avoid a proliferation of parameters, we will use $a_n$ and $b_n$ for the parameters in both (6.7) and (6.15), i.e. we will drop the tildes from now on. Which parameter set is meant will be clear from the context.

tion $B(k)$ captures the uncertainty in the damping scales $\Sigma_\parallel$ and $\Sigma_\perp$ in (6.11). This implies that our marginalization procedure will eliminate any cosmological information associated with the non-linear damping of the power spectrum, leaving the distinct information contained in the oscillating part of the spectrum $O_{\text{lin}}(k'(k, \mu))$. This type of procedure is used in the analysis of BAO data to correct for errors made in the modelling of $P^{\text{nw}}(k)$, see e.g. [28].

We will choose various levels of marginalization in our forecasts. This will help to distinguish the information encoded in the smooth shape of the spectrum, $P^{\text{nw}}(k)$, from that contained in the frequency and phase of the BAO spectrum, $P^{\text{w}}(k)$. In addition, these marginalizations also give a sense for the level of robustness of each type of information when accounting for the various uncertainties in modelling the data of a realistic galaxy survey.

**Extracting the BAO Signal**

In describing the power spectrum, we introduced the idea of marginalizing over polynomials to remove the information in $P_g(k)$ that is thought to be degenerate with non-linear evolution and galaxy biasing. The BAO spectrum is known to be robust to these effects and should therefore survive any such treatment. In principle, the BAO signal could be isolated with sufficient marginalization. However, in practice, it is more useful to extract the information associated with the BAO signal before any marginalization. The robustness of the BAO spectrum to non-linearities means we can be more aggressive with our choice of $k_{\text{max}}$ and less cautious with our marginalization. Consequently, it is convenient to treat the BAO signal and the broadband information independently.

The observed BAO spectrum is defined by

$$O_g(k, \mu) \equiv \frac{P_g^{\text{w}}(k, \mu)}{P_g^{\text{nw}}(k, \mu)} = D(k, \mu)\, O(k, \mu)\,, \tag{6.18}$$

where $D(k, \mu)$ and $O(k, \mu)$ were introduced in (6.6). To derive the new Fisher matrix for the BAO spectrum directly, we first write the derivatives of $P_g(k, \mu)$ as

$$\frac{\partial \ln P_g(k, \mu)}{\partial \theta_i} = \frac{1}{P_g^{\text{nw}} + P_g^{\text{w}}} \left( \frac{\partial P_g^{\text{nw}}}{\partial \theta_i} + \frac{\partial P_g^{\text{w}}}{\partial \theta_i} \right). \tag{6.19}$$

We then drop the term proportional to $\partial_{\theta_i} P_g^{\text{nw}}$ since it is degenerate with the marginalization over the broadband corrections. For the same reason, we write $\partial_{\theta_i} P_g^{\text{w}} \approx b^2 F^2 P^{\text{nw}} D\, \partial_{\theta_i} O$, i.e. we do not act with the derivatives on the functions $D(k, \mu)$ and $bF(k, \mu)$. The derivative in (6.19) therefore becomes

$$\frac{\partial \ln P_g(k, \mu)}{\partial \theta_i} \approx \frac{D(k, \mu)}{1 + D(k, \mu)\, O(k, \mu)} \frac{\partial O(k, \mu)}{\partial \theta_i}. \tag{6.20}$$

While the derivatives that we have dropped are non-zero, the marginalization procedure described above is designed to remove them and the forecasts for cosmic parameters should consequently be the same. Removing this information by hand (and marginalizing) ensures that our BAO-forecasts do not include these broadband effects, as we will show in Fig. 6.3. The resulting Fisher matrix is then given by

$$F_{ij} = \int_{-1}^{1} \frac{d\mu}{2} \int_{k_{min}}^{k_{max}} \frac{dk\, k^2}{(2\pi)^2} \frac{D(k, \mu)^2}{(1 + D(k, \mu)\, O(k, \mu))^2} \frac{\partial O(k, \mu)}{\partial \theta_i} \frac{\partial O(k, \mu)}{\partial \theta_j} V_{eff}(k, \mu).$$

(6.21)

We note that this Fisher matrix depends on $P_g^{nw}(k, \mu)$ only through $V_{eff}(k, \mu)$, which determines the signal-to-noise. For photometric surveys, we replace $V_{eff}(k, \mu) \rightarrow Z(k, \mu)^2 V_{eff}(k, \mu)$ to account for the redshift error and the associated reduction of power along the line-of-sight. In principle, we should model $P_g^{nw}(k, \mu)$ using the non-linear (galaxy) power spectrum, given that we will work close to the non-linear regime. However, non-linear evolution also correlates the modes and produces a non-Gaussian covariance matrix. Since most of the surveys under consideration in this chapter are limited by shot noise, using the non-linear power spectrum without taking into account the associated mode coupling in the covariance would artificially increase the number of signal-dominated modes. To be consistent with the use of a Gaussian covariance, our forecasts will therefore use the linear broadband spectrum.

## 6.2.2 Constraints from Planned Surveys

We are now ready to forecast the constraints of current and future CMB and LSS observations on the effective number of relativistic species $N_{eff}$. Unless stated otherwise, our baseline analysis assumes a $\Lambda CDM + N_{eff}$ cosmology in which the primordial helium fraction $Y_p$ is fixed by consistency with BBN. At the end of the section, we will also present results with $Y_p$ as a free parameter. We will further dissect the information content of the BAO spectrum in Sect. 6.3.

In Appendix C, we present detailed forecasts for current and future CMB experiments. The expected $1\sigma$ constraints for representative versions of the Planck satellite, a near-term CMB-S3 experiment and a future CMB-S4 mission are $\sigma(N_{eff}) = 0.18$, 0.054, 0.030, respectively.[3] We also show how these constraints depend on variations of the experimental configurations in the same appendix.

We would like to know how much these CMB constraints would improve with the addition of LSS data. A number of galaxy surveys are expected to take place over the next decade. The power of these surveys to constrain $N_{eff}$ is most sensitive to the survey volume, the number densities of galaxies and the redshift errors (spectroscopic versus photometric). The precise specifications of the surveys used in our analysis are given in Appendix C, where we also present more detailed forecasts for the full set of parameters.

---

[3]This precise CMB-S4 value of $\sigma(N_{eff})$ differs from that in Chap. 5 because we employ slightly different experimental specifications based on [29].

We will give the results of two types of forecasts based on $P(k)$ and $P^w(k)$. Our $P(k)$-forecasts apply the Fisher matrix (6.4) with $k_{max} = 0.2\,h\,\text{Mpc}^{-1}$ and marginalize over $b_{m \leq 1}$. To be conservative about non-linear biasing, we do not increase $k_{max}$ at large redshifts, despite the (near-)linearity of the matter power spectrum. Our BAO-forecasts use the Fisher matrix (6.21) with $k_{max} = 0.5\,h\,\text{Mpc}^{-1}$ and marginalize over $a_{n \leq 4}$, $b_{m \leq 3}$. We will also show how these forecasts depend on the choice of $k_{max}$ and the level of marginalization.

**Baseline results** In Table 6.1, we present the $1\sigma$ constraints on $N_{eff}$ for various combinations of current and future CMB and LSS experiments using the full $P(k)$-forecast. In Table 6.2, we compare these results to the same experiments using our BAO-forecasts. At BOSS levels of sensitivity and number densities, the BAO feature makes the most significant impact on constraints, particularly when combined with a CMB experiment like Planck. In contrast, with the larger volume and redshift range of DESI, the broadband shape carries most of the information and can lead to a significant improvement in the constraint on $N_{eff}$ both for Planck and a typical CMB-S3 experiment. Finally, photometric redshift surveys like DES and LSST generally perform worse than spectroscopic surveys because they are effectively two-dimensional for the scales of interest. However, the employed redshift error is conservative and we do not take the full potential of these surveys into account as we are only considering observations of galaxy clustering and have not included weak gravitational lensing measurements, for instance. We expect the constraints to improve with these additional LSS observables, but quantifying this is beyond the scope of this work.

**Table 6.1** Forecasted $1\sigma$ constraints on $N_{eff}$ for various combinations of current and future CMB and LSS experiments using $P(k)$-forecasts with $k_{max} = 0.2\,h\,\text{Mpc}^{-1}$

|        | CMB   | Spectroscopic | | | | Photometric | |
|--------|-------|------|-------|-------|--------|-------|-------|
|        |       | BOSS  | eBOSS | DESI  | Euclid | DES   | LSST  |
| Planck | 0.18  | 0.14  | 0.13  | 0.087 | 0.079  | 0.17  | 0.14  |
| CMB-S3 | 0.054 | 0.052 | 0.051 | 0.045 | 0.043  | 0.054 | 0.052 |
| CMB-S4 | 0.030 | 0.030 | 0.030 | 0.028 | 0.027  | 0.030 | 0.030 |

**Table 6.2** Forecasted $1\sigma$ constraints on $N_{eff}$ for various combinations of current and future CMB and LSS experiments using BAO-forecasts with $k_{max} = 0.5\,h\,\text{Mpc}^{-1}$

|        | CMB   | Spectroscopic | | | | Photometric | |
|--------|-------|------|-------|-------|--------|-------|-------|
|        |       | BOSS  | eBOSS | DESI  | Euclid | DES   | LSST  |
| Planck | 0.18  | 0.15  | 0.15  | 0.14  | 0.14   | 0.16  | 0.15  |
| CMB-S3 | 0.054 | 0.052 | 0.052 | 0.050 | 0.050  | 0.054 | 0.052 |
| CMB-S4 | 0.030 | 0.030 | 0.030 | 0.029 | 0.029  | 0.030 | 0.030 |

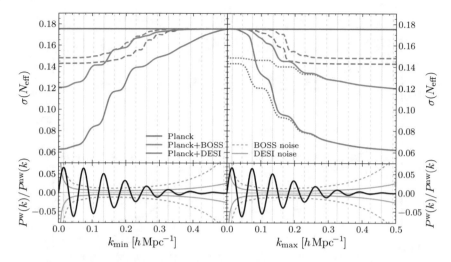

**Fig. 6.2** Forecasts for BOSS and DESI combined with Planck as a function of the smallest (*left*) and largest (*right*) Fourier modes used in the forecast, $k_{min}$ and $k_{max}$, with $k_{max} = 0.5\,h\,\text{Mpc}^{-1}$ in the left panel (from [1]). The solid and dashed lines indicate the constraints from the $P(k)$- and BAO-forecasts, respectively. Shown as the dotted lines are the "optimal constraints" as described in the main text. The lower panel displays the linear BAO spectrum and an estimate of the noise levels

**Sensitivity to $k_{max}$** The broadband signal is sensitive to non-linear effects and we should therefore understand how dependent these results are on the choice of $k_{max}$. In particular, we have chosen $k_{max} = 0.2\,h\,\text{Mpc}^{-1}$ in Table 6.1, but the usable range of scales is uncertain. Figure 6.2 shows how the constraints vary as a function of the maximal wavenumbers included in the analysis, $k_{max}$, for both the $P(k)$- and BAO-forecasts. For the BAO-forecasts, we see a clear plateau for $k_{max} > 0.2\,h\,\text{Mpc}^{-1}$. This behaviour is due to the damping of the oscillations at higher $k$ relative to the smooth power spectrum. Cosmic variance is ultimately determined by the amplitude of the smooth power spectrum and one cannot recover the high-$k$ oscillations even by lowering the shot noise. In contrast, the $P(k)$-forecasts show improvements out to $k_{max} > 0.3\,h\,\text{Mpc}^{-1}$.

Given that the BAO spectrum is robust to non-linear evolution, it is natural to consider an optimal combination of the $P(k)$ and BAO spectra that uses all the available information. This means using $P(k)$ up to a certain $k_{max}$ and adding BAO-only information for larger $k$. The $k_{max}$ of the $P(k)$ analysis then becomes the $k_{min}$ of the BAO analysis to avoid double counting the information. Results for this optimal combination are shown as the dotted line in Fig. 6.2.

**Sensitivity to marginalization** High-redshift galaxy surveys benefit significantly from measuring highly biased objects. These large biases can offset the growth function, $b(z)D_1(z) = \text{const}$, and keep the amplitude of the galaxy power spectrum effectively fixed at high redshift. This boost is important for maintaining a signal above the shot noise, which we have assumed is redshift-independent. As a

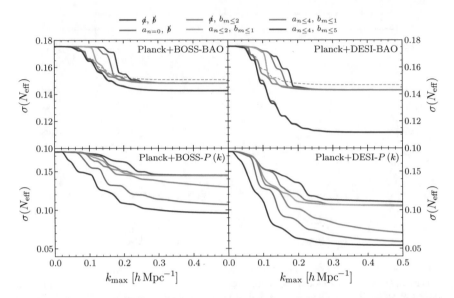

**Fig. 6.3** Forecasts for BOSS and DESI combined with Planck as a function of the largest Fourier modes used in the forecast, $k_{max}$, using various levels of both additive and multiplicative marginalization, cf. the $a_i$ and $b_i$-terms in (6.17) (from [1]). We have varied the number of parameters in the marginalization from none ($\cancel{a}$) to five ($a_{n\leq4}$) and none ($\cancel{b}$) to six ($b_{m\leq5}$), respectively. The dashed line shows the constraints from a standard isotropic BAO analysis for comparison

consequence, high-redshift and low-redshift galaxy power spectra are equally sensitive to uncertainties in the biasing coefficients. This is particularly significant when determining the largest wavenumbers that carry useful cosmological information. While taking $k_{max} > 0.2\,h\,\mathrm{Mpc}^{-1}$ is appealing to maximize the constraints on $N_{eff}$, we must also marginalize over successively more bias parameters. Figure 6.3 shows how the results depend on the marginalization scheme. While both the $P(k)$- and BAO-constraints degrade significantly when going from no marginalization to a few bias parameters, the BAO-forecasts quickly become robust to the marginalization. In contrast, the $P(k)$-forecasts weaken notably with additional biasing, but always lie below the BAO-only results, as one would expect. This confirms the intuition that the information that is primarily driving the constraints derived from $P(k)$ is present in the no-wiggle power spectrum, $P^{\mathrm{nw}}(k)$, instead of the BAO spectrum.

It is instructive to compare the results of our BAO-forecasts with those of a standard BAO analysis. Specifically, it is conventional to use the BAO signal to constrain only $q_i$, $i = \perp, \parallel$, defined in (6.8) and derive parameter constraints from them.[4] These derived limits on $N_{eff}$ are shown as the dashed lines in Fig. 6.3. The fact that the standard BAO constraints are slightly weaker than those of our full BAO-forecasts, even

---

[4]We also compared the constraints coming from the full anisotropic treatment (cf. Sect. 6.2.1) with the isotropic approximation. The BAO-forecasts only weaken at small wavenumbers depending on the marginalization procedure, but reach the same plateau values at large wavenumbers as our baseline analysis. In contrast, the constraints on $N_{eff}$ are systematically weaker in the isotropic $P(k)$-forecasts at the level of 15% for $k_{max} = 0.2\,h\,\mathrm{Mpc}^{-1}$.

**Table 6.3** Forecasted $1\sigma$ constraints on $N_{eff}$ and $Y_p$ for various combinations of current and future CMB and LSS experiments using $P(k)$-forecasts with $k_{max} = 0.2\,h\,\mathrm{Mpc}^{-1}$

|  | Parameter | CMB | Spectroscopic | | | | Photometric | |
|---|---|---|---|---|---|---|---|---|
|  |  |  | BOSS | eBOSS | DESI | Euclid | DES | LSST |
| Planck | $N_{eff}$ | 0.32 | 0.25 | 0.22 | 0.14 | 0.13 | 0.29 | 0.23 |
|  | $Y_p$ | 0.018 | 0.016 | 0.016 | 0.013 | 0.012 | 0.017 | 0.015 |
| CMB-S3 | $N_{eff}$ | 0.12 | 0.12 | 0.11 | 0.094 | 0.088 | 0.12 | 0.11 |
|  | $Y_p$ | 0.0069 | 0.0068 | 0.0067 | 0.0060 | 0.0058 | 0.0069 | 0.0066 |
| CMB-S4 | $N_{eff}$ | 0.081 | 0.079 | 0.078 | 0.070 | 0.067 | 0.081 | 0.078 |
|  | $Y_p$ | 0.0047 | 0.0046 | 0.0046 | 0.0043 | 0.0042 | 0.0047 | 0.0046 |

**Table 6.4** Forecasted $1\sigma$ constraints on $N_{eff}$ and $Y_p$ for various combinations of current and future CMB and LSS experiments using BAO-forecasts with $k_{max} = 0.5\,h\,\mathrm{Mpc}^{-1}$

|  | Parameter | CMB | Spectroscopic | | | | Photometric | |
|---|---|---|---|---|---|---|---|---|
|  |  |  | BOSS | eBOSS | DESI | Euclid | DES | LSST |
| Planck | $N_{eff}$ | 0.32 | 0.29 | 0.29 | 0.28 | 0.28 | 0.30 | 0.29 |
|  | $Y_p$ | 0.018 | 0.018 | 0.018 | 0.018 | 0.018 | 0.018 | 0.018 |
| CMB-S3 | $N_{eff}$ | 0.12 | 0.12 | 0.12 | 0.12 | 0.12 | 0.12 | 0.12 |
|  | $Y_p$ | 0.0069 | 0.0069 | 0.0069 | 0.0069 | 0.0069 | 0.0069 | 0.0069 |
| CMB-S4 | $N_{eff}$ | 0.081 | 0.080 | 0.080 | 0.079 | 0.079 | 0.081 | 0.080 |
|  | $Y_p$ | 0.0047 | 0.0047 | 0.0047 | 0.0046 | 0.0046 | 0.0047 | 0.0047 |

after marginalization, suggests there is information in the BAO spectrum beyond the BAO scale. Combined with our marginalization procedure, the analytic insights of Chap. 5 suggest that the improvement from the grey dashed line to the green line in the top row of Fig. 6.3 is carried by the phase shift. The more prominent enhancement in sensitivity indicated by the blue lines should then be related to the amplitude shift induced by free-streaming particles. We will explore this further in Sect. 6.3.

**Degeneracy with $Y_p$** To explore possible degeneracies between the effective number of relativistic species $N_{eff}$ and the primordial helium fraction $Y_p$, we now consider a $\Lambda$CDM+$N_{eff}$+$Y_p$ cosmology. In Tables 6.3 and 6.4, we present the $1\sigma$ constraints on $N_{eff}$ and $Y_p$ for various combinations of current and future CMB and LSS experiments using $P(k)$-forecasts and BAO-forecasts, respectively. As expected, the CMB-only constraint on $N_{eff}$ become worse due to the well-known degeneracy between $N_{eff}$ and $Y_p$ in the CMB damping tail. When broadband information is included, we find significant improvements in the constraints on both $N_{eff}$ and $Y_p$. However, this improvement cannot be attributed to the phase shift as we see only modest improvements in our BAO-forecasts. The broadband shape of the matter distribution is sensitive to the expansion history and to free-streaming neutrinos, but is not significantly affected by $Y_p$. As a result, the broadband information in $P(k)$ can break CMB degeneracies even without the phase shift information.

**Comments on reconstruction** In our baseline forecasts, we took $R \equiv 1$ in (6.9), which is equivalent to taking $\Sigma_s \to \infty$. A few comments are in order regarding the effect of a finite $\Sigma_s$. As discussed in [30], the optimal smoothing scale $\Sigma_s$ used in the BAO reconstruction depends on the noise levels of the experiment. Having said that, we have found only small changes in our results when going from $\Sigma_s = \infty$ to finite $\Sigma_s$. The constraints quoted in Tables 6.1–6.4 are basically unaffected, except for DESI and Euclid in the $P(k)$-forecasts, where the impact is also mild. Changing $\Sigma_s$ from $30\,h^{-1}$ Mpc to $15\,h^{-1}$ Mpc and $10\,h^{-1}$ Mpc, the constraint on $N_{\mathrm{eff}}$ slightly weakens from 0.090 to 0.093 and 0.096 for Planck+DESI (0.082, 0.086 and 0.090 for Planck+Euclid) in $\Lambda$CDM+$N_{\mathrm{eff}}$ compared to the quoted 0.087 (0.079) in Table 6.1. In practice, this roughly 10% effect has to be compared to the impact on the reconstruction efficiency.

## 6.2.3  Designer's Guide for Future Surveys

One of the main benefits of a Fisher forecast is that it can inform the design of future experiments. For spectroscopic surveys, the basic parameters are the total number of objects, $N_g$, the maximal redshift, $z_{\mathrm{max}}$, and the sky area in square degrees, $\Omega$. From these, we derive the survey volume, $V$, and the comoving number density, $\bar{n}_g$.[5] In this section, we will explore how the constraints on $N_{\mathrm{eff}}$ depend on these parameters.

Most of the survey characteristics are encoded in the effective survey volume,[6] $V_{\mathrm{eff}}$, cf. (6.5) and (6.21). The dependence of $V_{\mathrm{eff}}$ on the survey parameters is somewhat non-trivial. Increasing $V$ (by increasing $z_{\mathrm{max}}$ and/or $\Omega$), at fixed $N_g$, will also reduce $\bar{n}_g$. For signal-dominated modes, $\bar{n}_g P_g \gg 1$, this effect is not important and the effective volume scales approximately as $V_{\mathrm{eff}} \propto V$. However, for $\bar{n}_g P_g \ll 1$, the shot noise is important and the reduction in the comoving density is more important than the increase in the volume, so that the effective volume scales as $V_{\mathrm{eff}} \propto V^{-1}$. This means that we will only benefit from an increase in the volume as long as the modes of interest, $k \in [0.1, 0.3]\,h\,\mathrm{Mpc}^{-1}$, are signal dominated.

As mentioned before, the increased linearity of the matter distribution at high redshifts is undermined by the larger biasing. As a result, the main benefit of large $z_{\mathrm{max}}$ is the increased survey volume and hence the total number of modes. Unfortunately, the survey volume only grows slowly with redshift for $z > 2$ and the resulting improvements in parameters is relatively modest for large increases in $z_{\mathrm{max}}$. The situation is slightly different for the BAO spectrum as the non-linear damping factor $D(k, \mu)$ depends on the clustering of the matter directly and is therefore less important at high redshifts. However, the BAO signal alone has a relatively modest effect on

---

[5]For simplicity, we will assume that the comoving number density can be approximated by a constant over the complete survey volume. However, very similar results are obtained for BOSS and DESI when using the specific redshift-dependent number densities.

[6]The effective survey volume also depends on the linear bias parameter $b$ through $\bar{n}_g P_g \propto \bar{n}_g b^2$. This dependence is degenerate with a rescaling of $\bar{n}_g$, so we will take $b(z = 0) \equiv 1$ and vary $\bar{n}_g$. This ignores the impact that changes in $b$ may have on redshift space distortions.

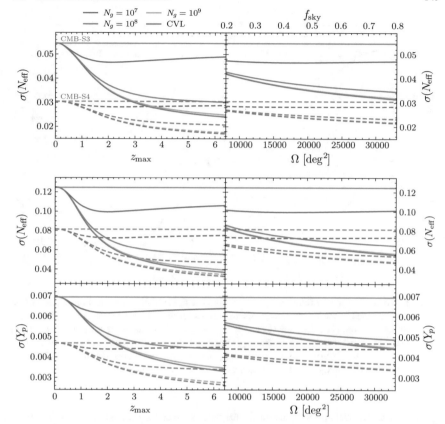

**Fig. 6.4** Future constraints for $\Lambda$CDM+$N_{\text{eff}}$ (*top*) and $\Lambda$CDM+$N_{\text{eff}}$+$Y_p$ (*bottom*) from the full galaxy power spectrum, $P_g(k)$, up to $k_{\max} = 0.2\,h\,\text{Mpc}^{-1}$ as a function of the total number of objects, $N_g$, at fixed survey area $\Omega = 20\,000\,\text{deg}^2$ (*left*) and as a function of the survey area $\Omega$ (or sky fraction $f_{\text{sky}}$) for fixed $z_{\max} = 2$ (*right*; from [1]). The comoving number density is assumed to be constant and given by the total volume of the survey. For "CVL" (red), all modes in the survey are assumed to be measured up to the limit set by cosmic variance. Solid and dashed lines correspond to combining the LSS data with CMB-S3 and CMB-S4 data, respectively. The gray lines indicate the level of sensitivity of the respective CMB experiments alone

$N_{\text{eff}}$-forecasts in general and the change to the damping factor consequently does not make a visible difference in our forecasts.

In the top panel of Fig. 6.4, we present $P(k)$-forecasts for $N_{\text{eff}}$ for a variety of survey configurations, assuming $Y_p$ is fixed by BBN consistency. We see that the largest improvement comes from increasing $N_g$ from $10^7$ to $10^8$. As we increase the number of objects further, we reach the cosmic variance limit for all modes of interest. We see that an optimistic future survey combined with a near-term CMB experiment can provide constraints that are comparable to (or slightly stronger than) those projected for CMB-S4 alone. Having said that, it does not appear that one can push the measurement of $N_{\text{eff}}$ well beyond the CMB-S4 target. Moreover, as in the case of the planned experiments, the improvements from the BAO signal alone are rather small.

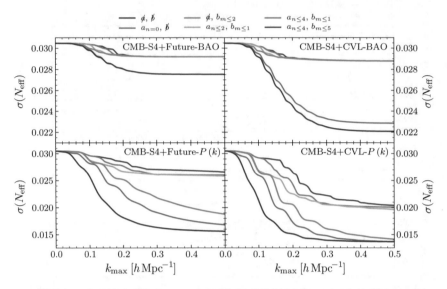

**Fig. 6.5** Forecasts for two future surveys combined with CMB-S4 as a function of the largest Fourier modes used in the forecast, $k_{\max}$, using various levels of both additive and multiplicative marginalization (from [1]). We have varied the number of parameters in the marginalization from none ($\phi$) to five ($a_{n \leq 4}$) and none ($\not b$) to six ($b_{n \leq 5}$), respectively. The employed experimental specifications for the "Future"-survey are $N_g = 10^8$, $z_{\max} = 3$ and $f_{\mathrm{sky}} = 0.5$, whereas the "CVL"-survey is cosmic variance limited for all $k$ up to $k_{\max}$ over $f_{\mathrm{sky}} = 0.5$ and $z_{\max} = 6$

The value of LSS becomes more significant as we expand the space of parameters. The bottom panel of Fig. 6.4 shows $P(k)$-forecasts for $\Lambda$CDM+$N_{\mathrm{eff}}$+$Y_p$. We again see that the most significant jump in sensitivity arises when $N_g$ increases from $10^7$ to $10^8$. We note that a factor of two improvement in $\sigma(N_{\mathrm{eff}})$ over CMB-S4 seems possible. We also see that the $P(k)$-forecasts for $N_{\mathrm{eff}}$ marginalized over $Y_p$ are competitive with CMB-only forecasts with $Y_p$ held fixed. In this sense, $P(k)$ adds robustness to the measurement of $N_{\mathrm{eff}}$ under broader extensions of $\Lambda$CDM. The improvement in $Y_p$ is slightly weaker, but shows the same general trend.

The range of accessible modes in near-term galaxy surveys is limited by their reliance on highly biased objects, but more futuristic surveys may not have the same limitations. Future surveys can also have high signal-to-noise beyond $k = 0.2\,h\,\mathrm{Mpc}^{-1}$, making it worth to consider the impact of increasing $k_{\max}$. In Fig. 6.5, we show the potential reach of two representative surveys. The first, denoted "Future", is characterised by $N_g = 10^8$, $f_{\mathrm{sky}} = 0.5$ and $z_{\max} = 3$, which is roughly the same as a spectroscopic follow-up to LSST. The second, denoted "CVL", is cosmic variance limited for all $k \leq k_{\max}$ over $f_{\mathrm{sky}} = 0.5$ and $z_{\max} = 6$. In principle, a 21 cm intensity mapping survey could achieve similar performance [10]. We see that $\sigma(N_{\mathrm{eff}}) \sim 0.015$ is achievable through the measurement of $P(k)$ in either survey for $k_{\max} = 0.5\,h\,\mathrm{Mpc}^{-1}$, although the improvement with CVL is more robust to marginalization.

## 6.3  Measurements of the Phase Shift

In the previous section, we showed how much the combination of future CMB and LSS measurements can improve the sensitivity to extra relativistic species. The dominant source of improvement came from the broadband shape of the power spectra, $P^{\mathrm{nw}}(k)$, rather than the BAO spectrum, $P^{\mathrm{w}}(k)$. Nevertheless, the shift of the acoustic peaks is a particularly robust signature of free-streaming, relativistic species [14] and is therefore an interesting observable in its own right. In this section, we will isolate the signal coming from the phase shift and forecast our ability to measure it in future surveys. Measuring the BAO phase shift provides an independent test of pre-recombination physics in a low-redshift observable. This could be used to shed light on possible discrepancies between low- and high-redshift measurements or as a discovery channel for exotic new physics.

### 6.3.1  Isolating the Phase Shift

The BAO feature in Fourier space can be written as

$$O_{\mathrm{lin}}(k) = A(k) \sin\left[\alpha^{-1} r_s k + \phi(k)\right], \tag{6.22}$$

where the parameter $\alpha$ represents changes in the BAO scale $r_s$, and the amplitude modulation $A(k)$ and the phase shift $\phi(k)$ encode a number of physical effects that alter the time evolution of the baryons. While $\alpha$ and $A(k)$ are implicit functions of redshift, $\phi(k)$ is redshift independent. Relativistic species are the unique source of a constant shift in the locations of the BAO peaks in the limit of large wavenumbers, i.e. $\phi(k \to \infty) = \phi_\infty$ (cf. Sect. 5.2.3). In practice, however, the measurement of the BAO spectrum occurs over a relatively small range of scales with a small number of (damped) acoustic oscillations. On these scales, the $k$-dependence of the shift can be relevant. Furthermore, additional $k$-dependent shifts from other cosmological parameters may also have to be taken into account [31].

To measure the phase shift $\phi(k)$, we will construct a template for the $k$-dependence as a function of the relevant parameters. For small variations around their fiducial values, it is a good approximation to treat the shifts arising from each cosmological parameter independently. By varying one parameter at a time and measuring the change in the peak locations, we can construct a template $\phi(k) = \sum_i \beta_i(\vec{\theta}) f_i(k)$. For $\Lambda\mathrm{CDM}+N_{\mathrm{eff}}$, the parameters $A_s$, $n_s$ and $\tau$ do not affect the evolution of the baryons prior to recombination and, therefore, do not change the phase of the oscillations. The parameters $\omega_b$ and $\theta_s$ do alter the BAO spectrum, but are effectively negligible for any realistic parameter range. The shifts induced by $\omega_c$ and $N_{\mathrm{eff}}$, on the other hand, can be significant.

The parameter that is most independent of $N_{\mathrm{eff}}$ is not the dark matter density $\omega_c$, but the scale factor at the time of matter-radiation equality, $a_{\mathrm{eq}}$. Since CMB data essentially fixes $a_{\mathrm{eq}}$, our template model can be reduced to

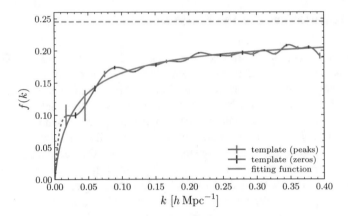

**Fig. 6.6** Template of the phase shift $f(k)$ as defined in (6.23) (from [1]). The numerical phase shifts (blue) were obtained by sampling from 100 different cosmologies with varying $N_{eff}$ and rescaling by $\beta(N_{eff})$ as defined in (6.27). The bars indicate the standard deviation in these measurements at the positions of the peaks (light blue) and zeros (dark blue) compared to the fiducial BAO spectrum. The red line shows the fitting function defined in (6.24). The dashed gray line is the analytic approximation (5.61) to the constant phase shift

$$\phi(k) = \beta(N_{eff})f(k), \qquad (6.23)$$

namely the shift induced by changing $N_{eff}$ at fixed $a_{eq}$. This is the same choice made by Follin et al. [32] in their CMB measurement of the phase shift. Fixing $a_{eq}$ also reproduces the expected constant phase shift at large wavenumbers. The template for the phase shift at fixed $\omega_c$, in contrast, does not approach a constant at large wavenumbers, which implies that the change of $a_{eq}$ to maintain constant $\omega_c$ is introducing a phase shift of comparable size to the constant shift induced by varying $N_{eff}$. For our applications, this additional shift plays no role, but it could be useful in future investigations.

We describe the measurement of the phase shift and the construction of the template in Appendix D. In short, we determine the shift in the locations of the peaks/troughs and zeros of the BAO spectrum compared to the fiducial cosmology with $N_{eff} = 3.046$ and sample 100 different cosmologies with varying $N_{eff}$ at fixed $a_{eq}$. It is convenient to normalize the template $f(k)$ such that $\beta = 0$ and 1 for $N_{eff} = 0$ and 3.046, respectively. In Fig. 6.6, we illustrate how the peaks/troughs and zeros of the BAO spectrum change in response to this variation in $N_{eff}$. We see that the phase shift created by $N_{eff}$ approaches a constant at large wavenumbers in line with physical expectations.

The measurement of the phase shift is challenging because it requires a very accurate model of the no-wiggle spectrum $P^{nw}(k)$ across a wide range of cosmological parameters. Errors in $P^{nw}(k)$ effectively change the functions $A(k)$ and $B(k)$ in (6.7) and lead to errors in the measurement of the BAO peaks and zeros, respectively. The small size of the phase shift in Fig. 6.6 only exacerbates this problem. Fortunately, while the template is difficult to generate, our forecasts using the template are very stable. Furthermore, the template is well approximated by a simple fitting function,

$$f(k) = \frac{\phi_\infty}{1 + (k_\star/k)^\xi}, \tag{6.24}$$

where $\phi_\infty = 0.227$, $k_\star = 0.0324\,h\,\mathrm{Mpc}^{-1}$ and $\xi = 0.872$ were obtained by a weighted fitting procedure. From the analytic treatment in Sect. 5.2.3, we expect $\phi_\infty = 0.191\pi\,\epsilon_{\mathrm{fid}} + \mathcal{O}(\epsilon_{\mathrm{fid}}^2) \approx 0.245$ to linear order, where $\epsilon(N_{\mathrm{eff}}) = N_{\mathrm{eff}}/(a_\nu + N_{\mathrm{eff}})$ is a measure of the excess radiation density, $(\rho_r - \rho_\gamma)/\rho_r$, with $a_\nu \approx 4.40$ as introduced in (3.16). This approximation overestimates the value obtained using the fitting formula by about 8%, which is consistent with the expected corrections from higher orders in $\epsilon_{\mathrm{fid}} = \epsilon_\nu \approx 0.41$. Around $k \sim 0.1\,h\,\mathrm{Mpc}^{-1}$, where BOSS and DESI have the largest signal-to-noise ratio, the relative difference is almost 50%, which makes it evident that the offset from the analytic approximation has to be taken into account in an analysis such as the one proposed below, whereas the precise shape of the template plays a sub-dominant role. We also note that this template is basically independent of changes to the BAO scale $r_s$, for example due to changes in the dark matter density.

We use the measured phase template to write the BAO spectrum in terms of the spectrum in the fiducial cosmology:

$$O(k) = O_{\mathrm{fid}}\big(\alpha^{-1}k + (\beta - 1)\,f(k)/r_s^{\mathrm{fid}}\big), \tag{6.25}$$

where $\alpha \equiv \alpha(z_i)$ takes an independent value in each redshift bin centred around $z_i$ and $\beta$ is a single parameter for the entire survey. A measurement of $\alpha(z_i)$ and $\beta$ can then be translated into constraints on cosmological parameters using

$$\alpha(\vec{\theta}; z) \equiv q\,r_s^{\mathrm{fid}}/r_s = [D_V(z)/r_s] / [D_V(z)/r_s]_{\mathrm{fid}}, \tag{6.26}$$

$$\beta(N_{\mathrm{eff}}) \equiv \epsilon/\epsilon_{\mathrm{fid}}, \tag{6.27}$$

where the parameters $q$ and $D_V$ were introduced in Sect. 6.2.1. With this normalization, the largest possible phase shift due to $N_{\mathrm{eff}}$ is given by $\beta(N_{\mathrm{eff}} \to \infty) = 2.45$.

In Sect. 7.2.1, we will show that the forecasts produced using only these templates are in agreement with the forecasts using the full BAO spectrum. From a measurement of $\beta > 0$, one gets a constraint on $N_{\mathrm{eff}}$ that is only associated to the size of the phase shift. This approach is analogous to the template-based measurement of the phase shift in the CMB by Follin et al. [32]. The measurement of $N_{\mathrm{eff}}$ from the phase alone ignores the effects of $N_{\mathrm{eff}}$ on $\alpha$, but has the advantage that any detection is unambiguously[7] a measurement of free-streaming relativistic particles.

---

[7]We have explicitly checked that our template gives an unbiased measurement of $\beta$, cf. Sect. 7.2.2. In particular, we have verified that we reproduce $\beta \approx 0$ for a cosmology with $N_{\mathrm{eff}} = 0$.

We will also be interested in the measurement of $\beta$ when a prior on $\alpha$ is included, e.g. from the CMB.[8] In a given cosmological model, the parameter $\alpha$ is fully determined by the set of cosmological parameters, $\alpha = \alpha(\vec{\theta})$. As the $\alpha(z_i)$ inferred from the CMB are correlated between the $n$ redshift bins of a galaxy survey and $n$ is in general larger than the number of cosmological parameters, we compute the $n$-dimensional inverse covariance matrix according to $C_\alpha^{-1} = A^T F A$, where $F$ is the Fisher matrix and $A$ is the pseudo-inverse of $\nabla_{\vec{\theta}}\, \vec{\alpha}$. We use the CMB Fisher matrices for the $\Lambda$CDM+$N_{\mathrm{eff}}$ cosmology as in Sect. 6.2. We can then impose the $\alpha(z_i)$-prior on the redshift-binned likelihood function $\mathcal{L}(\alpha, \beta; z_i)$ according to $\mathcal{L}(\beta) \propto \int \prod_{z_i} \mathrm{d}\alpha_i \prod_{z_i} \mathcal{L}(\alpha_i, \beta; z_i)\, \pi(\alpha_1, \ldots, \alpha_n)$, where $\alpha_i \equiv \alpha(z_i)$ and $\pi$ is the $n$-dimensional Gaussian prior with covariance matrix $C_\alpha$. The observed posterior distribution for $\alpha(z_i)$ could also be constructed by evaluating $\alpha(z_i)$ for each point in a given CMB Markov chain.

## 6.3.2  Constraints from Planned Surveys

We will now show how well the phase shift can be measured in planned galaxy surveys. It is useful to first understand the parameter space $\alpha$-$\beta$ without imposing a prior on $\alpha$. Both parameters affect the locations of the acoustic peaks and are therefore quite degenerate. We will use likelihood-based forecasts to ensure accuracy. We will confirm that the posterior distributions[9] of $\alpha$ and $\beta$ are Gaussian, while the constraints on $N_{\mathrm{eff}}$ derived from this parametrization are significantly non-Gaussian. This suggests that a Fisher matrix forecast in terms of $\alpha$ and $\beta$ would be more reliable than one that starts directly from $N_{\mathrm{eff}}$.

We define the phase shift relative to the fiducial model with $N_{\mathrm{eff}} = 3.046$. The broadband spectrum for the fiducial model can be isolated by using the method in Appendix D or through the use of a fitting function along the lines of [28]. These methods generate the BAO spectrum $O_{\mathrm{fid}}(k)$ and hence $O(k)$ via (6.25). We compute the log-likelihood using the same noise and modelling as in the Fisher matrix (6.21).

Forecasts for the one- and two-dimensional posteriors are shown in Fig. 6.7 for both BOSS and DESI. We see that for both surveys the posterior distributions are Gaussian. The best-fit Gaussian for BOSS and DESI has $\sigma(\beta) = 1.3$ and $0.47$, respectively, which corresponds to a rejection of $\beta = 0$ at 77 and 98% confidence. Clearly, BOSS cannot exclude $\beta = 0$ (and hence $N_{\mathrm{eff}} = 0$) without any prior information from the CMB. Since the weakness of the constraint on $\beta$ is driven by the degeneracy with $\alpha$ (see the left panel in Fig. 6.7), we expect to get significant improvements in the constraints on $\beta$ after imposing a CMB prior on $\alpha$. Inspection of the two-dimensional contours already shows that we will sizeably limit the range of $\beta$.

---

[8] We also indirectly use the CMB data to constrain other cosmological parameters, in particular the scale factor at matter-radiation equality $a_{\mathrm{eq}}$, so that we can ignore any additional phase shifts not associated with $N_{\mathrm{eff}}$.

[9] Since we assume flat priors for the parameters, we can identify the posteriors with the likelihoods.

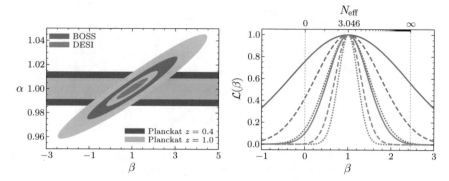

**Fig. 6.7** *Left:* Contours showing $1\sigma$ and $2\sigma$ exclusions in the $\alpha$-$\beta$ plane for BOSS and DESI (from [1]). For purpose of illustration, we have reduced these surveys to a single redshift bin (and therefore a single $\alpha$-parameter). The gray bands indicate Planck priors for $\alpha$ assuming the median redshift is $z = 0.4$ and 1.0 for BOSS and DESI, respectively. *Right:* One-dimensional posterior distributions of $\beta$ for BOSS and DESI. The dashed and dotted lines indicate the use of a redshift-dependent CMB prior on $\alpha$ from Planck and CMB-S4, respectively

The posterior distribution with the prior from Planck (CMB-S4) is shown in the right panel of Fig. 6.7. For BOSS, we find $\sigma(\beta) = 0.76$ (0.50) which implies that $\beta > 0$ at 81% (95%) confidence. For DESI, we should find strong evidence for a phase shift with $\sigma(\beta) = 0.30$ (0.22) which excludes $\beta = 0$ at $3.5\sigma$ ($4.6\sigma$).

To translate these results into constraints on $N_{\rm eff}$, we use the relationship between $\beta$ and $N_{\rm eff}$ given in (6.27). This map is non-linear over the measured range of $\beta$ and we therefore anticipate the posteriors to be non-Gaussian. The derived $N_{\rm eff}$-posteriors in Fig. 6.8 indeed show a highly non-Gaussian distribution. As anticipated from the $\beta$-posterior for BOSS, the constraints on $N_{\rm eff}$ are relatively weak without imposing a Planck prior on $\alpha$.

We also see that the constraining power is significantly weaker at bounding large values of $N_{\rm eff}$ than small ones. This asymmetry is simply a reflection of the fact that increasing $N_{\rm eff}$ does not produce proportionally larger phases shifts. The same asymmetry was also seen in the CMB constraints of Follin et al. [32], likely for the same reason. Recall that we have an upper limit on the phase shift of $\beta < 2.45$, which is saturated for $N_{\rm eff} \to \infty$. In practice, this means that for $N_{\rm eff} \gg a_\nu \approx 4.40$, we will have an equal likelihood[10] for every value of $N_{\rm eff}$ because they produce identical spectra. As a result, a flat prior on $N_{\rm eff}$ (rather than $\beta$) will lead to ill-defined results because the integral $\int^\infty {\rm d}N_{\rm eff}\, \mathcal{L}(N_{\rm eff})$ will diverge. On the other hand, for highly-significant detections of $\beta > 0$, a flat prior over any reasonable range of $N_{\rm eff}$ will produce stable results. We are not quite in this regime with BOSS, which is why we will only quote constraints on $\beta$.

---

[10]Realistic values of $N_{\rm eff}$ are not quite in the asymptotic regime, but still show the weakened distinguishing power for larger $N_{\rm eff}$.

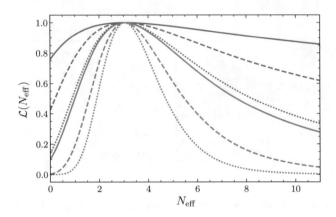

**Fig. 6.8** Posterior distributions of $N_{\mathrm{eff}}$ for BOSS (blue) and DESI (red) derived from the phase shift in the BAO spectrum, i.e. via the measurement of $\beta$ (from [1]). The dashed and dotted lines indicate that a redshift-dependent CMB prior on $\alpha$ has been imposed using Planck and CMB-S4, respectively

**Table 6.5** Forecasted $1\sigma$ constraints on the amplitude of the phase shift $\beta$ for current and future LSS experiments. We also show the constraints on $\beta$ after imposing a redshift-dependent prior on the BAO parameter $\alpha$ from Planck and CMB-S4

|                | Spectroscopic | | | | Photometric | |
|----------------|------|-------|------|--------|------|------|
|                | BOSS | eBOSS | DESI | Euclid | DES  | LSST |
| BAO            | 1.3  | 1.0   | 0.47 | 0.40   | 2.6  | 1.0  |
| + Planck prior | 0.76 | 0.70  | 0.30 | 0.26   | 1.1  | 0.50 |
| + CMB-S4 prior | 0.50 | 0.48  | 0.22 | 0.19   | 1.0  | 0.42 |

Table 6.5 shows the projected constraints on $\beta$ for a variety of planned surveys with and without priors from the CMB. We see that roughly a factor of three improvement can be achieved in spectroscopic surveys going from BOSS to Euclid. Both DESI and Euclid should have sufficient sensitivity to reach a more than $5\sigma$ exclusion of $\beta = 0$ when imposing a Planck prior. As before, galaxy clustering measurements in photometric surveys do not lead to competitive constraints as they are effectively two-dimensional on the relevant scales.

## 6.3.3   Constraints from Future Surveys

Given the robustness of the phase shift as a probe of light relics, a high-significance detection of the phase shift in LSS would be a valuable piece of cosmological information. We have seen that current and planned surveys can detect the phase shift, but are not expected to produce constraints on $N_{\mathrm{eff}}$ that are competitive with those from the CMB. It is natural to ask if future surveys can reach this level of sensitivity.

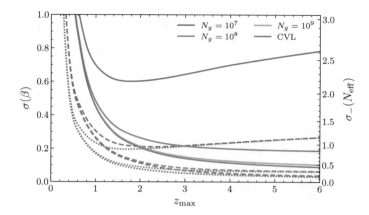

**Fig. 6.9** Future constraints on the amplitude of the phase shift $\beta$ as a function of $z_{max}$ and $N_g$, assuming $f_{sky} = 0.5$ (from [1]). The dashed and dotted lines indicate that a CMB prior on $\alpha$ has been imposed using Planck and CMB-S4, respectively. The corresponding $1\sigma$ lower limit on $N_{eff}$, which is $N_{eff} = 3.046 - \sigma_-(N_{eff})$, is computed by inverting (6.27) and indicated by the right axis

Like the measurement of the BAO scale, the measurement of the phase requires large signal-to-noise for $0.1\,h\,\mathrm{Mpc}^{-1} \lesssim k \lesssim 0.3\,h\,\mathrm{Mpc}^{-1}$. As long as the number density is sufficiently large to keep the shot noise below cosmic variance, we gain primarily by increasing $z_{max}$ to achieve larger survey volumes. At larger levels of the shot noise, we only measure a few peak locations well which increases the degeneracy between $\alpha$ and $\beta$. Figure 6.9 shows results for a variety of possible survey configurations. As before, the constraints on $\beta$ can be mapped into constraints on $N_{eff}$ using (6.27). We see that with $10^8$ objects and $z_{max} > 3$, we consistently obtain $\sigma(N_{eff}) < 0.5$ (1.0) with (without) a prior on $\alpha$ included.

To put these results into context, the measurement of Follin et al. [32] of $N_{eff}^{\phi} = 2.3^{+1.1}_{-0.4}$ from the Planck TT spectrum is comparable to a survey with $N_g = 10^9$ objects out to redshift $z_{max} = 3$. Follin et al. also forecasted $\sigma(N_{eff}^{\phi}) = 0.41$ for Planck TT+TE+EE which is near the sensitivity of future LSS surveys when increasing the redshift range to $z_{max} = 6$. Reaching this level of sensitivity will be extremely challenging with an optical survey, but could potentially be achieved with 21 cm intensity mapping [10].

## 6.4 Summary

Large-scale structure surveys are an untapped resource in the search for light relics of the hot big bang. The growing statistical power of these surveys will make them competitive with the CMB in terms of the constraints they will provide on a broad range of cosmological parameters. Moreover, the combination of CMB and LSS observations will allow powerful and robust tests of the physical laws that determined the structure and evolution of the early universe.

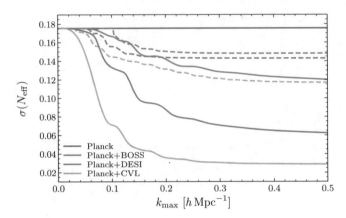

**Fig. 6.10** Sensitivity of planned and future LSS surveys to $N_{\mathrm{eff}}$ using the galaxy power spectrum (solid) and the BAO spectrum (dashed) marginalized over two bias parameters, $b_{m \le 1}$ (from [1])

In this chapter, we have explored the potential impact of LSS surveys on measurements of the parameter $N_{\mathrm{eff}}$. We have found that the dominant statistical impact of future surveys lies in the shape of the galaxy power spectrum. The distribution of dark matter in the universe is altered through the gravitational influence of the free-streaming radiation, leading to changes in the shape of the power spectrum that can be detected at high significance. A summary of the reach of selected planned and future surveys is given in Fig. 6.10. We see that BOSS and DESI can extend results significantly beyond the current CMB constraints. Futuristic surveys combined with a future CMB-S4 mission could achieve $\sigma(N_{\mathrm{eff}}) \sim 0.015$, which is close to reaching the target of $\Delta N_{\mathrm{eff}} = 0.027$ at a significance of $2\sigma$.

Future LSS surveys will also be able to detect the coherent shift in the peak locations of the BAO spectrum. This would be an intriguing measurement as this phase shift is a highly robust and unambiguous probe of light relics and the cosmic neutrino background [14]. The fact that the phase shift should agree between the CMB and BAO measurements is a highly non-trivial consequence of physics both before and after recombination, and could be an interesting test of exotic extensions of $\Lambda$CDM without requiring the CMB as an anchor. Being a new low-redshift observable, improved measurements of the BAO phase shift may therefore play a useful role in elucidating apparent low-$z$/high-$z$ discrepancies in some cosmological data [33].

In the future, the combination of different cosmological observations might significantly advance our insights into fundamental physics. In this chapter, we observed that CMB and LSS observations can complement each other by providing tighter as well as more robust constraints on $N_{\mathrm{eff}}$. In addition, we demonstrated that the BAO spectrum encodes cosmological information beyond the acoustic scale which can be extracted reliably. A broader exploration will likely reveal more targets that benefit from this complementarity.

# References

1. D. Baumann, D. Green, B. Wallisch, Searching for light relics with large-scale structure. JCAP **08**, 029 (2018). arXiv:1712.08067 [astro-ph.CO]
2. K. Dawson et al. (BOSS Collaboration), The Baryon Oscillation Spectroscopic Survey of SDSS-III. Astron. J. **145**, 10 (2013). arXiv:1208.0022 [astro-ph.CO]
3. K. Dawson et al., The SDSS-IV extended Baryon Oscillation Spectroscopic Survey: overview and early data. Astron. J. **151**, 44 (2016). arXiv:1508.04473 [astro-ph.CO]
4. T. Abbott et al. (DES Collaboration), The Dark Energy Survey. arXiv:astro-ph/0510346 [astro-ph]
5. A. Aghamousa et al. (DESI Collaboration), The DESI experiment part I: science, targeting and survey design. arXiv:1611.00036 [astro-ph.IM]
6. Ž. Ivezić et al. (LSST Collaboration), LSST: from science drivers to reference design and anticipated data products. arXiv:0805.2366 [astro-ph]
7. R. Laureijs et al. (Euclid Collaboration), Euclid definition study report. arXiv:1110.3193 [astro-ph.CO]
8. A. Font-Ribera, P. McDonald, N. Mostek, B. Reid, H.-J. Seo, A. Slosar, DESI and other dark energy experiments in the era of neutrino mass measurements. JCAP **05**, 023 (2014). arXiv:1308.4164 [astro-ph.CO]
9. S. Dodelson, K. Heitmann, C. Hirata, K. Honscheid, A. Roodman, U. Seljak, A. Slosar, M. Trodden, Cosmic visions dark energy: science. arXiv:1604.07626 [astro-ph.CO]
10. A. Obuljen, E. Castorina, F. Villaescusa-Navarro, M. Viel, High-redshift post-reionization cosmology with 21 cm intensity mapping. JCAP **05**, 004 (2018). arXiv:1709.07893 [astro-ph.CO]
11. D. Eisenstein, H.-J. Seo, M. White, On the robustness of the acoustic scale in the low-redshift clustering of matter. Astrophys. J. **664**, 660 (2007). arXiv:astro-ph/0604361 [astro-ph]
12. M. Crocce, R. Scoccimarro, Non-linear evolution of baryon acoustic oscillations. Phys. Rev. D **77**, 023533 (2008). arXiv:0704.2783 [astro-ph]
13. N. Sugiyama, D. Spergel, How does non-linear dynamics affect the baryon acoustic oscillation? JCAP **02**, 042 (2014). arXiv:1306.6660 [astro-ph.CO]
14. D. Baumann, D. Green, M. Zaldarriaga, Phases of new physics in the BAO spectrum. JCAP **11**, 007 (2017). arXiv:1703.00894 [astro-ph.CO]
15. B. Bassett, R. Hložek, Baryon acoustic oscillations. arXiv:0910.5224 [astro-ph.CO]
16. L. Verde, Statistical methods in cosmology. Lect. Notes Phys. **800**, 147 (2010). arXiv:0911.3105 [astro-ph.CO]
17. M. Tegmark, Measuring cosmological parameters with galaxy surveys. Phys. Rev. Lett. **79**, 3806 (1997). arXiv:astro-ph/9706198 [astro-ph]
18. D. Blas, M. Garny, M. Ivanov, S. Sibiryakov, Time-sliced perturbation theory II: baryon acoustic oscillations and infrared resummation. JCAP **07**, 028 (2016). arXiv:1605.02149 [astro-ph.CO]
19. N. Hand, U. Seljak, F. Beutler, Z. Vlah, Extending the modelling of the anisotropic galaxy power spectrum to $k = 0.4\, h\,\mathrm{Mpc}^{-1}$. JCAP **10**, 009 (2017). arXiv:1706.02362 [astro-ph.CO]
20. Z. Ding, H.-J. Seo, Z. Vlah, Y. Feng, M. Schmittfull, F. Beutler, Theoretical systematics of future baryon acoustic oscillation surveys. Mon. Not. Roy. Astron. Soc. **479**, 1021 (2018). arXiv:1708.01297 [astro-ph.CO]
21. W. Ballinger, J. Peacock, A. Heavens, Measuring the cosmological constant with redshift surveys. Mon. Not. Roy. Astron. Soc. **282**, 877 (1996). arXiv:astro-ph/9605017 [astro-ph]
22. N. Kaiser, Clustering in real space and in redshift space. Mon. Not. Roy. Astron. Soc. **227**, 1 (1987)
23. H.-J. Seo, D. Eisenstein, Improved forecasts for the baryon acoustic oscillations and cosmological distance scale. Astrophys. J. **665**, 14 (2007). arXiv:astro-ph/0701079 [astro-ph]
24. M. White, Shot noise and reconstruction of the acoustic peak. arXiv:1004.0250 [astro-ph.CO]
25. H.-J. Seo, D. Eisenstein, Probing dark energy with baryonic acoustic oscillations from future large galaxy redshift surveys. Astrophys. J. **598**, 720 (2003). arXiv:astro-ph/0307460 [astro-ph]

26. H. Zhan, L. Knox, Baryon oscillations and consistency tests for photometrically-determined redshifts of very faint galaxies. Astrophys. J. **644**, 663 (2006). arXiv:astro-ph/0509260 [astro-ph]
27. J. Gleyzes, R. de Putter, D. Green, O. Doré, Biasing and the search for primordial non-Gaussianity beyond the local type. JCAP **04**, 002 (2017). arXiv:1612.06366 [astro-ph.CO]
28. F. Beutler et al. (BOSS Collaboration), The clustering of galaxies in the completed SDSS-III Baryon Oscillation Spectroscopic Survey: baryon acoustic oscillations in Fourier space. Mon. Not. Roy. Astron. Soc. **464**, 3409 (2017). arXiv:1607.03149 [astro-ph.CO]
29. K. Abazajian et al. (CMB-S4 Collaboration), CMB-S4 Science Book, 1st edn. arXiv: 1610.02743 [astro-ph.CO]
30. H.-J. Seo, F. Beutler, A. Ross, S. Saito, Modelling the reconstructed BAO in Fourier space. Mon. Not. Roy. Astron. Soc. **460**, 2453 (2016). arXiv:1511.00663 [astro-ph.CO]
31. Z. Pan, L. Knox, B. Mulroe, A. Narimani, Cosmic microwave background acoustic peak locations. Mon. Not. Roy. Astron. Soc. **459**, 2513 (2016). arXiv:1603.03091 [astro-ph.CO]
32. B. Follin, L. Knox, M. Millea, Z. Pan, First detection of the acoustic oscillation phase shift expected from the cosmic neutrino background. Phys. Rev. Lett. **115**, 091301 (2015). arXiv:1503.07863 [astro-ph.CO]
33. W. Freedman, Cosmology at a crossroads: tension with the Hubble constant. Nat. Astron. **1**, 0121 (2017). arXiv:1706.02739 [astro-ph.CO]

# Chapter 7
# Measurement of Neutrinos in the BAO Spectrum

The existence of the cosmic neutrino background is a remarkable prediction of the hot big bang model. These neutrinos were a dominant component of the energy density in the early universe and, therefore, played an important role in the evolution of cosmological perturbations. In particular, fluctuations in the neutrino density produced a distinct shift in the temporal phase of sound waves in the primordial plasma, which has recently been detected in the cosmic microwave background [1]. In the previous chapter, we proposed a new analysis of the baryon acoustic oscillation signal which extends the conventional BAO analysis presented in [2, 3] by including the amplitude of the neutrino-induced phase shift as a free parameter in addition to the BAO scale. In this chapter, we report on the first measurement of this neutrino-induced phase shift in the spectrum of baryon acoustic oscillations of the BOSS DR12 data, based on this novel approach. Constraining the acoustic scale using Planck data while marginalizing over the effects of neutrinos in the CMB, we find evidence for a non-zero phase shift at greater than 95% confidence. We also demonstrate the robustness of this result in simulations and forecasts. Besides providing a new measurement of the cosmic neutrino background, this work is the first application of the BAO feature beyond its application as a standard ruler and to early universe physics.

The outline of this chapter, which is based on the research presented in [4], is as follows. In Sect. 7.1, we briefly review the phase shift induced by cosmic neutrinos and lay out the modified BAO analysis pipeline which employs the proposed template-based approach. In Sect. 7.2, which is partly based on [5], we check the new method in likelihood-based forecasts and on mock catalogues. In Sect. 7.3, we measure the amplitude of the phase shift in Fourier space and find statistically significant evidence for the presence of the phase shift in the BOSS DR12 dataset, in line with expectations from the mocks and forecasts. In Sect. 7.4, we describe and perform our alternative analysis in configuration space with results that are broadly compatible with the main Fourier-space constraints. We conclude, in Sect. 7.5, with a brief summary of our results and an outlook on future improvements of our measurement.

© Springer Nature Switzerland AG 2019                    161
B. Wallisch, *Cosmological Probes of Light Relics*, Springer Theses,
https://doi.org/10.1007/978-3-030-31098-1_7

## 7.1   Modified BAO Analysis

A variety of experiments have been proposed to observe the C$\nu$B directly [6–8]. However, these experiments are very challenging because neutrino interactions at low energies are extremely weak. Cosmological observations of the radiation density in the early universe, on the other hand, are making an increasingly strong case that the C$\nu$B has already been detected indirectly (cf. e.g. Sect. 3.3.1). As we showed in Chaps. 5 and 6, the effect of neutrinos on the perturbations in the primordial plasma, which have been detected in the CMB, is a particularly robust probe of the C$\nu$B. An interesting feature of the phase shift in the BAO spectrum is the fact that it is robust to the effects of non-linear gravitational evolution [9]. This provides the rare opportunity of extracting a signature of primordial physics which is immune to many of the uncertainties that inflict the modelling of non-linear effects in large-scale structure observables. The BAO phase shift therefore presents a clean imprint of early universe physics and the C$\nu$B. In this chapter, we will provide its first measurement.

The analysis of the (isotropic) BAO signal is usually reduced to the measurement of a single parameter, the BAO scale. In this chapter, we consider the extension of the conventional BAO analysis that we proposed in Chap. 6 and which takes the information contained in the phase of the spectrum into account. Since we only observe a finite number of modes, some of which evolved primarily during matter domination, we cannot simply search for a constant phase shift in the data. This means that recovering all of the accessible information requires an accurate momentum-dependent template for the phase shift that applies to the modes of interest.

The phase shift (relative to $N_{\text{eff}} = 0$) can be written as $\phi(k) \equiv \beta(N_{\text{eff}})f(k)$, where $\beta$ is the amplitude of the phase shift and $f(k)$ is a function that encodes its momentum dependence (see Sect. 6.3 and Appendix D for further details). The amplitude is proportional to the fractional neutrino density, $\epsilon_\nu(N_{\text{eff}}) \approx N_{\text{eff}}/(4.4 + N_{\text{eff}})$, and we have chosen the normalization so that $\beta = 0$ and 1 correspond to $N_{\text{eff}} = 0$ and 3.046, respectively. We note that the parameter $\beta$ is a non-linear function of $N_{\text{eff}}$ that asymptotes to $\beta \rightarrow 2.45$ for $N_{\text{eff}} \rightarrow \infty$ because adding more neutrinos does not change the phase shift when neutrinos become the dominant source of energy density in the universe. The template $f(k)$ is shown in Fig. 7.1 and is well approximated by the fitting function (6.24).

The observed BAO spectrum receives various non-linear corrections. We model these contributions as in the standard BAO analysis, e.g. [2], but now introduce the amplitude of the phase shift $\beta$ as an additional free parameter. Following Eq. (6.25), we write the non-linear BAO spectrum as

$$O_g(k) \equiv O_{\text{lin}}^{\text{fid}}\big(k/\alpha + (\beta - 1)f(k)/r_s^{\text{fid}}\big)\,\mathrm{e}^{-k^2\Sigma_{\text{nl}}^2/2}\,,  \tag{7.1}$$

where $O_{\text{lin}}^{\text{fid}}(k)$ and $r_s^{\text{fid}}$ are the linear BAO spectrum and the BAO scale in the fiducial

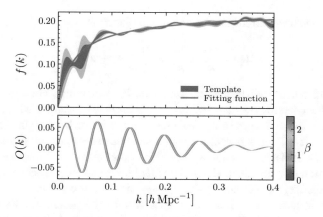

**Fig. 7.1** Template of the phase shift $f(k)$ as defined in (6.23), with the fitting function (6.24) shown as the red curve (after [4]). The bottom panel displays the linear BAO spectrum $O(k)$ as a function of the amplitude of the phase shift $\beta$

cosmology, which is chosen to be the same as in [2].[1] The exponential factor in (7.1) describes the non-linear damping of the BAO signal after reconstruction [10, 11]. The parameter $\alpha$ captures the change in the apparent location of the BAO peak due to changes in the acoustic scale and the angular projection,

$$\alpha(N_{\text{eff}}) = \frac{D_V(z)\, r_s^{\text{fid}}}{D_V^{\text{fid}}(z)\, r_s}, \quad \text{with} \quad D_V(z) = \left[ (1+z)^2 D_A^2(z) \frac{cz}{H(z)} \right]^{1/3}. \tag{7.2}$$

In Sect. 7.2, we will show that this model is effectively unbiased in the sense that we recover $\beta \approx 0$ for a universe with $N_{\text{eff}} = 0$ even when we assume a fiducial model with $N_{\text{eff}} = 3.046$. Moreover, given the template (6.24), the modelling is robust to the precise method for extracting $O_{\text{lin}}^{\text{fid}}(k)$ and we will therefore use the same method as [2].

We model the non-linear broadband spectrum in each redshift bin as

$$P^{\text{nw}}(k) = B^2 P_{\text{lin}}^{\text{nw}}(k) F(k, \Sigma_s) + A(k). \tag{7.3}$$

This includes two physical parameters: a linear bias parameter, $B$, and a velocity damping term arising from the non-linear velocity field ("Fingers of God"),

$$F(k, \Sigma_s) = \frac{1}{(1 + k^2 \Sigma_s^2/2)^2}. \tag{7.4}$$

---

[1]The fiducial cosmology is described by the following parameters: $\Omega_b h^2 = 0.022$, $\Omega_c h^2 = 0.119$, $h = 0.676$, $A_s = 2.18 \times 10^{-9}$, $n_s = 0.96$, $\tau = 0.08$, $N_{\text{eff}} = 3.046$ and $\sum m_\nu = 0.06\,\text{eV}$.

In addition, we have introduced the polynomial function

$$A(k) = \frac{a_1}{k^3} + \frac{a_2}{k^2} + \frac{a_3}{k} + a_4 + a_5 k^2 \,, \tag{7.5}$$

whose coefficients $a_n$ will be marginalized over. This polynomial does not represent a physical effect, but removes any residual information that is not encoded in the locations of the peaks and zeros of the BAO spectrum. With such a marginalization over broadband effects, our $\alpha$-$\beta$ parametrization contains essentially all of the information of the $\Lambda$CDM+$N_{\rm eff}$ cosmology available in the BAO spectrum (cf. Sect. 7.2.1). The free parameters in this model will be fitted independently in each redshift bin.

To summarize, the measured galaxy power spectrum is described by two cosmological parameters, $\alpha$ and $\beta$, and a number of nuisance parameters. Except for $\beta$, all parameters are redshift dependent, and we will fit our model to the signal in two independent redshift bins, $(0.2 < z_1 < 0.5)$ and $(0.5 < z_3 < 0.75)$.[2] In total, our fit to the power spectrum in the range $0.01\,h\,Mpc^{-1} < k < 0.3\,h\,Mpc^{-1}$ therefore has 21 free parameters:

$$\beta, \ \alpha_{z_1}, \ \alpha_{z_3}; \ \{B_{{\rm NGC},z}, \ B_{{\rm SGC},z}, \ \Sigma_{s,z}, \ \Sigma_{{\rm nl},z}, \ a_{n,z}\}_{z_1,z_3} \,, \tag{7.6}$$

where we have allowed for independent bias parameters in the North Galactic Cap (NGC) and the South Galactic Cap (SGC) as in [2]. We generally employ flat priors for all parameters, in particular $\beta$.[3] We require the $\alpha_z$ parameters to be between 0.8 and 1.2, and the damping scales, $\Sigma_{s,z}$ and $\Sigma_{{\rm nl},z}$, to be between 0 and $20\,h^{-1}$ Mpc, while no explicit priors are imposed on the bias parameters $B_{i,z}$, the phase parameter $\beta$ or the polynomial terms $a_{n,z}$. Our goal is to measure the new parameter $\beta$, while marginalizing over all other parameters.

## 7.2  Validation of the Method

Before applying our analysis pipeline to the BOSS data, we perform several checks and show that the new method provides reliable and consistent results. First, we explicitly demonstrate that the advocated template-based approach captures most of the information on neutrinos in the BAO spectrum (Sect. 7.2.1). Second, we establish in forecasts that the phase shift inferred in analyses employing the BAO spectrum (7.1) does not depend on the employed BAO extraction and is unbiased, in the sense that it correctly recovers the input value of the phase amplitude even if a different fiducial

---

[2]The middle redshift bin $(0.4 < z_2 < 0.6)$, which was used in the BOSS DR12 analysis, carries little additional information on the BAO signal since it overlaps with the other two bins.

[3]We note that the choice of a flat prior on $\beta$, rather than $N_{\rm eff}$, weakens the statistical significance of the $\beta > 0$ measurement compared to the analyses in the CMB, which use $N_{\rm eff}$. In other words, a flat prior on $N_{\rm eff}$ would lead to stronger constraints on the phase shift and, therefore, the C$\nu$B.

cosmology is assumed (Sect. 7.2.2). Finally, we validate our modified BAO analysis using mock catalogues created for the BOSS DR12 analysis (Sect. 7.2.3).

## 7.2.1 Comparison to Parameter-Based Approach

We have suggested the use of a phase template to characterise the effect of neutrinos. This is a natural choice as the phase shift is the physical effect we wish to isolate. One might however worry that the template (6.23) does not capture the entire relevant information in the BAO spectrum. For this purpose, it is instructive to compare the results of our template-based forecasts of Sect. 6.3 to a more direct parameter-based approach to isolating the phase shift. In the parameter-based approach, we define two new parameters $\tilde{\theta}_s$ and $\tilde{N}_{\mathrm{eff}}$ that play the role of $\theta_s$ and $N_{\mathrm{eff}}$ in the BAO signal, but are taken to be independent of the same parameters in the CMB. We will then fix all remaining cosmological parameters in the BAO spectrum using the CMB, except the physical cold dark matter density $\omega_c$ which we traded for the scale factor at matter-radiation equality $a_{\mathrm{eq}}$. As with our template extraction, holding $a_{\mathrm{eq}}$ fixed ensures that the phase shift approaches a constant at large wavenumbers, whose value is determined by $\tilde{N}_{\mathrm{eff}}$. Beside measuring the phase of the BAO signal, the parameter $\tilde{N}_{\mathrm{eff}}$ also contributes to the scale parameter $\alpha$ and could therefore be constrained by the standard BAO-scale measurement if all the other cosmological parameters are fixed to their Planck best-fit values. Introducing the additional parameter $\tilde{\theta}_s$ gives enough freedom to remove this effect and any constraint on $\tilde{N}_{\mathrm{eff}}$ must be coming from the phase shift alone. This is analogous to isolating the phase shift in the CMB by marginalizing over $Y_p$ or any other parameters that are degenerate with the $N_{\mathrm{eff}}$ -induced change to the damping tail. We will confirm this expectation in our forecasts.

Typically, the advantage of the parameter-based approach is that it is easy to implement. However, in this case, we found it more difficult to set up reliably. The phase shift ultimately controls the breaking of the degeneracy between $\tilde{\theta}_s$ and $\tilde{N}_{\mathrm{eff}}$ and, as we discussed in Sect. 6.3.1, $P^{\mathrm{nw}}(k)$ must therefore be determined sufficiently accurately to not produce errors in this shift. To compute the likelihood directly, we must re-compute $P^{\mathrm{nw}}(k)$ for every value of the cosmological parameters. Producing stable results for the BAO spectrum across a wide range of parameters can be very computationally expensive and technically challenging. Simpler and faster methods can work well near the fiducial cosmology (such as the use of a fitting function), but often produce noisy results as the parameters vary significantly and typically underestimate the likelihood as we depart from the fiducial cosmology (and, hence, overestimate the constraining power).

Despite the challenge presented by a parameter-based approach, it has the advantage that it should capture all of the cosmological information available. For this reason, it is useful to compare the results of the parameter-based and template-based approaches to see if the template is missing information. Fortunately, we will see that the posterior distributions for $\tilde{N}_{\mathrm{eff}}$ and $\tilde{\theta}_s$ can be largely reproduced as a derived consequence of the template-based forecasts. From our results of Sect. 6.3.2, we should

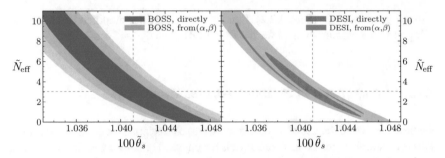

**Fig. 7.2** BOSS (*left*) and DESI (*right*) two-dimensional $1\sigma$ and $2\sigma$ contours for $\tilde{N}_{\text{eff}}$ and $\tilde{\theta}_s$, determined ('directly') from the likelihood for the BAO spectrum for each value of the parameters and derived ('from $(\alpha, \beta)$') from the redshift-binned likelihood for $\alpha$ and $\beta$ (from [5]). We find good agreement between both methods, suggesting that the two-dimensional parametrization is capturing most of the relevant information. The dashed lines indicate the fiducial values

anticipate that the posteriors for $\tilde{N}_{\text{eff}}$ and $\tilde{\theta}_s$ will be non-Gaussian, and will therefore require the calculation of the likelihood for $\tilde{N}_{\text{eff}}$ and $\tilde{\theta}_s$ directly (and not only the Fisher matrix). We will follow the same approach as described in Sect. 6.3.2. Computing the full likelihood is quite involved, which is the reason why we will assume that the CMB data fixes the other cosmological parameters to their fiducial values, except for $\tilde{N}_{\text{eff}}$ and $\tilde{\theta}_s$.

Results of the likelihood analysis in terms of these parameters for both BOSS and DESI are shown in Fig. 7.2. We see that the results are similar, which establishes that our templates are capturing most of the information available in the BAO spectrum, in particular for BOSS. This is an important observation because it allows us to simplify the analysis to a two-parameter template without much loss of information. In fact, the conclusion that these likelihoods are the same is not easily reproduced with any method of BAO extraction, but requires the robustness and stability of a method such as the one we use (see Appendix D). Given instead our phase shift template, one can reliably compute Fisher matrices or likelihoods for $\alpha$ and $\beta$, and derive the implications for cosmological parameters from them. Future surveys, such as DESI, show somewhat larger differences between the two methods, which suggests that more information could potentially be extracted by using additional and/or alternative templates.

The doubling of cosmological parameters to treat the CMB and LSS independently, like in the case of $\tilde{N}_{\text{eff}}$ and $\tilde{\theta}_s$, has useful conceptual advantages even if we derive constraints on these parameters from the posterior of $\alpha$ and $\beta$. Growing tensions between the CMB and certain low-$z$ measurements have garnered much attention, but lack a compelling explanation. Measuring $\tilde{\theta}_s$ and $\tilde{N}_{\text{eff}}$ in the BAO spectrum may provide a new perspective on this issue without the need for a CMB anchor.

## 7.2.2 Validation of the Modified Analysis

We have seen in the left panel of Fig. 7.2 that our approach captures essentially all of the information in the BAO spectrum at the sensitivity levels of the BOSS experiment. However, one may still worry that the mapping

$$O_{\text{lin}}(k) \to O_{\text{fid}}^{\text{lin}}\big(k/\alpha + (\beta - 1)f(k)/r_s^{\text{fid}}\big) \tag{7.7}$$

introduces additional unphysical changes to the BAO spectrum. Since we use $N_{\text{eff}} = 3.046$, corresponding to $\beta = 1$, as the fiducial model, a poor modelling for $\beta \neq 1$ could lead to artificially strong evidence for a phase shift and could bias the measurement of $\beta$ if $N_{\text{eff}} \neq 3.046$.

Our interest lies mostly in the exclusion of $\beta = 0$. A straightforward check that our method is reliable is to compute the posterior distribution for $\beta$ in a cosmology with $N_{\text{eff}} = 0$ to see that the result is effectively unbiased. We use the same likelihood-based forecasts as in Sect. 6.3 and the resulting posterior for $\beta$ is shown in Fig. 7.3. The expected values for $\alpha$ and $\beta$ are retrieved reliably in both cases. We also find good agreement when imposing the CMB prior from Planck with the respective input values of $N_{\text{eff}}$. This test demonstrates that even though the fiducial model with $N_{\text{eff}} = 3.046$ is used for constructing the template, the model with $N_{\text{eff}} = 0$ is correctly recovered, especially for BOSS.[4] At higher levels of sensitivity, e.g. for DESI, the expected values for $\beta$ are recovered even more accurately for both $N_{\text{eff}} = 0$ and 3.046. However, due to the smaller error bars and the slight difference between the parameter-based and template-based approaches around $N_{\text{eff}} = 0$ for DESI, the mean $\bar{\beta}$ is found about $0.8\,\sigma(\beta)$ too high, whereas it is excellent for the fiducial $N_{\text{eff}} = 3.046$.

One may also be concerned that these results could depend sensitively on the method of BAO extraction. Indeed, as we have discussed, the phase shift template $f(k)$ is quite sensitive to the BAO extraction and demands a method that is accurate across a wide range in $N_{\text{eff}}$. In contrast, the model in (7.7) only requires an accurate BAO extraction for the fiducial cosmology. We have verified that the results in Fig. 7.3 do not depend on the BAO extraction method being used.

## 7.2.3 Tests on Mock Catalogues

Finally, we test our entire analysis pipeline including the phase shift template using 999 MultiDark-Patchy mock catalogues [12], which have been created for the

---

[4]In detail, the solid red curve in Fig. 7.3 shows a mean of $\bar{\beta} = 0.27$ rather than zero for a $N_{\text{eff}} = 0$ cosmology. This level of bias is acceptably small given the much larger statistical error of $\sigma(\beta) = 0.97$. Of course, this bias should be accounted for when determining the precise statistical significance of the exclusion of $\beta = 0$, but it does not affect our main conclusion in this chapter that $\beta > 0$ at 95% confidence in the BOSS DR12 dataset.

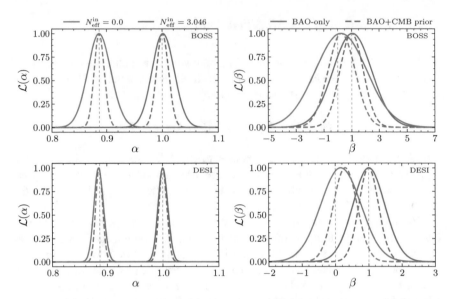

**Fig. 7.3** Posterior distributions for the BAO scale parameter $\alpha$ and the amplitude of the phase shift $\beta$ when the mock data for BOSS (*top*) and DESI (*bottom*) were generated using $N_{\mathrm{eff}}^{\mathrm{in}} = 3.046$ (blue) and 0 (red), corresponding to $\beta = 1$ and 0. In both cases, the model in (7.7) used a fiducial cosmology with $N_{\mathrm{eff}} = 3.046$ and the posteriors for $\alpha$ are shown for one redshift bin similar to $z_1$. The dotted lines show the posterior distributions after imposing a prior from a Planck-type CMB experiment. We see that the posteriors reproduce the expected behaviour, which indicates that the estimation of $\alpha_i$ and $\beta$ is essentially unbiased

BOSS DR12 analysis. The Patchy mock catalogues have been calibrated to an N-body simulation-based reference sample using analytical-statistical biasing models. The reference catalogue is extracted from one of the BigMultiDark simulations [13]. The mock catalogues have a known issue with overdamping of the BAO, making the signal for the traditional BAO approximately 30% weaker [2]. We therefore forecast the mocks and the real data separately, taking these differences into account. For the mock forecasts, we use $\Sigma_{\mathrm{nl}} = 7\,h^{-1}$ Mpc as the fiducial value of the non-linear damping scale.

An appealing feature of using the mock catalogues is that we can check that the performance expected from our forecasts is reproduced by the distribution of maximum-likelihood points across the catalogue. Figure 7.4 confirms that the distributions for the parameters $\alpha$ and $\beta$ are indeed in good agreement with the fiducial value of $\beta = 1$. A Gaussian fit to the distribution of maximum-likelihood values yields $\beta = 1.0 \pm 2.4$ ($\alpha_{z_1} = 1.000 \pm 0.035$, $\alpha_{z_3} = 1.000 \pm 0.035$), which is consistent with the value found from a likelihood-based forecast as in Sect. 6.3, $\sigma(\beta) = 2.1$.

As seen in the left panel of Fig. 7.4, there is a strong degeneracy between the effects of the parameters $\alpha$ and $\beta$. The origin of this degeneracy is easy to understand. If the only well-determined quantity in the data were the position of the first peak in the BAO spectrum, there would be a perfect degeneracy between phase and frequency

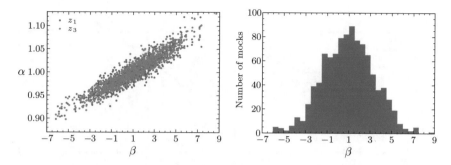

**Fig. 7.4** Distribution of maximum-likelihood values in the $\alpha$-$\beta$ plane for the two BOSS redshift bins $z_1$ and $z_3$ (*left*), and for $\beta$ (*right*) in 999 MultiDark-Patchy mock catalogues [12], yielding $\beta = 1.0 \pm 2.4$ (after [4])

determination. In reality, several peaks and troughs are present in the data which breaks the perfect degeneracy and allows the parameters $\alpha$ and $\beta$ to be constrained independently. However, one still expects them to remain significantly correlated, partly because the peaks are measured with decreasing accuracy due to damping. Since this degeneracy is a limiting factor in the measurement of $\beta$, we anticipate a significant improvement in the constraint on $\beta$ when the degeneracy with $\alpha$ is broken with additional data. Below we will see that this is indeed the case.

## 7.3   Analysis of BOSS Data

Having demonstrated that our new method passes all the reported checks, we can apply it to the BAO signal of the final data release (DR12) of the Baryon Oscillation Spectroscopic Survey; see [14]. As described in detail in [15], this dataset contains 1 198 006 galaxies with spectroscopic redshifts in the range $0.2 < z < 0.75$ and covers 10 252 deg$^2$ of the sky.

### 7.3.1   BAO-Only Analysis

First, we analyse the two BOSS redshift bins without any additional data or external prior. To explore the BAO likelihood function, we use the Python-based, affine-invariant ensemble sampler emcee [16] for Markov chain Monte Carlo. The convergence is determined with the Gelman–Rubin criterion [17] by comparing eight separate chains and requiring all scale-reduction parameters to be smaller than $R - 1 = 0.01$. Figure 7.5 shows the posterior distribution for the parameters $\beta$ and $\alpha_{z_1}$, $\alpha_{z_3}$. The measured $\alpha$-values are in good agreement with those found in [2], but the errors have increased due to the degeneracy with $\beta$. We find

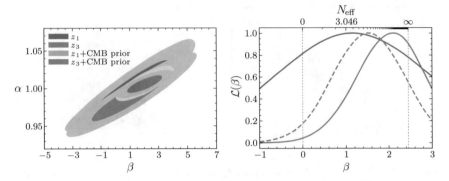

**Fig. 7.5** Constraints on the neutrino-induced phase shift inferred from the BOSS DR12 dataset (after [4]). *Left*: Contours showing $1\sigma$ and $2\sigma$ exclusions in the $\alpha$-$\beta$ plane for the two redshift bins $z_1$ and $z_3$, both from the BAO data alone and after imposing a CMB prior on $\alpha$. *Right*: One-dimensional posterior distributions of $\beta$ without (blue) and with (red) the $\alpha$-prior for the combined redshift bins. The dashed line is the result after marginalizing over the lensing amplitude $A_L$

$\alpha_{z_1} = 1.001 \pm 0.025$, $\alpha_{z_3} = 0.991 \pm 0.022$ and $\beta = 1.2 \pm 1.8$. These results are in good agreement with likelihood-based forecasts for the data,[5] $\sigma(\alpha_{z_1}) = 0.021$, $\sigma(\alpha_{z_3}) = 0.019$ and $\sigma(\beta) = 1.5$. A similar level of agreement between forecasts and actual performance was obtained for the measurement of $\alpha$ in the conventional BAO analysis of BOSS DR12 [2].

## 7.3.2    Adding a CMB Prior

The BAO-only measurement of $\beta$ is limited by the degeneracy with $\alpha(z)$. However, in a given cosmology, the allowed range of $\alpha(z)$ has to be consistent with constraints on the cosmological parameters. Our interest is to measure the neutrino-induced phase shift in the BAO signal assuming a background cosmology that is consistent with the Planck CMB constraints. We infer the prior on $\alpha(z)$ from the Planck 2015 temperature and polarization data[6] [18], while marginalizing over any additional cosmological information (including all effects of $N_{\text{eff}}$). If available, we directly employ the Markov chains supplied by the Planck collaboration, which were computed using CAMB [19] and CosmoMC [20] with the publicly released priors and settings. In particular for the $\Lambda$CDM+$N_{\text{eff}}$+$A_L$ prior cosmology, we sample the data using the same codes and priors. At each point in the Monte Carlo Markov chains obtained from the Planck likelihood for a certain background cosmology, we compute the values of $\alpha_{z_1}$ and $\alpha_{z_3}$ associated with the given set of cosmological parameters.

---

[5]These forecasted values are slightly larger than those in Sect. 6.3.2 because we accounted for the roughly 30% smaller galaxy bias measured in [2].

[6]We use the low-$\ell$ ($2 \le \ell \le 29$) temperature and LFI polarization data, and the high-$\ell$ ($30 \le \ell \le 2508$) plik cross half-mission temperature and polarization spectra. In "TT-only", we omit the high-$\ell$ polarization spectra.

**Table 7.1** Constraints on the amplitude of the phase shift $\beta$ with and without a Planck prior on the BAO scale, assuming various underlying cosmologies. Our baseline result uses the $\Lambda$CDM+$N_{eff}$ prior, marginalizing over all of the effects of $N_{eff}$ in the CMB. We see that this result is robust to including or excluding $N_{eff}$ and $A_L$ in the prior cosmology. Finally, we show the large central value of $\beta$ also appears when using TT-only spectra and is therefore not solely a consequence of the polarization data

| Prior cosmology | $\beta$ |
| --- | --- |
| None (BAO-only) | $1.2 \pm 1.8$ |
| $\Lambda$CDM+$N_{eff}$ | $2.05 \pm 0.81$ |
| $\Lambda$CDM | $1.97 \pm 0.73$ |
| $\Lambda$CDM+$N_{eff}$ (TT-only) | $1.6 \pm 1.1$ |
| $\Lambda$CDM (TT-only) | $1.87 \pm 0.89$ |
| $\Lambda$CDM+$N_{eff}$+$A_L$ | $1.53 \pm 0.83$ |
| $\Lambda$CDM+$A_L$ | $1.49 \pm 0.76$ |

In this way, we obtain the two-dimensional (Gaussian) posterior for $\alpha_{z_1}$-$\alpha_{z_3}$. We confirmed on the mock catalogues that a Gaussian prior with the expected mean values and the Planck $\Lambda$CDM+$N_{eff}$ covariance matrix results in an unbiased measurement of $\beta = 1.00 \pm 0.85$ (see also Sect. 7.2.2 for the equivalent forecasts with unbiased posterior distributions). On the data, we impose the Planck posterior on $\alpha$ by importance-sampling our BAO-only Monte Carlo Markov chains.

The right panel of Fig. 7.5 shows the marginalized posterior distributions for the parameter $\beta$. We see that including the $\alpha$-posterior from the Planck $\Lambda$CDM+$N_{eff}$ chains as a prior sharpens the distribution significantly. Having obtained a constraint of $\beta = 2.05 \pm 0.81$ on the phase shift amplitude, we want to evaluate the statistical significance of an exclusion of $\beta = 0$, corresponding to no phase shift and no free-streaming neutrinos. For this purpose, we extract the fraction of Monte Carlo samples which have $\beta > \beta_0$. To be cautious about the small bias found in Sect. 7.2.2, we employ $\beta_0 = 0.27$ instead of $\beta_0 = 0$.[7] In this and other aspects of the analysis, we have therefore made intentionally conservative choices in stating our statistical significance. The measurement of $\beta = 2.05 \pm 0.81$ consequently corresponds to an exclusion of $\beta = 0$ at greater than 99% confidence. The statistical error of the measurement is in good agreement with the forecasted value of $\sigma(\beta) = 0.77$. On the other hand, the central value is more than a $1\sigma$ fluctuation away from the expected Standard Model value $\beta = 1$. Any upward fluctuation adds to the confidence of our exclusion, provided that it is simply a statistical fluctuation. We tested the stability of this upward fluctuation to changes in the cosmological model and the CMB likelihood (see Table 7.1).

The statistical significance of the result is largely insensitive to the choice of cosmology and likelihood. The largest deviation from $\Lambda$CDM within the Planck data alone is the preference for a larger lensing amplitude $A_L$ [21]. To estimate the im-

---

[7]We explicitly checked that the computation based on likelihood ratios leads to essentially the same confidence levels, which is expected since the posterior distributions are very close to Gaussian.

pact of this upward fluctuation on our analysis, we marginalized over $A_L$ in the implementation of the $\alpha$-prior. The dashed posterior curve in Fig. 7.5 shows the result obtained from the $\Lambda$CDM+$N_{\text{eff}}$+$A_L$ prior cosmology, which corresponds to $\beta = 1.53 \pm 0.83$. We see that marginalizing over $A_L$ indeed brings the central value of $\beta$ into closer agreement with $\beta = 1$, suggesting that part of our large central value is due to a known upward fluctuation of the Planck data. Having said that, even with this marginalization, we find evidence for a positive phase shift, $\beta > 0$, at greater than 94% confidence.[8] In summary, while the precise significance of the phase shift measurement depends on the implementation of the CMB prior, the exclusion of $\beta = 0$ at greater than 95% confidence is stable to all choices of the prior that we have considered.

## 7.4  Analysis in Configuration Space

The neutrino-induced phase shift is characteristically a Fourier-space (FS) quantity. By contrast, the BAO frequency is more commonly described in configuration space (CS) as the scale of the BAO feature in the two-point correlation function. Consequently, the measurement of the BAO scale is often depicted as the determination of the BAO peak location in CS [22, 23]. The phase shift modifies the shape of the BAO peak and manifests itself in CS as a transfer of correlations around the peak position from small to large scales (see Fig. 7.6). Given that the BAO scale measurement is known to give compatible results in CS and FS (see e.g. [3]), we anticipate the same to be true of the phase shift. We will therefore implement a modified version of the CS method used in [23] as a cross-check of our main FS analysis.

In the following, we first describe how we have incorporated the phase shift into the configuration-space analysis of the BAO signal (Sect. 7.4.1). Similar to the validation in FS, we also check the modified method on mock catalogues. We then apply the analysis pipeline to the BOSS DR12 data and impose the same Planck priors on $\alpha$ as in FS, and find results that are consistent with those of the Fourier-space analysis (Sect. 7.4.2).

### 7.4.1  Modified Pipeline and Cross-Checks

Our non-linear model for the correlation function starts from the processed matter power spectrum

$$P(k) = F(k, \Sigma_s) P_{\text{lin}}^{\text{nw}}(k) \left[ 1 + O_g(k) \right], \tag{7.8}$$

---

[8]Note that we marginalized over $A_L$ because it experiences a large fluctuation in the Planck data. The statistical significance of the corresponding result should therefore not be compared to the results of our blind analysis.

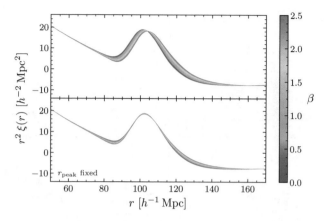

**Fig. 7.6** Rescaled linear correlation function $r^2\xi(r)$ as a function of the amplitude of the phase shift $\beta$ (after [4]). The upper panel keeps $\alpha = 1$ fixed, while $\alpha$ is varied in the lower panel to fix the position of the peak, $r_{\rm peak}$. This illustrates the degeneracy between $\alpha$ and $\beta$ in configuration space

where $O_g(k)$ is the template-based non-linear BAO spectrum defined in (7.1) and $F(k, \Sigma_s)$ is given by (7.4). The two-point galaxy correlation function is then modelled as

$$\xi_g(r) = B^2 \int d\log k \, \frac{k^3}{2\pi^2} P(k) \, j_0(kr) + A(r) \,, \tag{7.9}$$

where $j_0(kr)$ is a spherical Bessel function. We introduced the constant bias parameter $B$ and the polynomial function $A(r)$, taken to have the same form as in [23],

$$A(r) = \frac{a_1}{r^2} + \frac{a_2}{r} + a_3 \,, \tag{7.10}$$

where the coefficients $a_n$ are marginalized over. While the constant bias matches the same parameter in the FS analysis, the polynomial $A(r)$ is *not* equivalent to the polynomial $A(k)$ in (7.5). This is one of the notable differences between the FS and CS analyses. Except for the amplitude of the phase shift $\beta$, all parameters are redshift dependent. Since the scale $\Sigma_s$ is held fixed to the best-fit value obtained on the mock catalogues, we fit the following 13 parameters to the correlation function in the range $r \in [55 - 160] \, h^{-1}\,\mathrm{Mpc}$:[9]

$$\beta, \alpha_{z_1}, \alpha_{z_3}; \{B_z, \Sigma_{\mathrm{nl},z}, a_{n,z}\}_{z_1,z_3} \,, \tag{7.11}$$

for the same two redshift bins as in Fourier space.

---

[9]We employ flat priors on the cosmological parameters, requiring $\beta$ to be between $-10$ and $10$, and $\alpha_z$ to be between 0.5 and 1.5, but do not impose explicit priors for the other ten parameters. On the data, we speed up the analysis by analytically marginalizing over the broadband parameters $a_{n,z}$ in each step.

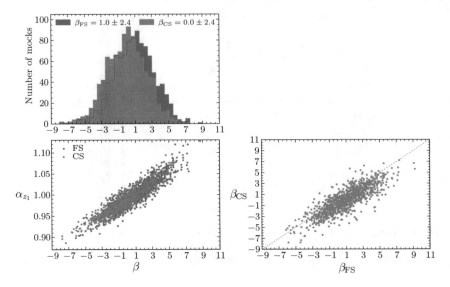

**Fig. 7.7** Comparison of the distribution of maximum-likelihood values in 999 mock catalogues [12] for the Fourier-space (FS, blue) and configuration-space (CS, red) analyses (after [4]). We also show the correlation between the inferred phase shift amplitudes in the two analyses (green)

We apply the same pipeline as in [23] to the MultiDark-Patchy mock catalogues [12] and determine the distributions of maximum-likelihood values for the parameters $\alpha$ and $\beta$. The results are shown in Fig. 7.7 and correspond to $\beta_{CS} = 0.0 \pm 2.4$ ($\alpha_{z_1} = 0.989 \pm 0.033$, $\alpha_{z_3} = 0.990 \pm 0.034$). Comparing these distributions with the FS mock catalogue analysis of Sect. 7.2.3, we observe a strong correlation with correlation coefficient $r = 0.84$, but a statistically significant bias of about 1/3 of a standard deviation for both $\alpha_i$ and $\beta$, albeit with approximately the same standard deviations. When including the CMB prior, the mean shifts upwards and gives $\beta_{CS} = 0.75 \pm 0.89$, corresponding to a bias of about 1/4 of a standard deviation which is also slightly larger than in FS. These values demonstrate good statistical agreement between the CS and FS analyses, and demonstrate that CS provides a useful cross-check of the FS analysis. While CS does show larger biases, they are sufficiently small that they should not meaningfully affect the statistical significance of our results. On the other hand, we noticed that the precise choice of the broadband polynomial $A(r)$ altered both the mean and standard deviation, while being consistent with the fiducial cosmology. These features of the CS analysis will be explored in future work. The shifts seen in CS further highlight the remarkable robustness of the phase shift in FS.

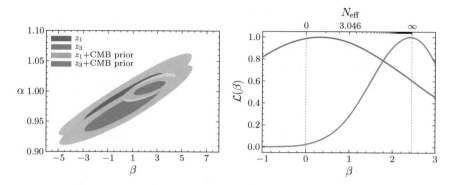

**Fig. 7.8** Constraints on the neutrino-induced phase shift inferred from the BOSS DR12 data in configuration space (after [4]). *Left*: Contours showing $1\sigma$ and $2\sigma$ exclusions in the $\alpha$-$\beta$ plane for the two redshift bins $z_1$ and $z_3$ in configuration space, both from the BAO data alone and after imposing a CMB prior on $\alpha$. *Right*: One-dimensional posterior distributions of $\beta$ without (blue) and with (red) the $\alpha$-prior for the combined redshift bins

### 7.4.2 Analysis of BOSS Data

With these caveats in mind, we apply the CS pipeline to the BOSS DR12 dataset. The posterior distributions for the parameters $\alpha_{z_1}$, $\alpha_{z_3}$ and $\beta$ are presented in Fig. 7.8, and correspond to measurements of $\alpha_{z_1} = 0.991 \pm 0.027$, $\alpha_{z_3} = 0.973 \pm 0.026$ and $\beta_{CS} = 0.4 \pm 2.1$. These mean values of $\alpha_i$ are about $1/4$ of a standard deviation lower than the ones found in the standard BAO analysis [23]. In addition, the error bars increased, mainly related to the degeneracy between $\alpha$ and $\beta$ discussed in Sect. 7.2.3. The value of $\bar{\beta}$ is $0.3\sigma$ lower than in FS with a 16% larger error. When adding a Planck prior to break the degeneracy, we measure $\beta_{CS} = 2.36 \pm 0.86$ which is larger than in FS because of the mentioned bias in $\alpha_i$ towards lower values. Nevertheless, these CS measurements are statistically consistent with the main FS results, with similar shifts in the mean values as observed in the mock analysis. Given that the broadband modelling and peak isolation in configuration and Fourier space are distinct, an agreement between the two analyses was not guaranteed, although the change to the BAO peak is simply the inverse Fourier transform of the phase shift. Having said that, this analysis confirms that a measurement can also be made in configuration space and, despite the discussed differences, is comparable to the main analysis in Fourier space.

## 7.5 Summary

In this chapter, we have reported on the first measurement of the neutrino-induced phase shift in the BAO spectrum. This is the first evidence for the cosmic neutrino background in the clustering of galaxies and the first application of the BAO signal beyond its use as a standard ruler.

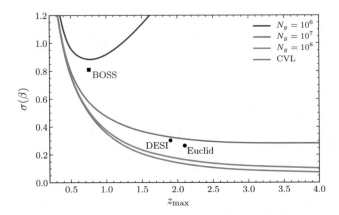

**Fig. 7.9** Constraints on the amplitude of the phase shift $\beta$ including a CMB prior on the BAO scale parameter $\alpha$ from Planck (after [4]). The lines are forecasted constraints as a function of the maximum redshift $z_{max}$ and the number of objects $N_g$ of a cosmological survey observing a sky fraction of $f_{sky} = 0.5$ (see Sect. 6.3.3 for details). Shown is also the cosmic variance limit. The square indicates the result obtained in this work. The dots mark projected constraints for DESI and Euclid assuming $z_{max}$ to be given by the largest redshift bin used to define the survey in [24]

To extract the phase information, we modified the conventional BAO data analysis by allowing the amplitude of the phase shift to be an additional free parameter. We determined this new parameter to be non-zero at greater than 95% confidence, even allowing for very conservative marginalization over corrections to the broadband spectrum. Our measurement is a nontrivial confirmation of the standard cosmological model at low redshifts and a proof of principle that there is additional untapped information in the phase of the BAO spectrum. Since this phase information is protected from the effects of non-linear gravitational evolution [9], it is a particularly robust probe of early universe physics.

A number of galaxy surveys are planned over the next decade which have the potential to significantly improve on our measurement of the neutrino background (see Fig. 7.9). The Dark Energy Spectroscopic Instrument, for example, should be sensitive to the C$\nu$B at more than $3\sigma$, making the BAO phase shift measurement more comparable to current limits from the CMB [1]. Combining Euclid with a prior from a next-generation CMB experiment would allow a $5\sigma$ detection of the C$\nu$B. Moreover, having shown that there is valuable information in the phase of the BAO spectrum, we should ask what else can be learned from it beyond the specific application to light relics. As the observed BAO feature is the result of the combined dynamics of the dark matter and baryons, it is broadly sensitive to new physics in these sectors. The BAO phase shift is one particularly clean probe of this physics and we hope that our work will inspire new ideas for exploring the early universe at low redshifts.

# References

1. B. Follin, L. Knox, M. Millea, Z. Pan, First detection of the acoustic oscillation phase shift expected from the cosmic neutrino background. Phys. Rev. Lett. **115**, 091301 (2015). arXiv:1503.07863 [astro-ph.CO]
2. F. Beutler et al. (BOSS collaboration), The clustering of galaxies in the completed SDSS-III Baryon Oscillation Spectroscopic Survey: baryon acoustic oscillations in Fourier space. Mon. Not. Roy. Astron. Soc. **464**, 3409 (2017). arXiv:1607.03149 [astro-ph.CO]
3. S. Alam et al. (BOSS Collaboration), The clustering of galaxies in the completed SDSS-III Baryon Oscillation Spectroscopic Survey: cosmological analysis of the DR12 galaxy sample. Mon. Not. Roy. Astron. Soc. **470**, 2617 (2017). arXiv:1607.03155 [astro-ph.CO]
4. D. Baumann, F. Beutler, R. Flauger, D. Green, A. Slosar, M. Vargas-Magaña, B. Wallisch, C. Yèche, First constraint on the neutrino-induced phase shift in the spectrum of baryon acoustic oscillations. Nat. Phys. **15**, 465 (2019). arXiv:1803.10741 [astro-ph.CO]
5. D. Baumann, D. Green, B. Wallisch, Searching for light relics with large-scale structure. JCAP **08**, 029 (2018). arXiv:1712.08067 [astro-ph.CO]
6. S. Weinberg, Universal neutrino degeneracy. Phys. Rev. **128**, 1457 (1962)
7. A. Ringwald, Prospects for the direct detection of the cosmic neutrino background. Nucl. Phys. A **827**, 501C (2009). arXiv:0901.1529 [astro-ph.CO]
8. S. Betts et al., Development of a relic neutrino detection experiment at PTOLEMY: Princeton Tritium Observatory for Light, Early-Universe, Massive-Neutrino Yield. arXiv:1307.4738 [astro-ph.IM]
9. D. Baumann, D. Green, M. Zaldarriaga, Phases of new physics in the BAO spectrum. JCAP **11**, 007 (2017). arXiv:1703.00894 [astro-ph.CO]
10. D. Eisenstein, H.-J, Seo, E. Sirko, D. Spergel, Improving cosmological distance measurements by reconstruction of the baryon acoustic peak. Astrophys. J. **664**, 675 (2007). arXiv:astro-ph/0604362 [astro-ph]
11. N. Padmanabhan, X. Xu, D. Eisenstein, R. Scalzo, A. Cuesta, K. Mehta, E. Kazin, A two-percent distance to z = 0.35 by reconstructing baryon acoustic oscillations – I. Methods and application to the Sloan Digital Sky Survey. Mon. Not. Roy. Astron. Soc. **427**, 2132–2145 (2012). arXiv:1202.0090 [astro-ph.CO]
12. F.-S. Kitaura et al., The clustering of galaxies in the SDSS-III Baryon Oscillation Spectroscopic Survey: mock galaxy catalogues for the BOSS final data release. Mon. Not. Roy. Astron. Soc. **456**, 4156 (2016). arXiv:1509.06400 [astro-ph.CO]
13. A. Klypin, G. Yepes, S. Gottlober, F. Prada, S. Hess, MultiDark simulations: the story of dark matter halo concentrations and density profiles. Mon. Not. Roy. Astron. Soc. **457**, 4340 (2016). arXiv:1411.4001 [astro-ph.CO]
14. S. Alam et al. (SDSS-III Collaboration), The eleventh and twelfth data releases of the Sloan Digital Sky Survey: final data from SDSS-III. Astrophys. J. Suppl. **219**, 12 (2015). arXiv:1501.00963 [astro-ph.IM]
15. B. Reid et al., SDSS-III Baryon Oscillation Spectroscopic Survey data release 12: galaxy target selection and large-scale structure catalogues. Mon. Not. Roy. Astron. Soc. **455**, 1553 (2016). arXiv:1509.06529 [astro-ph.CO]
16. D. Foreman-Mackey, D. Hogg, D. Lang, J. Goodman, emcee: the MCMC hammer. Publ. Astron. Soc. Pac. **125**, 306 (2013). arXiv:1202.3665 [astro-ph.IM]
17. A. Gelman, D. Rubin, Inference from iterative simulation using multiple sequences. Statist. Sci. **7**, 457 (1992)
18. P.A.R. Ade et al. (Planck Collaboration), Planck 2015 results. XIII. Cosmological parameters. Astron. Astrophys. **594**, A13 (2016). arXiv:1502.01589 [astro-ph.CO]
19. A. Lewis, A. Challinor, A. Lasenby, Efficient computation of CMB anisotropies in closed FRW models. Astrophys. J. **538**, 473 (2000). arXiv:astro-ph/9911177 [astro-ph]
20. A. Lewis, S. Bridle, Cosmological parameters from CMB and other data: a Monte Carlo approach. Phys. Rev. D **66**, 103511 (2002). arXiv:astro-ph/0205436 [astro-ph]

21. N. Aghanim et al. (Planck Collaboration), Planck intermediate results. LI. Features in the cos-
    mic microwave background temperature power spectrum and shifts in cosmological parameters.
    Astron. Astrophys. **607**, A95 (2017). arXiv:1608.02487 [astro-ph.CO]
22. A. Ross et al. (BOSS Collaboration), The clustering of galaxies in the completed SDSS-
    III Baryon Oscillation Spectroscopic Survey: observational systematics and baryon acous-
    tic oscillations in the correlation function. Mon. Not. Roy. Astron. Soc. **464**, 1168 (2017).
    arXiv:1607.03145 [astro-ph.CO]
23. M. Vargas-Magaña et al. (BOSS Collaboration), The clustering of galaxies in the completed
    SDSS-III Baryon Oscillation Spectroscopic Survey: theoretical systematics and baryon acous-
    tic oscillations in the galaxy correlation function. Mon. Not. Roy. Astron. Soc. **477**, 1153
    (2018). arXiv:1610.03506 [astro-ph.CO]
24. A. Font-Ribera, P. McDonald, N. Mostek, B. Reid, H.-J. Seo, A. Slosar, DESI and other
    dark energy experiments in the era of neutrino mass measurements. JCAP **05**, 023 (2014).
    arXiv:1308.4164 [astro-ph.CO]

# Chapter 8
# Conclusions and Outlook

The wealth of cosmic microwave background and large-scale structure data has transformed the field of cosmology. Remarkably, these observations have not only become precise enough to answer questions about the universe at large, but also to start addressing puzzles in the microscopic description of Nature. Cosmological measurements are particularly well suited to study Standard Model neutrinos and to shed light on the possible existence of other light relics beyond the Standard Model. In this thesis, we have contributed to this endeavour by uncovering new constraints and identifying robust signatures of these particles in cosmological observables. We established the free-streaming nature of cosmic neutrinos in both the cosmic microwave background anisotropies and the clustering of galaxies, and paved the way to more efficiently using the wealth and precision of cosmological datasets in the future. In this final chapter, we summarize the main results of this work and make a few remarks on future directions.

## Summary

We employed cosmological observations to probe fundamental physics in two domains: within and beyond the Standard Model of particle physics. We focussed on providing new insights into the neutrino sector of the Standard Model as well as some of its extensions containing extra light and weakly-coupled species. In particular, we presented the first measurement of the cosmic neutrino background in the clustering of galaxies, obtained new CMB constraints on additional forms of radiation and derived novel bounds on light scalar particles, such as axions. Along the way, we highlighted the power of a subtle phase shift in the acoustic oscillations of the primordial plasma as a robust probe of neutrinos and other free-streaming relativistic species.

Based on a detailed analytical understanding of the phase shift, we presented new evidence for the cosmic neutrino background. We achieved this by establishing the free-streaming nature of these particles separately in cosmic microwave background observations of the Planck satellite and in the large-scale structure mapped by the Baryon Oscillation Spectroscopic Survey. In the BAO analysis, we established a new

© Springer Nature Switzerland AG 2019                                          179
B. Wallisch, *Cosmological Probes of Light Relics*, Springer Theses,
https://doi.org/10.1007/978-3-030-31098-1_8

method to extract the neutrino-induced phase shift. This allowed us to perform the first measurement of the imprint of neutrinos in a low-redshift observable. At the same time, this investigation marked the first application of the BAO signal to early universe physics and illustrated that the spectrum of baryon acoustic oscillations carries more accessible cosmological information than only the acoustic scale. The gravitational effect of the cosmic neutrino background has now been observed in the damping and the phase shift of both the CMB anisotropies and the BAO spectrum of galaxy clustering. Cosmological observations have therefore been able to provide new tests of the least understood part of the Standard Model.

Our forecasts indicate that future observations are guaranteed to explore regimes of BSM physics which have so far been inaccessible. We showed that future CMB measurements have the potential to probe the energy density in neutrinos and other relativistic species at the one-percent level, corresponding to an order of magnitude improvement over current bounds. In addition, we established that there is further, currently untapped information in the LSS data which will help us to push constraints on the effective number of relativistic species, $N_{\text{eff}}$, below well-motivated theoretical targets. This might have far-reaching consequences for new light particles, such as axions, which are predicted in many interesting SM extensions. These BSM species are hard to detect in terrestrial experiments due to their weak couplings, but the large number densities in the early universe make it possible to measure their gravitational effects. We demonstrated that reaching the sensitivity of the minimal thermal contribution of one scalar particle, $\Delta N_{\text{eff}} = 0.027$, would have important implications: We could either detect any light particles that have ever been in thermal equilibrium with the Standard Model, or put strong bounds on their SM interactions. We exemplified this for axions and other scalar BSM particles, and found that, in many cases, existing constraints from astrophysical and terrestrial searches could be surpassed by orders of magnitude. This result and the target $\Delta N_{\text{eff}} = 0.027$ have now been adopted by the CMB Stage-4 collaboration as one of their main science targets [1].

## Outlook

We have shown that future CMB and LSS observations could have a significant impact on our understanding of fundamental physics. In order to harvest all of the hidden clues stored in the observables, our ability to extract information from cosmological measurements should however further increase in this age of data-driven cosmology. In addition, we have argued that, in the case of $N_{\text{eff}}$, these observations can play complementary roles by both enhancing the raw sensitivity and adding to the robustness of the measurement. Our forecasts also suggest that the constraints on certain cosmological parameters, such as the effective number of relativistic species or the matter density, may be tightened when moving beyond the standard BAO analysis. A broader exploration, including additional observables such as weak gravitational lensing and the cross-correlation of different probes, will likely reveal more aspects of the early and/or late universe that could be discovered in this way. It would also be intriguing in this respect if we could uncover further quantities like the phase shift that are robust against non-linear gravitational evolution.

With the increasing precision of cosmological surveys, we are not only measuring the homogeneous background evolution of the universe, but have become sensitive to differences in the evolution of perturbations. This motivates revisiting the predictions of specific models of BSM physics. When we studied the impact of future $N_{eff}$ measurements on the SM couplings of axions, we provided one such avenue within an effective field theory of light scalar species. Since the theoretical threshold for $\Delta N_{eff}$ is larger for particles with spin and might soon be reached by CMB polarization experiments together with LSS surveys, the EFT of light species should be revisited. Moreover, these effective models could also include massive particles which are abundant in well-motivated SM extensions and might decay. The type of energy injection from such decays may then be observable over a wide range of times, even in the absence of new light fields. Including other cosmological observables, such as CMB spectral distortions or further late-time LSS probes, could therefore jointly constrain these models. Furthermore, the detailed predictions for $N_{eff}$ may be significantly altered from the minimal case, which can change the impact of future constraints. Finally, the investigation of light dark matter models, whose SM interactions must be mediated by new light fields to satisfy current thermal abundance limits, could link the search for light relics with the hunt for the (particle) nature of dark matter.

The prospects of probing the early universe with the future influx of cosmological data are very bright. At the same time, the possible implications on particle physics cannot be overstated. These observations will provide an opportunity to probe physics beyond the Standard Model at a much more precise level than was previously possible and in a regime that is inaccessible to terrestrial experiments. We are optimistic that this will teach us something interesting. We will either discover a whole new world of dark physics, or learn to what remarkable degree it is decoupled from the rest of physics. In the meantime, we remain curious what future cosmological measurements will tell us about the universe and the underlying laws of Nature.

# Reference

1. K. Abazajian et al. (CMB-S4 Collaboration), CMB-S4 Science Book, 1st edn. arXiv: 1610.02743 [astro-ph.CO]

# Appendix A
# Goldstone Production and Decay

In this appendix, we provide supplemental material to Chap. 4, which is based on [1]. We compute the production rates of axions, familons and majorons via their Standard Model interactions (Sect. A.1), and discuss the effects of the possible decays of these Goldstone bosons to photons and neutrinos (Sect. A.2).

## A.1 Production Rates

In this section, we derive the rates of Goldstone boson production used in the main text. We consider separately the couplings to gauge fields and to matter fields.

### A.1.1 Couplings to Gauge Fields

Above the scale of electroweak symmetry breaking, the coupling of the Goldstone boson to the Standard Model gauge sector is

$$\mathcal{L}_{\phi\text{EW}} = -\frac{1}{4}\frac{\phi}{\Lambda}\left(c_1\, B_{\mu\nu}\tilde{B}^{\mu\nu} + c_2\, W^a_{\mu\nu}\tilde{W}^{\mu\nu,a} + c_3\, G^a_{\mu\nu}\tilde{G}^{\mu\nu,a}\right). \tag{A.1}$$

The dominant processes leading to the production of the axion $\phi$ are illustrated in Fig. A.1. In the limit of massless gauge bosons, the cross sections for some of these processes have infrared (IR) divergences. The results therefore depend slightly on how these divergences are regulated; see e.g. [2–6]. The most detailed analysis has been performed in [6], where the total production rate was found to be

$$\Gamma = \frac{T^3}{8\pi\Lambda^2}\left[c_1^2\, F_1(T) + 3c_2^2\, F_2(T) + 8c_3^2\, F_3(T)\right], \tag{A.2}$$

© Springer Nature Switzerland AG 2019
B. Wallisch, *Cosmological Probes of Light Relics*, Springer Theses,
https://doi.org/10.1007/978-3-030-31098-1

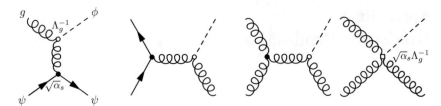

(a) Primakoff process.  (b) Fermion annihilation.  (c) Gluon fusion (representative diagrams).

**Fig. A.1** Feynman diagrams for the dominant Goldstone production via the gluon coupling (from [1]). For gluon fusion, there are $t$- and $u$-channel diagrams in addition to the presented $s$-channel diagram. Similar diagrams apply for the couplings to the electroweak gauge bosons

where the functions $F_n(T)$ were derived numerically. We extracted $F_n(T)$ from Fig. 1 of [6], together with the one-loop running of the gauge couplings $\alpha_i(T)$.

## Coupling to Gluons

To isolate the effect of the coupling to gluons, we write $c_1 = c_2 \equiv 0$ and define $\Lambda_g \equiv \Lambda/c_3$. In this case, the production rate (A.2) becomes

$$\Gamma_g(T) = \frac{F_3(T)}{\pi} \frac{T^3}{\Lambda_g^2} \equiv \gamma_g(T) \frac{T^3}{\Lambda_g^2}, \tag{A.3}$$

with $\gamma_g(10^{10}\,\text{GeV}) = 0.41$. The function $\gamma_g(T)$ is presented in the left panel of Fig. A.2. The freeze-out bound on the gluon coupling then is

$$\Lambda_g > \left(\frac{\pi^2}{90} g_{*,R}\right)^{-1/4} \sqrt{\gamma_{g,R}\, T_R\, M_{\text{pl}}} \equiv \lambda_g(T_R) \left(\frac{T_R}{10^{10}\,\text{GeV}}\right)^{1/2}, \tag{A.4}$$

where $g_{*,R} \equiv g_*(T_R)$ and $\gamma_{g,R} \equiv \gamma_g(T_R)$. The bound in (A.4) is illustrated in the right panel of Fig. A.2. In the main text, we used $\lambda_g(10^{10}\,\text{GeV}) = 5.4 \times 10^{13}\,\text{GeV}$.

## Coupling to Photons

To isolate the coupling to the electroweak sector, we set $c_3 = 0$. In this case, the Lagrangian (A.1) can be written as

$$\mathcal{L}_{\phi\text{EW}} = -\frac{1}{4} \frac{\phi}{\Lambda} \left( c_a\, B_{\mu\nu} \tilde{B}^{\mu\nu} + s_a\, W_{\mu\nu}^a \tilde{W}^{\mu\nu,a} \right), \tag{A.5}$$

where we have defined

$$\Lambda \rightarrow \frac{\Lambda}{\sqrt{c_1^2 + c_2^2}} \quad \text{and} \quad c_a \equiv \frac{c_1}{\sqrt{c_1^2 + c_2^2}}, \quad s_a \equiv \frac{c_2}{\sqrt{c_1^2 + c_2^2}}. \tag{A.6}$$

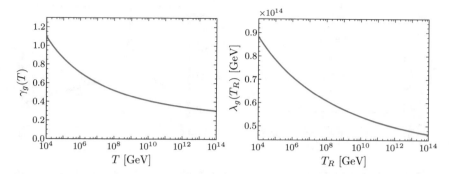

**Fig. A.2** *Left*: Axion production rate associated with the coupling to gluons as parametrized by $\gamma_g(T)$ in (A.3) (after [1, 7]). *Right*: Constraint on the axion-gluon coupling $\Lambda_g$ as parametrized by $\lambda_g(T_R)$ in (A.4) (after [1, 7])

Note that $c_a^2 + s_a^2 = 1$, so we can use $\Lambda$ and $c_a$ as the two free parameters. The production rate (A.2) is then given by

$$\Gamma = \frac{[c_a^2 F_1(T) + 3s_a^2 F_2(T)]\, T^3}{8\pi \,\Lambda^2} \equiv \gamma(T, c_a) \frac{T^3}{\Lambda^2}. \tag{A.7}$$

The function $\gamma(T, c_a)$ is shown in the left panel of Fig. A.3. In the main text, we employed $\gamma(10^{10}\,\mathrm{GeV}, 1) = 0.017$. The freeze-out bound on the coupling then is

$$\Lambda(c_a) > \left(\frac{\pi^2}{90} g_{*,R}\right)^{-1/4} \sqrt{\gamma_R(c_a)\, T_R\, M_{\mathrm{pl}}}, \tag{A.8}$$

with $\gamma_R(c_a) \equiv \gamma(T_R, c_a)$. We wish to relate this bound to the couplings below the EWSB scale.

At low energies, the axion couplings to the electroweak sector become

$$\mathcal{L}_{\phi\mathrm{EW}} = -\frac{1}{4}\left(\frac{\phi}{\Lambda_\gamma} F_{\mu\nu}\tilde{F}^{\mu\nu} + \frac{\phi}{\Lambda_Z} Z_{\mu\nu}\tilde{Z}^{\mu\nu} + \frac{\phi}{\Lambda_{Z\gamma}} Z_{\mu\nu}\tilde{F}^{\mu\nu} + \frac{\phi}{\Lambda_W} W^+_{\mu\nu}\tilde{W}^{-\mu\nu}\right), \tag{A.9}$$

where $F_{\mu\nu}$, $Z_{\mu\nu}$ and $W^\pm_{\mu\nu}$ are the field strengths for the photon, $Z$ and $W^\pm$, respectively. Here, we have dropped additional (non-Abelian) terms proportional to $c_2$ which are cubic in the gauge fields. In order to match the high-energy couplings in (A.5) to the low-energy couplings in (A.9), we define

$$\Lambda_\gamma^{-1} = \left(c_w^2 c_a + s_w^2 s_a\right)\Lambda^{-1}, \tag{A.10}$$

$$\Lambda_Z^{-1} = \left(c_w^2 s_a + s_w^2 c_a\right)\Lambda^{-1}, \tag{A.11}$$

$$\Lambda_{Z\gamma}^{-1} = 2 s_w c_w \left(s_a - c_a\right)\Lambda^{-1}, \tag{A.12}$$

$$\Lambda_W^{-1} = s_a \Lambda^{-1}, \tag{A.13}$$

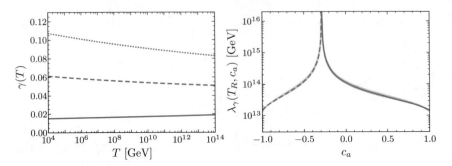

**Fig. A.3** *Left*: Axion production rate associated with the coupling to the electroweak gauge bosons as parametrized by $\gamma(T, c_a)$ in (A.7) for $c_a = 0$ (dotted line), $1/\sqrt{2}$ (dashed line) and 1 (solid line; after [1, 7]). *Right*: Constraint on the axion-photon coupling $\Lambda_\gamma$ as parametrized by $\lambda_\gamma(T_R, c_a)$ in (A.14) (after [1]). The solid and dashed lines correspond to bounds on positive and negative $\Lambda_\gamma$ for $T_R = 10^{10}$ GeV. The band displays the change for reheating temperatures between $10^4$ GeV (upper edge) and $10^{15}$ GeV (lower edge)

where $\{c_w, s_w\} \equiv \{\cos\theta_w, \sin\theta_w\}$, with Weinberg's mixing angle $\theta_w \approx 30°$. Using (A.10), we can write (A.8) as a bound on the photon coupling,

$$
\Lambda_\gamma(c_a) > \left(c_w^2 c_a + s_w^2 s_a\right)^{-1} \times \left(\frac{\pi^2}{90} g_{*,R}\right)^{-1/4} \sqrt{\gamma_R(c_a) T_R M_{\text{pl}}}
$$

$$
\equiv \lambda_\gamma(T_R, c_a) \left(\frac{T_R}{10^{10} \text{ GeV}}\right)^{1/2}. \tag{A.14}
$$

This bound is illustrated in the right panel of Fig. A.3. We see that we get the most conservative constraint by setting $s_a \equiv 0$, for which we have $\lambda_\gamma(10^{10} \text{ GeV}, 1) = 1.4 \times 10^{13}$ GeV.

## A.1.2 Couplings to Matter Fields

The calculation of the Goldstone production rates associated with the couplings to the SM fermions is somewhat less developed. In this section, we calculate the relevant rates following the procedure outlined in [4].

**Preliminaries**

The integrated Boltzmann equation for the evolution of the number density of the Goldstone boson takes the form

$$
\frac{dn_\phi}{dt} + 3Hn_\phi = \Gamma(n_\phi^{\text{eq}} - n_\phi), \tag{A.15}
$$

where $n_\phi^{eq} = \zeta(3)T^3/\pi^2$ is the equilibrium density of a relativistic scalar. In order to simplify the analysis, we will replace the integration over the phase space of the final states with the centre-of-mass cross section, $\sigma_{cm}$, or the centre-of-mass decay rate, $\Gamma_{cm}$. While this approach is not perfectly accurate, it has the advantage of relating the vacuum amplitudes to the thermal production rates in terms of relatively simple integrals.

- For a two-to-two process, $1 + 2 \to 3 + 4$, we have

$$\Gamma_{2\to2} \simeq \frac{1}{n_\phi^{eq}} \int \frac{d^3 p_1}{(2\pi)^3} \frac{d^3 p_2}{(2\pi)^3} \frac{f_1(p_1)}{2E_1} \frac{f_2(p_2)}{2E_2} [1 \pm f_3][1 \pm f_4] 2s\sigma_{cm}(s), \quad (A.16)$$

where $f_{1,2}$ are the distribution functions of the initial states and $s \equiv (p_1 + p_2)^2$ is the Mandelstam variable. We have included simplified Bose enhancement and Pauli blocking terms, $[1 \pm f_3][1 \pm f_4] \to \frac{1}{2}([1 \pm f_3(p_1)][1 \pm f_4(p_2)] + \{p_1 \leftrightarrow p_2\})$, which is applicable in the centre-of-mass frame where the initial and final momenta are all equal.[1] For $s \gg m_i^2$, the centre-of-mass cross section is given by

$$\sigma_{cm}(s) \simeq \frac{1}{32\pi} \int d\cos\theta \frac{\sum |\mathcal{M}|^2(s, \theta)}{s}, \quad (A.17)$$

where $\sum |\mathcal{M}|^2$ is the squared scattering amplitude including the sum over spins and charges, and $\theta$ is the azimuthal angle in the centre-of-mass frame. For all models of freeze-out considered in the main text, the centre-of-mass cross section is independent of $s$. Moreover, we will only encounter fermion-boson scattering or fermion annihilation in this section. With the enhancement/blocking terms, one finds that the numerical pre-factors in both cases agree to within 10%. To simplify the calculations, we will therefore use the fermion annihilation rate throughout,

$$\Gamma_{2\to2} \simeq \sigma_{cm} T^3 \left(\frac{7}{8}\right)^2 \frac{\zeta(3)}{\pi^2} \approx 0.093\, \sigma_{cm} T^3. \quad (A.18)$$

The advantage of this approach is that we can relate the centre-of-mass cross section directly to the production rate with minimal effort and reasonable accuracy.
- For a one-to-two process, $1 \to 2 + 3$, the decay rate in the centre-of-mass frame is

$$\Gamma_{cm} \simeq \frac{1}{32\pi m_1} \int d\cos\theta \sum |\mathcal{M}|^2, \quad (A.19)$$

where we have taken the two final particles to be massless. Since $\Gamma_{cm}$ is independent of energy, the rate only depends on whether the initial state is a fermion or boson. Transforming this rate to a general frame gives

---

[1] These Pauli blocking and Bose enhancement terms were not included in [4], as they complicate the rate calculations. We have included them to ensure that the rates computed for both the forward and backward processes give the same results.

$$\Gamma_{1\to 2} \simeq \frac{1}{n_\phi^{eq}} \int \frac{d^3 p_1}{(2\pi)^3} f_1(p_1)\left[1 \pm f_2(p_1/2)\right]\left[1 \pm f_3(p_1/2)\right]\frac{m_1}{E_1}\Gamma_{cm}, \quad (A.20)$$

where $f_1$ is the distribution function of the decaying particle (not necessarily $\phi$). We are mostly interested in the limit $T \gg m_1$, in which case the rate (A.20) reduces to

$$\Gamma_{1\to 2} \simeq \frac{m_1}{T}\frac{\pi^2}{16\zeta(3)}\Gamma_{cm} \times \begin{cases} 1 - \dfrac{4}{\pi^2} & \text{fermion,} \\ 1 & \text{boson,} \end{cases} \quad (A.21)$$

where the dependence on the number of degrees of freedom of the decaying particle has been absorbed into $\Gamma_{cm}$ through the sum over spins and charges. Note that, in equilibrium, the rates for decay and inverse decay are equal.

## Coupling to Charged Fermions

We consider the following coupling between a Goldstone boson and charged fermions:

$$\mathcal{L}_{\phi\psi} = \frac{\phi}{\Lambda_\psi}\left(iH\,\bar{\psi}_{L,i}\left[(\lambda_i - \lambda_j)g_V^{ij} + (\lambda_i + \lambda_j)g_A^{ij}\right]\psi_{R,j} + \text{h.c.}\right), \quad (A.22)$$

where $H$ is the Higgs doublet, $\psi_{L,R} \equiv \frac{1}{2}(1 \mp \gamma^5)\psi$, and the $SU(2)_L$ and $SU(3)_C$ structures have been left implicit. Distinct processes dominate in the various limits of interest:

- **Freeze-out**   At high energies, the Goldstone boson is produced through the following two processes (see Fig. A.4): (a) $\psi_i + \bar{\psi}_j \to H + \phi$ and (b) $\psi_i + H \to \psi_j + \phi$. Summing over the spins and charges, we get

$$\sum |\mathcal{M}|^2_{(a)} = 4N_\psi\, s\, \frac{(\lambda_i - \lambda_j)^2 (g_V^{ij})^2 + (\lambda_i + \lambda_j)^2 (g_A^{ij})^2}{\Lambda_\psi^2}, \quad (A.23)$$

$$\sum |\mathcal{M}|^2_{(b)} = 4N_\psi\, s(1 - \cos\theta)\, \frac{(\lambda_i - \lambda_j)^2 (g_V^{ij})^2 + (\lambda_i + \lambda_j)^2 (g_A^{ij})^2}{\Lambda_\psi^2}, \quad (A.24)$$

where we have combined fermion and anti-fermion scattering in the sum over charges and introduced

$$N_\psi \equiv \begin{cases} 1 & \psi = \text{lepton,} \\ 3 & \psi = \text{quark.} \end{cases} \quad (A.25)$$

We also find it convenient to define $\Lambda_{ij}^I \equiv \Lambda_\psi/g_I^{ij}$, with $I \in \{V, A\}$. Using (A.17) and (A.18), and treating the vector and axial-vector couplings separately, we obtain

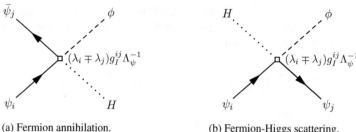

(a) Fermion annihilation.                              (b) Fermion-Higgs scattering.

**Fig. A.4** Feynman diagrams for the dominant Goldstone production via the coupling to charged fermions above the electroweak scale (from [1, 8]). For the vector and axial vector couplings, $I \in \{V, A\}$, the '$-$' and '$+$' signs apply, respectively

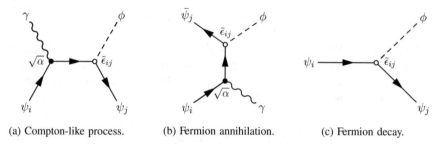

(a) Compton-like process.        (b) Fermion annihilation.        (c) Fermion decay.

**Fig. A.5** Feynman diagrams for the dominant Goldstone production via the coupling to charged fermions below the electroweak scale (from [1, 8]). For quarks, the coupling to photons is replaced by that to gluons. In addition to the displayed $s$- and $t$-channel diagrams for the Compton-like process and fermion annihilation, there are $u$-channel diagrams which are not shown

$$\Gamma_{ij}^I = N_\psi \left(\frac{7}{8}\right)^2 \frac{4\zeta(3)}{\pi^2} \frac{(\lambda_i \mp \lambda_j)^2}{8\pi} \frac{T^3}{(\Lambda_{ij}^I)^2} \simeq 0.19 N_\psi \frac{(\lambda_i \mp \lambda_j)^2}{8\pi} \frac{T^3}{(\Lambda_{ij}^I)^2},$$
(A.26)

where the '$-$' and '$+$' signs apply to $I = V$ and $I = A$, respectively.

- **Freeze-in**  Below the scale of EWSB, the Lagrangian (A.22) becomes

$$\mathcal{L}_{\phi\psi} = i\frac{\phi}{\Lambda_\psi} \bar{\psi}_i \left[(m_i - m_j)g_V^{ij} + (m_i + m_j)g_A^{ij}\gamma^5\right] \psi_j,$$
(A.27)

with $m_i \equiv \sqrt{2}\lambda_i/v$. The Goldstone production processes associated with these couplings are shown in Fig. A.5.

*Diagonal couplings.*—We first consider the diagonal part of the interaction, which takes the form $i\tilde{\epsilon}_{ii} \, \phi\bar{\psi}_i\gamma^5\psi_i$, with $\tilde{\epsilon}_{ii} \equiv 2m_i g_A^{ii}/\Lambda_\psi$. Kinematical constraints require us to include at least one additional particle in order to get a non-zero amplitude. The two leading processes are (a) $\psi_i + \{\gamma, g\} \to \psi_i + \phi$ (cf. Fig. A.5a) and

(b) $\psi_i + \bar{\psi}_i \to \phi + \{\gamma, g\}$ (cf. Fig. A.5b), where $\{\gamma, g\}$ is either a photon or gluon depending on whether the fermion is a lepton or quark, respectively. Summing over spins and charges, we obtain

$$\sum |\mathcal{M}|^2_{(a)} = 16\pi A_\psi |\tilde{\epsilon}_{ii}|^2 \frac{s^2}{(m_i^2 - t)(m_i^2 - u)}, \tag{A.28}$$

$$\sum |\mathcal{M}|^2_{(b)} = 16\pi A_\psi |\tilde{\epsilon}_{ii}|^2 \frac{t^2}{(s - m_i^2)(m_i^2 - u)}, \tag{A.29}$$

where $s$, $t$ and $u$ are the Mandelstam variables and

$$A_\psi \equiv \begin{cases} \alpha & \psi = \text{lepton,} \\ 4\alpha_s & \psi = \text{quark.} \end{cases} \tag{A.30}$$

In the massless limit, the cross section has IR divergences in the $t$- and $u$-channels from the exchange of a massless fermion. The precise production rate therefore depends on the treatment of the soft modes. Regulating the IR divergence with the fermion mass and taking the limit $s \gg m_i^2$, we find

$$\sigma_{\text{cm}}(s) \simeq \frac{1}{s} A_\psi |\tilde{\epsilon}_{ii}|^2 \left[ 3 \log \frac{s}{m_i^2} - \frac{3}{2} \right]. \tag{A.31}$$

At high temperatures, the fermion mass is controlled by the thermal mass $m_i^2 \to m_T^2 = \frac{1}{2}\pi A_\psi T^2$ and the production rate becomes

$$\tilde{\Gamma}_{ii} = \frac{3\pi^3}{64\zeta(3)} A_\psi \frac{|\tilde{\epsilon}_{ii}|^2}{8\pi} T \left[ \log \frac{2}{\pi A_\psi} + 2 \log 2 - \frac{3}{2} \right]. \tag{A.32}$$

This formula is expected to break down at $T \lesssim m_i$, but will be sufficient at the level of approximation being used in this work. A proper treatment of freeze-in at $T \sim m_i$ should go beyond $\Gamma = H$ and fully solve the Boltzmann equations. However, this level of accuracy is not needed for estimating the constraint on the coupling $\tilde{\epsilon}_{ii}$.

The result (A.32) will be of limited utility for the coupling to quarks. This is due to the fact that, for $T \lesssim 30\,\text{GeV}$, the QCD coupling becomes large and our perturbative calculation becomes unreliable.[2] In fact, we see that the production rate (A.32) becomes negative in this regime. While the top quark is sufficiently heavy to be still at weak coupling, its mass is close to the electroweak phase transition and, therefore, the assumption $s \gg m_i^2$ is not applicable. For these reasons, we will not derive bounds on the quark couplings from these production rates.

---

[2]These effects are computable using the techniques of [6], but this is beyond the scope of the present work.

*Off-diagonal couplings.*—When the coupling of $\phi$ is off-diagonal in the mass basis, the dominant process at low energies is the decay $\psi_i \to \psi_j + \phi$, cf. Fig. A.5c. Since the mass splittings of the SM fermions are large and $m_\phi \ll m_\psi$, the centre-of-mass decay rate is well approximated by

$$\Gamma_{\rm cm} = \frac{N_\psi}{8\pi} \frac{m_i^3}{\Lambda_{ij}^2}, \tag{A.33}$$

where $\Lambda_{ij} \equiv \left[(g_V^{ij})^2 + (g_A^{ij})^2\right]^{-1/2} \Lambda_\psi$. Using (A.21), we get

$$\tilde{\Gamma}_{ij} = \frac{(\pi^2 - 4)}{16\zeta(3)} \frac{N_\psi}{8\pi} \frac{1}{T} \frac{m_i^4}{\Lambda_{ij}^2} \simeq 0.31 N_\psi \frac{|\tilde{\epsilon}_{ij}|^2}{8\pi} \frac{m_i^2}{T}, \tag{A.34}$$

with $\tilde{\epsilon}_{ij} \approx m_i/\Lambda_{ij}$. In addition to this decay, we also have production with a photon or gluon, given by (A.32) with $\tilde{\epsilon}_{ii} \to \tilde{\epsilon}_{ij}$. We will neglect this contribution as it is suppressed by a factor of $\alpha$ or $\alpha_s$ for $T \sim m_i$.

**Coupling to Neutrinos**

The coupling between the Goldstone boson and neutrinos is

$$\mathcal{L}_{\phi\nu} = -\frac{1}{2}\left(i\tilde{\epsilon}_{ij}\phi\nu_i\nu_j - \frac{1}{2\Lambda_\nu}\epsilon_{ij}\phi^2\nu_i\nu_j + \cdots\right) + \text{h.c.}, \tag{A.35}$$

where we have written the Majorana neutrinos in two-component notation. The first term in (A.35) will control freeze-in and the second will determine freeze-out:

- **Freeze-out**   At high energies, the dominant production mechanism is $\nu_i + \nu_j \to \phi + \phi$ (cf. Fig. A.6a) through the second term in $\mathcal{L}_{\nu\phi}$. The spin-summed amplitude squared is

$$\sum |\mathcal{M}|^2 = |\epsilon_{ij}|^2 \frac{2s}{\Lambda_\nu^2}, \tag{A.36}$$

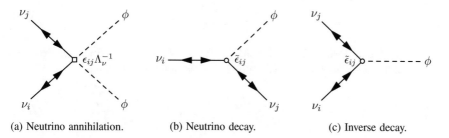

(a) Neutrino annihilation.        (b) Neutrino decay.        (c) Inverse decay.

**Fig. A.6** Feynman diagrams for the dominant Goldstone production via the coupling to neutrinos (from [1]). The double arrows denote the sum over the spinor index structure for two-component fermions [9]

which results in the production rate

$$\Gamma_{ij} = \frac{1}{2} s_{ij} \left(\frac{7}{8}\right)^2 \frac{\zeta(3)}{\pi^2} \frac{|\epsilon_{ij}|^2}{8\pi} \frac{T^3}{\Lambda_\nu^2} \simeq 0.047 \, s_{ij} \frac{|\epsilon_{ij}|^2}{8\pi} \frac{T^3}{\Lambda_\nu^2} \,, \tag{A.37}$$

where the factors of $\frac{1}{2}$ and $s_{ij} \equiv 1 - \frac{1}{2}\delta_{ij}$ are the symmetry factors for identical particles in the initial and final states, respectively. The contribution to the rate from higher-order terms in (A.35) is suppressed by further powers of $T^2/\Lambda_\nu^2$.

• **Freeze-in**   Unlike for charged fermions, the freeze-in abundance from the coupling to neutrinos arises only through decays. Below the scale of EWSB, the couplings of neutrinos to the rest of the SM are suppressed by the weak scale and are irrelevant. The only freeze-in processes that are allowed by kinematics are therefore three-body decays.

*Low-mass regime.*—For $m_\phi \ll m_i - m_j$, with $m_i > m_j$, the off-diagonal linear coupling allows the decay $\nu_i \to \nu_j + \phi$, cf. Fig. A.6b. The decay rate in the centre-of-mass frame is

$$\Gamma_{\text{cm}} = \frac{1}{8\pi} \frac{m_i^2 - m_j^2}{m_i^3} \left(|\tilde{\epsilon}_{ij}|^2 (m_i^2 + m_j^2) + 2 \, \text{Re}\big[(\tilde{\epsilon}_{ij})^2\big] m_i m_j\right). \tag{A.38}$$

In order to simplify the calculations in the main text, we take $m_i \gg m_j$ which is guaranteed for the minimal mass normal hierarchy. Since the decaying particle is a fermion, the thermal production rate in (A.21) becomes

$$\tilde{\Gamma}_{ij} = \frac{\pi^2 - 4}{16\zeta(3)} \frac{|\tilde{\epsilon}_{ij}|^2}{8\pi} \frac{m_i^2}{T} \simeq 0.31 \frac{|\tilde{\epsilon}_{ij}|^2}{8\pi} \frac{m_i^2}{T}. \tag{A.39}$$

Notice that the off-diagonal decay rate is the same for charged leptons and neutrinos even though the neutrinos have a Majorana mass.

*High-mass regime.*—For $m_\phi \gg m_i \geq m_j$, the Goldstone boson decays to fermions, $\phi \to \nu_i + \nu_j$, both through the diagonal and off-diagonal couplings. The inverse decay $\nu_i + \nu_j \to \phi$ (see Fig. A.6c) is therefore a production channel. The decay rate is given by

$$\Gamma_{\text{cm}} = \frac{|\tilde{\epsilon}_{ij}|^2}{8\pi} m_\phi \,, \tag{A.40}$$

which, in equilibrium, is equal to the rate for the inverse decay. Since the decaying particle is a boson, the thermal production rate in (A.21) becomes

$$\tilde{\Gamma}_{ij} = s_{ij} \frac{\pi^2}{16\zeta(3)} \frac{|\tilde{\epsilon}_{ij}|^2}{8\pi} \frac{m_\phi^2}{T} \simeq 0.51 \, s_{ij} \frac{|\tilde{\epsilon}_{ij}|^2}{8\pi} \frac{m_\phi^2}{T}. \tag{A.41}$$

The rate is somewhat enhanced compared to the decay in the low-mass regime because the decaying particle is a boson.

The dominant Goldstone production mechanism through the couplings to neutrinos is quite sensitive to kinematics. For $m_\phi \lesssim m_\nu$, the diagonal decay is forbidden and the dominant Goldstone production is through freeze-out. In addition, when $m_\phi \sim m_\nu$ there are additional kinematic constraints for both diagonal and off-diagonal couplings. As a result, the limits on the interaction scale $\Lambda_\nu$ (or the dimensionless couplings $\epsilon_{ij}$ and $\tilde{\epsilon}_{ij}$) are sensitive to $m_\phi$.

## A.2 Comments on Decays

Throughout Chap. 4, we treated each of the operators which couple the pseudo-Nambu-Goldstone bosons to the Standard Model independently. For computing the production rates, this is justified since the amplitudes for the different processes that we consider do not interfere and the couplings therefore add in quadrature. One may still ask, however, if the interplay between several operators can affect the cosmological evolution after the production. In particular, one might worry that some operators would allow for the decay of the pNGBs and that this might evade the limits on $N_{\rm eff}$. In the following, we will address this concern. We are assuming that $m_\phi < 1$ MeV, so that the only kinematically allowed decays are to photons and neutrinos.

### A.2.1 Decay to Photons

If the decay occurs after recombination, then the pNGBs are effectively stable as far as the CMB is concerned and our treatment in the main text applies directly. To see when this is the case, we computed the decay temperature $T_D$ associated with the decay mediated by the coupling to photons (4.6):

$$\frac{T_D}{T_{\rm rec}} \approx 9.5 \times 10^{-10} \left( \frac{\Lambda_\gamma}{10^{10}\,{\rm GeV}} \right)^{-4/3} \left( \frac{m_\phi}{T_{\rm rec}} \right)^2 . \tag{A.42}$$

Recalling the stellar cooling bound, $\Lambda_\gamma > 1.3 \times 10^{10}$ GeV [10], we see that the pNGBs are effectively stable as long as $m_\phi \lesssim 10$ keV. For comparison, a stable particle with $m_\phi \gtrsim 100$ eV produces $\Omega_m > 1$ and is therefore excluded by constraints on the dark matter abundance. For $m_\phi > 10$ keV, the decay to photons does affect the phenomenology and must be considered explicitly. Nevertheless, in the regime of interest, the pNGBs are non-relativistic and, therefore, carry a large energy density, $\rho_\phi \simeq m_\phi n_\phi$. As a result, this region is highly constrained by current cosmological observations [11, 12].

### A.2.2  Decay to Neutrinos

Depending on the mass of the pNGB, the decay to neutrinos leads to the following three scenarios:

- For $m_\phi < T_{\text{rec}}$, the implications of the decays are relatively easy to characterise. As discussed in Sect. 4.4.2, the phenomenology is only modified if $T_{\text{fluid}} > T_{\text{rec}}$. In this case, strong interactions between the pNGBs and the neutrinos imply that the neutrinos are no longer free-streaming particles, which is ruled out by current CMB observations, cf. Sect. 5.3.

- For $T_D > m_\phi > T_{\text{rec}}$, the pNGBs are brought into equilibrium with the neutrinos at $T \sim T_D$ and then become Boltzmann suppressed for $T \lesssim m_\phi$. This process leads to a contribution to $N_{\text{eff}}$, even if the pNGBs have negligible energy density to begin with. To estimate the size of the effect, we first note that the freeze-in at $T_D$ conserves the total energy density in neutrinos and pNGBs,

$$(g_{*,\nu} + g_{*,\phi})(a_1 T_1)^4 = g_{*,\nu}(a_0 T_0)^4 , \qquad (A.43)$$

where $T_0$ and $T_1$ are the initial and final temperatures during the equilibration, and $g_{*,\nu}$ and $g_{*,\phi} = 1$ are the effective numbers of degrees of freedom in $\nu$ and $\phi$, respectively. When the temperature drops below the mass of the pNGBs, their energy density is converted to neutrinos. This process conserves the comoving entropy density,

$$(g_{*,\nu} + g_{*,\phi})(a_1 T_1)^3 = g_{*,\nu}(a_2 T_2)^3 , \qquad (A.44)$$

where $T_2 \ll m_\phi$ is some temperature after the pNGB population has decayed. The final energy density of the neutrinos becomes

$$a_2^4 \rho_{\nu,2} = \left( \frac{g_{*,\nu} + g_{*,\phi}}{g_{*,\nu}} \right)^{1/3} a_0^4 \rho_{\nu,0} , \qquad (A.45)$$

where $\rho_{\nu,i} \equiv \rho_\nu(a_i)$. Using the definition of $N_{\text{eff}}$ in (3.19) together with (2.20) as well as (3.16), and $\rho_\gamma a^4 = \text{const}$, we find

$$N_{\text{eff}} = \left( \frac{g_{*,\nu} + g_{*,\phi}}{g_{*,\nu}} \right)^{1/3} N_{\text{eff},0} . \qquad (A.46)$$

Considering the coupling to a single neutrino flavour (rather than all three), i.e. $N_{\text{eff},0} \simeq 1$ and $g_{*,\nu} = 7/4$, we then get

$$\Delta N_{\text{eff}} = \left( 1 + \frac{4}{7} \right)^{1/3} - 1 \simeq 0.16 , \qquad (A.47)$$

where $\Delta N_{\text{eff}} \equiv N_{\text{eff}} - N_{\text{eff},0}$. Coupling to more than one neutrino flavour and including a non-zero initial temperature for the pNGBs would increase this number slightly, so that we will use $\Delta N_{\text{eff}} \geq 0.16$.

- The production of pNGBs through the freeze-in process is avoided if $m_\phi > T_D > T_{\text{rec}}$, in which case the pNGBs decay to neutrinos out of equilibrium. To a good approximation, this decay conserves the energy density, which is therefore simply transferred from $\phi$ to $\nu$ at the time of the decay. The contribution to $\Delta N_{\text{eff}}$ is enhanced by the amount of time that $\phi$ is non-relativistic before its decay, which may be a large effect for $m_\phi \gg 1$ eV (see e.g. [13] for a related discussion).

In summary, operators that allow the Goldstone bosons to decay do not substantially alter the predictions presented in the main text. On the one hand, decays to photons cannot occur early enough to impact the CMB. On the other hand, decays to neutrinos typically increase the contributions to $\Delta N_{\text{eff}}$ and would therefore strengthen our bounds.

# References

1. D. Baumann, D. Green, B. Wallisch, New target for cosmic axion searches. Phys. Rev. Lett. **117**, 171301 (2016). arXiv:1604.08614 [astro-ph.CO]
2. E. Braaten, T. Yuan, Calculation of screening in a hot plasma. Phys. Rev. Lett. **66**, 2183 (1991)
3. M. Bolz, A. Brandenburg, W. Buchmüller, Thermal production of gravitinos. Nucl. Phys. B **606**, 518 (2001) [Erratum: Nucl. Phys. B **790**, 336 (2008)]. arXiv:hep-ph/0012052 [hep-ph]
4. E. Masso, F. Rota, G. Zsembinszki, On axion thermalization in the early universe. Phys. Rev. D **66**, 023004 (2002). arXiv:hep-ph/0203221 [hep-ph]
5. P. Graf, F. Steffen, Thermal axion production in the primordial quark-gluon plasma. Phys. Rev. D **83**, 075011 (2011). arXiv:1008.4528 [hep-ph]
6. A. Salvio, A. Strumia, W. Xue, Thermal axion production. JCAP **01**, 011 (2014). arXiv:1310.6982 [hep-ph]
7. Adapted from [1] with permission. Copyright by the American Physical Society
8. Reproduced from [1] with permission. Copyright by the American Physical Society
9. H. Dreiner, H. Haber, S. Martin, Two-component spinor techniques and Feynman rules for quantum field theory and supersymmetry. Phys. Rep. **494**, 1 (2010). arXiv:0812.1594 [hep-ph]
10. A. Friedland, M. Giannotti, M. Wise, Constraining the axion-photon coupling with massive stars. Phys. Rev. Lett. **110**, 061101 (2013). arXiv:1210.1271 [hep-ph]
11. D. Cadamuro, J. Redondo, Cosmological bounds on pseudo-Nambu-Goldstone bosons. JCAP **02**, 032 (2012). arXiv:1110.2895 [hep-ph]
12. M. Millea, L. Knox, B. Fields, New bounds for axions and axion-like particles with keV-GeV masses. Phys. Rev. D **92**, 023010 (2015). arXiv:1501.04097 [astro-ph.CO]
13. W. Fischler, J. Meyers, Dark radiation emerging after big bang nucleosynthesis? Phys. Rev. D **83**, 063520 (2011). arXiv:1011.3501 [astro-ph.CO]

# Appendix B
# Further Aspects of the Phase Shift

In this appendix, we provide supplemental material to Chap. 5, which is based on [1]. We estimate the effects of matter domination on the neutrino-induced phase shift (Sect. B.1) and derive that the phase shift in the CMB polarization spectrum is the same as in the temperature spectrum (Sect. B.2).

## B.1  Comments on Matter

We calculated the phase shift of the photon density fluctuations under the assumption of a radiation-dominated background in the main text. This simplified approach allowed an analytic treatment, but we may wonder if we missed any important effects in this way. While focussing on the contributions from free-streaming radiation, in this section, we will bridge this gap to the degree that is possible without resorting to numerics.

 We are interested in understanding the contributions to the phase shift from modes in the matter era for several reasons. First, recombination occurs during matter domination which could, in principle, be important for every mode in the CMB.[1] Second, horizon entry of modes which correspond to large angular scales (small $\ell$) happens during (or near) matter domination and the complete evolution of these modes is consequently dictated by the physics in the matter era. Finally, a logarithmic dependence of the phase shift on the multipoles $\ell$ was found in [2] for observable modes. This could potentially be interpreted as an effect of the finite matter density. By studying the limits $\ell \to \infty$ and $\ell \to 0$ while accounting for the contributions from matter, we aim to further clarify these effects in the following. We will study these two limits one after another:

---

[1] We derive the effects of matter on the phase shift in the acoustic peaks of the CMB. However, it should be clear from our discussion in Chap. 6 that this treatment equally applies to the phase shift in the BAO spectrum.

© Springer Nature Switzerland AG 2019
B. Wallisch, *Cosmological Probes of Light Relics*, Springer Theses,
https://doi.org/10.1007/978-3-030-31098-1

- We start with those modes that entered the horizon during the radiation era and therefore correspond to small angular scales in the CMB anisotropy spectra. Let us first decompose (5.8) as follows:

$$
B(y) = \underbrace{\int_0^{y_{eq}} dy'\, \Phi_+(y') \cos y'}_{\equiv\, B_{rad}} + \underbrace{\int_{y_{eq}}^{y} dy'\, \Phi_+(y') \cos y'}_{\equiv\, B_{mat}}, \tag{B.1}
$$

with $y_{eq} = c_\gamma k \tau_{eq}$ marking the moment of matter-radiation equality. We now consider the two terms separately. Modes that entered the horizon long before $\tau_{eq}$ correspond to $y_{eq} \gg 1$. For these modes, the first term in (B.1) can be approximated as

$$
B_{rad} \simeq \int_0^{\infty} dy'\, \Phi_+^{(rad)}(y') \cos y', \tag{B.2}
$$

which is exactly what we obtain in Sect. 5.2. The second term in (B.1) therefore constitutes the main correction from the matter era,

$$
B_{mat} \simeq \int_{y_{eq}}^{y} dy'\, \Phi_+^{(mat)}(y') \cos y'. \tag{B.3}
$$

In consequence, we can simply repeat the discussion of Sect. 5.2.3 for the matter era in order to estimate the size of this effect. While $\epsilon_X \equiv \bar\rho_X/\bar\rho$ was a constant in the main text, the important difference is that it now scales as $a^{-1} \propto \tau^{-2}$. Since $\epsilon_X = \epsilon_{X,in}\,\tau_{in}^2/\tau^2$ by setting $\tau_{in}$ in the matter era, Eq. (5.50) turns into

$$
\Phi_-(y) = -\frac{8k^2}{y^4}\epsilon_{X,in}\, y_{in}^2\, \sigma_X(y) = -\frac{16}{3}\frac{1}{y^4}\epsilon_{X,in}\, y_{in}^2\, D_{X,2}(y). \tag{B.4}
$$

As before, we only need the quadrupole moment $D_{X,2}(y)$ to zeroth order to compute $\Phi_-(y)$ to first order in $\epsilon_X$. Starting from (5.52), we find

$$
D_{X,2}^{(0)}(y) = d_{X,in}\, j_2\big[c_\gamma^{-1}(y - y_{in})\big]
$$
$$
+ \frac{3}{c_\gamma}\Phi_{+,in} \int_{y_{in}}^{y} dy' \left\{ \frac{2}{5} j_1\big[c_\gamma^{-1}(y - y')\big] - \frac{3}{5} j_3\big[c_\gamma^{-1}(y - y')\big] \right\}, \tag{B.5}
$$

where we have used that $\Phi_+^{(0)} = $ const during the matter era. Taking the appropriate limit of $y \gg 1$, we obtain

$$
\Phi_-^{(1)}(y) \simeq 16\epsilon_{X,in}\, y_{in}^2\, \frac{\sin(c_\gamma^{-1} y)}{c_\gamma^{-1} y^5}\left( \Phi_{+,in} + \frac{1}{3} d_{X,in} \right). \tag{B.6}
$$

As a result, we have $\Phi_- \propto y^{-5} \to 0$ in the limit $y \to \infty$. Consequently, $\Phi_+$ is no longer sourced by $\Phi_-$ at late times and will be given by the homogeneous solution (with coefficients that may depend on $\epsilon_X$). Importantly, the value of $B$ will be the same as that predicted in Sect. 5.2. To conclude, the phase shift of the acoustic oscillations will therefore be equal to the value in a radiation-dominated universe at high wavenumbers $k$ (and thus multipoles $\ell$).

- Now, we turn to the modes that entered the horizon during the matter era and correspond to large angular scales in the CMB. In this case, it is more challenging to cleanly separate the result into a correction to the amplitude of oscillations and a phase shift. This is why we primarily aim to understand the scaling of the result with wavenumber $k$ in the limit $k \to 0$. It is fortunate that this scaling is both easy to understand analytically, and the same for the amplitude correction and the phase shift.

Our intuition is that the contributions from dark radiation (including neutrinos) vanish as $k \to 0$. The reason for this is that the time of horizon entry increases compared to the time of matter-radiation equality as we lower $k$. The radiation energy density should therefore be diluted relative to the matter. The role of this radiation in the evolution of the modes should then become negligible because it only affects observations through its gravitational influence.

By returning to (B.4) and noticing that $y_{in} = c_\gamma k \tau_{in}$, we can confirm this intuition. (Here, the time $\tau_{in}$ is fixed and independent of $k$, e.g. we may choose $\tau_{in}$ to be the time of matter-radiation equality.) We consequently have $\Phi_-^{(1)} = \epsilon_{X,in} c_\gamma^2 k^2 \tau_{in}^2 g(y)$ and conclude that the correction to $d_\gamma$ at linear order in $\epsilon_X$ will be of the following form:

$$d_\gamma^{(1)}(\tau) = \epsilon_{X,in} \, c_\gamma^2 k^2 \tau_{in}^2 \int_{y_{in}}^{y} dy' f(y, y') \,. \tag{B.7}$$

It is obvious that $d_\gamma^{(1)} \propto k^2 \to 0$ if the integral converges as $y \to \infty$. As a matter of fact, due to the scaling of the upper limit of integration ($y = c_\gamma k \tau$ at fixed $\tau$), the result will be suppressed by additional powers of $k$ if the integral diverges as $\tau \to \infty$. We therefore deduce that both the amplitude and phase corrections from neutrinos (or any dark radiation) will approach zero *at least* as fast as $k^2$.

We can draw the following conclusions from these asymptotic scaling arguments:

- For $k \to \infty$, the phase shift due to free-streaming particles approaches a constant.
- For $k \to 0$, the phase and amplitude corrections scale at least as $k^2$.

These results correspond approximately to $\ell \simeq k(\tau_0 - \tau_{rec})$ in the flat-sky limit, where $\tau_0$ is the conformal time today. We therefore expect a constant phase shift at high $\ell$. We do however not expect our asymptotic formula for $k \to 0$ to be more than a rough guide since matter-radiation equality translates to relatively low multipoles. The primary purpose of this discussion was to highlight that a power law is the likely behaviour, simply because of the power-law decay of the energy density of the extra radiation. Consequently, the phase shift per $\ell$ should be some function that interpolates between a power law and a constant. Although the logarithmic dependence of the ansatz of [2] appears to work well enough on intermediate scales, it is unlikely to follow it in detail.

## B.2  Comments on Polarization

In the main text, we phrased the analytic discussions in terms of the temperature anisotropy. As we observed in Sect. 5.3, CMB polarization however plays a very important role in present and especially future data analyses. We show in this section that the phase shift of the polarization spectrum and that of the temperature spectrum are in fact the same.

We first write the Boltzmann equation for the amplitude of polarized anisotropies, $\Theta_P$, according to [3] as

$$\dot{\Theta}_P + ik\mu\Theta_P = -\dot{\kappa}\left[-\Theta_P + \frac{1}{2}(1 - P_2(\mu))\,\Pi\right], \tag{B.8}$$

with $\Pi \equiv \Theta_2 + \Theta_{P,0} + \Theta_{P,2}$. To avoid confusion with the conformal time $\tau$, we denote the optical depth by $\kappa$ and its time derivative is given by $\dot{\kappa} = -n_e\sigma_T a$. The photon anisotropic stress determines the temperature quadrupole, $\Theta_2 \equiv \frac{1}{2}k^2\sigma_\gamma$. The solution of (B.8) can be written in terms of a line-of-sight integral,

$$\Theta_P(\tau_0) = \int_{\tau_{\text{in}}}^{\tau_0} d\tau\, e^{ik\mu(\tau-\tau_0)-\kappa(\tau)}\left(\frac{3}{4}\dot{\kappa}(\tau)\,(\mu^2 - 1)\,\Pi(\tau)\right). \tag{B.9}$$

Since the integral in (B.9) is proportional to the visibility function $-\dot{\kappa}e^{-\kappa}$, it peaks at the surface of last-scattering. We therefore find

$$\Theta_P(\tau_0) \simeq e^{ik\mu(\tau_{\text{rec}}-\tau_0)}\frac{3}{4}(1 - \mu^2)\Pi(\tau_{\text{rec}}), \tag{B.10}$$

where we took the limit of instantaneous recombination, $-\dot{\kappa}e^{-\kappa} \simeq \delta_D(\tau - \tau_{\text{rec}})$. Using the collision term in the Boltzmann equation for temperature and solving for $\Pi$ to leading order in $\dot{\kappa} \gg 1$, we get $\Pi \simeq \frac{5}{2}\Theta_2 \simeq -\frac{10}{9}k\,\dot{\kappa}^{-1}\,\Theta_1$. We finally obtain

$$\Theta_{P,\ell}(\tau_0) \simeq \frac{5}{18}\dot{d}_\gamma(k, \tau_{\text{rec}})\,\dot{\kappa}^{-1}(\tau_{\text{rec}})\left(1 + \frac{\partial^2}{\partial(k\tau_0)^2}\right)j_\ell(k\tau_0), \tag{B.11}$$

where we applied the continuity equation, $\dot{d}_\gamma = -3k\Theta_1$, and performed a multipole expansion. To conclude, we highlight two facts about this result:

- $\Theta_{P,\ell} \propto \dot{d}_\gamma$.—This implies that the locations of the acoustic peaks in the polarization spectrum are affected by the fluctuations in the dark radiation in the same way as in the temperature spectrum because the time derivative will not affect the phase shift from dark radiation.
- $\Theta_{P,\ell} \propto \dot{\kappa}^{-1} \propto n_e^{-1}$.—Since the damping tail in the temperature spectrum scales as $(n_e H)^{-1}$ (cf. Sect. 3.4.1), this importantly breaks the degeneracy between $H$ and $n_e$ (or $Y_p$).

# References

1. D. Baumann, D. Green, J. Meyers, B. Wallisch, Phases of new physics in the CMB. JCAP **01**, 007 (2016). arXiv:1508.06342 [astro-ph.CO]
2. B. Follin, L. Knox, M. Millea, Z. Pan, First detection of the acoustic oscillation phase shift expected from the cosmic neutrino background. Phys. Rev. Lett. **115**, 091301 (2015). arXiv:1503.07863 [astro-ph.CO]
3. M. Zaldarriaga, D. Harari, Analytic approach to the polarization of the cosmic microwave background in flat and open universes. Phys. Rev. D **52**, 3276 (1995). arXiv:astro-ph/9504085 [astro-ph]

# Appendix C
# Details of the Fisher Forecasts

In this appendix, we provide supplemental material to Chap. 6, which is based on [1]. We first describe our CMB Fisher forecasts and present results for a range of experimental configurations (Sect. C.1). We then provide details of our LSS forecasts, define the specifications for the employed galaxy surveys, and report results for a range of data combinations and cosmologies (Sect. C.2). Finally, we show a few of the convergence tests that we performed to establish the stability of our numerical computations (Sect. C.3).

## C.1 Forecasting CMB Constraints

Forecasting the sensitivities of future CMB observations is by now a standard exercise; see e.g. [2–5]. For completeness, this appendix collects the basic elements of our CMB Fisher analysis as well as the specifications of the CMB experiments that were used in our forecasts of Sect. 6.2.

### C.1.1 Fisher Matrix

The Fisher matrix for CMB experiments can be written as

$$F_{ij} = \sum_{X,Y} \sum_{\ell=\ell_{\min}}^{\ell_{\max}} \frac{\partial C_\ell^X}{\partial \theta_i} \left[ \mathbf{C}_\ell^{XY} \right]^{-1} \frac{\partial C_\ell^Y}{\partial \theta_j} . \tag{C.1}$$

The covariance matrix $\mathbf{C}_\ell^{XY}$ for each multipole $\ell$ and $X = ab$, $Y = cd$, with $a, b, c, d = T, E, B$, is defined by

© Springer Nature Switzerland AG 2019
B. Wallisch, *Cosmological Probes of Light Relics*, Springer Theses,
https://doi.org/10.1007/978-3-030-31098-1

$$C_\ell^{abcd} = \frac{1}{(2\ell+1)f_{\text{sky}}} \left[ (C_\ell^{ac} + N_\ell^{ac})(C_\ell^{bd} + N_\ell^{bd}) + (C_\ell^{ad} + N_\ell^{ad})(C_\ell^{bc} + N_\ell^{bc}) \right],$$
$$(\text{C.2})$$

where $C_\ell^X$ are the theoretical CMB power spectra, $N_\ell^X$ are the (Gaussian) noise spectra of a given experiment and $f_{\text{sky}}$ is the effective sky fraction that is used in the cosmological analysis. We employ perfectly delensed power spectra and omit the lensing convergence for simplicity as it is sufficient for our purposes. We however comment on the effects of these assumptions below. The noise power spectra are

$$N_\ell^X = (\Delta X)^2 \exp\left\{ \frac{\ell(\ell+1)\,\theta_b^2}{8\ln 2} \right\}, \tag{C.3}$$

with the map sensitivities for temperature and polarization spectra $\Delta X = \Delta T, \Delta P$, respectively, and the beam width $\theta_b$ (taken to be the full width at half maximum). Note that we set $N_\ell^{TE} \equiv 0$ as we assume the noise in temperature and polarization to be uncorrelated. For a multi-frequency experiment, the noise spectrum (C.3) applies for each frequency channel separately. The effective noise after combining all channels is

$$N_\ell^X = \left[ \sum_\nu \left( N_\ell^{X,\nu} \right)^{-1} \right]^{-1}, \tag{C.4}$$

where $N_\ell^{X,\nu}$ are the noise power spectra for the separate frequency channels $\nu$.

### C.1.2  Experimental Specifications

Our specifications for the Planck satellite are collected in Table C.1. The adopted configuration is the same as that used in the CMB-S4 Science Book [5]. For the low-$\ell$ data, we use the unlensed TT spectrum with $\ell_{\min} = 2$, $\ell_{\max} = 29$ and $f_{\text{sky}} = 0.8$. We do not include low-$\ell$ polarization data, but instead impose a Gaussian prior on the optical depth, with $\sigma(\tau) = 0.01$. For the high-$\ell$ data, we use the unlensed TT, TE, EE spectra with $\ell_{\min} = 30$, $\ell_{\max} = 2500$ and $f_{\text{sky}} = 0.44$. Since the low-$\ell$ and high-$\ell$ modes are independent, we simply add the corresponding Fisher matrices.

**Table C.1**  Specifications for the Planck-like experiment used in [4] and in the CMB-S4 Science Book [5]. The dashes in the first two columns for $\Delta P$ indicate that those frequency channels are not sensitive to polarization

| Frequency (GHz) | 30 | 44 | 70 | 100 | 143 | 217 | 353 |
|---|---|---|---|---|---|---|---|
| $\theta_b$ (arcmin) | 33 | 23 | 14 | 10 | 7 | 5 | 5 |
| $\Delta T$ (μK arcmin) | 145 | 149 | 137 | 65 | 43 | 66 | 200 |
| $\Delta P$ (μK arcmin) | – | – | 450 | 103 | 81 | 134 | 406 |

We parametrize future CMB experiments in terms of a single effective frequency with noise level $\Delta T$, beam width $\theta_b$ and sky fraction $f_{sky}$. We will present constraints as a function of these three parameters. We take $\theta_b = 3\prime$, $\Delta T = 5\,\mu K$ arcmin and $f_{sky} = 0.3$ as the fiducial configuration of a CMB-S3-like experiment. For a representative CMB-S4 mission, we adopt the same configuration as in the CMB-S4 Science Book [5]: $\theta_b = 2\prime$, $\Delta T = 1\,\mu K$ arcmin and $f_{sky} = 0.4$. For both experiments, we use unlensed temperature and polarization spectra with $\ell_{min} = 30$, $\ell_{max}^T = 3000$, $\ell_{max}^P = 5000$. We add the low-$\ell$ Planck data as described above, include high-$\ell$ Planck data with $f_{sky} = 0.3$ and $f_{sky} = 0.2$ for CMB-S3 and CMB-S4, respectively, and impose the same Gaussian prior on the optical depth $\tau$ as for Planck. In addition, we forecast a cosmic variance-limited experiment with $\ell_{min} = 2$, $\ell_{max}^T = 3000$, $\ell_{max}^P = 5000$ and $f_{sky} = 0.75$ to estimate the potential reach.

Unlike the CMB-S4 Science Book, we do not include delensing of the T- and E-modes. For $N_{eff}$-forecasts, this was shown to have a negligible impact [6], while using unlensed spectra overestimates the constraining power of the CMB by roughly 30% for $N_{eff}+Y_p$. We are primarily interested in the improvement in parameters from adding LSS data, which should be robust to these relatively small differences. We also ignore the lensing convergence as it basically does not impact the constraints on these parameters.

## C.1.3 Future Constraints

As a point of reference and for comparison with the results of Sect. 5.3, we present constraints derived from CMB observations alone. In Table C.2, we show the $1\sigma$ constraints for Planck, the described representative configurations of CMB-S3 and CMB-S4, and the mentioned CVL experiment. In Table C.3, we display how these constraints vary when we allow the helium fraction $Y_p$ to be an additional free parameter. The differences in the forecasted sensitivities for Planck compared to the constraints published in [7] can be attributed entirely to the improvement in $\sigma(\tau)$ which arises from the imposed prior on the optical depth $\tau$. The forecast of $N_{eff}$ for CMB-S3 is a rough estimate and will be subject to the precise specifications of the respective experiment. While the precise design of CMB-S4 is also undetermined at this point, $\sigma(N_{eff}) = 0.03$ is a primary science target and is therefore more likely to be a reliable estimate of the expected performance. For a CVL experiment, some improvement is expected when including the lensing convergence, with constraints possibly reaching $\sigma(N_{eff}) \lesssim 0.008$.

In Fig. C.1, we demonstrate how the constraints on $N_{eff}$ depend on the sky fraction $f_{sky}$, for three different values of $\theta_b$ and fixed noise level $\Delta T = 1\,\mu K$ arcmin. When varying the total sky fraction, we also appropriately change the contribution of the included high-$\ell$ Planck data. In Fig. C.2, we illustrate the constraint on $N_{eff}$ as a function of the beam size $\theta_b$ and the noise level $\Delta T$, for fixed sky

**Table C.2** Forecasted sensitivities of Planck, CMB-S3, CMB-S4 and a CVL experiment for the parameters of $\Lambda$CDM and $\Lambda$CDM+$N_{\mathrm{eff}}$

| Parameter | Planck | CMB-S3 | CMB-S4 | CVL | Planck | CMB-S3 | CMB-S4 | CVL |
|---|---|---|---|---|---|---|---|---|
| $10^5\,\omega_b$ | 16 | 5.1 | 2.7 | 0.97 | 26 | 8.3 | 3.8 | 1.3 |
| $10^4\,\omega_c$ | 16 | 8.3 | 7.1 | 4.5 | 26 | 10 | 7.9 | 4.5 |
| $10^7\,\theta_s$ | 29 | 9.4 | 5.9 | 3.5 | 44 | 13 | 6.7 | 3.5 |
| $\ln(10^{10}A_s)$ | 0.020 | 0.020 | 0.020 | 0.0041 | 0.021 | 0.020 | 0.020 | 0.0041 |
| $n_{\mathrm{S}}$ | 0.0040 | 0.0023 | 0.0020 | 0.0012 | 0.0093 | 0.0040 | 0.0030 | 0.0020 |
| $\tau$ | 0.010 | 0.010 | 0.010 | 0.0020 | 0.010 | 0.010 | 0.010 | 0.0020 |
| $N_{\mathrm{eff}}$ | – | – | – | – | 0.18 | 0.054 | 0.030 | 0.011 |

**Table C.3** Forecasted sensitivities of Planck, CMB-S3, CMB-S4 and a CVL experiment for the parameters of $\Lambda$CDM+$Y_p$ and $\Lambda$CDM+$N_{\mathrm{eff}}$+$Y_p$

| Parameter | Planck | CMB-S3 | CMB-S4 | CVL | Planck | CMB-S3 | CMB-S4 | CVL |
|---|---|---|---|---|---|---|---|---|
| $10^5\,\omega_b$ | 24 | 8.2 | 3.8 | 1.4 | 26 | 8.4 | 3.8 | 1.4 |
| $10^4\,\omega_c$ | 17 | 8.6 | 7.2 | 5.0 | 49 | 21 | 14 | 8.4 |
| $10^7\,\theta_s$ | 33 | 9.9 | 6.3 | 3.8 | 89 | 27 | 15 | 6.9 |
| $\ln(10^{10}A_s)$ | 0.020 | 0.020 | 0.020 | 0.0041 | 0.022 | 0.020 | 0.020 | 0.0042 |
| $n_{\mathrm{S}}$ | 0.0082 | 0.0038 | 0.0029 | 0.0019 | 0.0093 | 0.0040 | 0.0030 | 0.0020 |
| $\tau$ | 0.010 | 0.010 | 0.010 | 0.0020 | 0.010 | 0.010 | 0.010 | 0.0020 |
| $N_{\mathrm{eff}}$ | – | – | – | – | 0.32 | 0.12 | 0.081 | 0.045 |
| $Y_p$ | 0.012 | 0.0037 | 0.0021 | 0.0008 | 0.018 | 0.0069 | 0.0047 | 0.0026 |

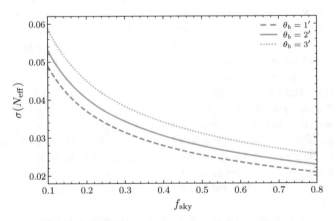

**Fig. C.1** Marginalized constraints on $N_{\mathrm{eff}}$ as a function of the sky fraction $f_{\mathrm{sky}}$ for three values of the beam width $\theta_b$ and fixed noise level $\Delta T = 1\,\mu\mathrm{K}$ arcmin (from [1])

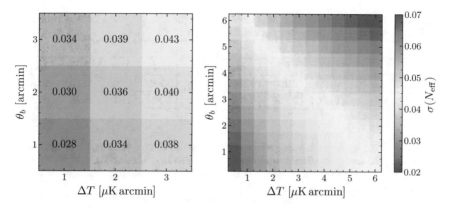

**Fig. C.2** Marginalized constraints on $N_{\mathrm{eff}}$ as a function of the beam size $\theta_b$ and the temperature noise level $\Delta T$, for fixed sky fraction $f_{\mathrm{sky}} = 0.4$ (from [1])

fraction $f_{\mathrm{sky}} = 0.4$. Comparing Fig. C.2 to the equivalent figure in the CMB-S4 Science Book [5] (Fig. 22), we see that the difference between the two forecasts is $\Delta\sigma(N_{\mathrm{eff}}) \approx 0.002$. This can be attributed to the effects of imperfect delensing and is completely negligible for our purposes.

## C.2    Forecasting LSS Constraints

In this appendix, we collect the specific information regarding the planned LSS surveys which we used in our Fisher and likelihood forecasts. We also provide the full set of constraints on all of the cosmological parameters and cosmologies that are studied in Chap. 6.

### C.2.1    Survey Specifications

Below, we provide the experimental specifications for the galaxy surveys used in our forecasts. We have slightly simplified the details compared to [8], for example. In particular, we group different types of tracers (e.g. luminous red galaxies, emission line galaxies or quasars) into a single effective number density and bias. We find our results to be fairly insensitive to many of these details and well approximated by a fixed number of objects distributed with a constant comoving number density over the same redshift range.

The employed parametrization of the spectroscopic redshift surveys BOSS, eBOSS, DESI and Euclid are provided in Tables C.4, C.5, C.6 and C.7. For eBOSS, we combine BOSS and the two eBOSS configurations of Table C.5 into one survey

**Table C.4** Basic specifications for BOSS derived from [8] with a sky area of $\Omega = 10\,000 \deg^2$ resulting in roughly $1.4 \times 10^6$ objects in a volume of about $7.5\,h^{-3}\mathrm{Gpc}^3$

| $\bar{z}$ | 0.05 | 0.15 | 0.25 | 0.35 | 0.45 | 0.55 | 0.65 | 0.75 |
|---|---|---|---|---|---|---|---|---|
| $b$ | 1.79 | 1.90 | 1.98 | 2.09 | 2.32 | 2.26 | 2.38 | 3.09 |
| $10^3\,\bar{n}_g\ (h^3\mathrm{Mpc}^{-3})$ | 0.289 | 0.290 | 0.300 | 0.304 | 0.276 | 0.323 | 0.120 | 0.0100 |
| $V\ (h^{-3}\,\mathrm{Gpc}^3)$ | 0.0255 | 0.164 | 0.402 | 0.704 | 1.04 | 1.38 | 1.72 | 2.04 |

**Table C.5** Basic specifications for eBOSS derived from [8]. The redshift range is covered twice, first showing the survey covering $\Omega = 1500 \deg^2$ that will include emission line galaxies (resulting in roughly $3.8 \times 10^5$ objects in a volume of about $8.0\,h^{-3}\mathrm{Gpc}^3$), and then the survey with $\Omega = 6000 \deg^2$ that will not (resulting in roughly $7.2 \times 10^5$ objects in a volume of about $32\,h^{-3}\mathrm{Gpc}^3$)

| $\bar{z}$ | 0.55 | 0.65 | 0.75 | 0.85 | 0.95 | 1.05 | 1.15 | 1.25 |
|---|---|---|---|---|---|---|---|---|
| $b$ | 3.07 | 2.07 | 1.57 | 1.57 | 1.61 | 3.51 | 1.98 | 2.35 |
| $10^5\,\bar{n}_g\ (h^3\,\mathrm{Mpc}^{-3})$ | 0.463 | 21.3 | 35.5 | 23.6 | 5.40 | 0.563 | 1.53 | 1.48 |
| $V\ (h^{-3}\mathrm{Gpc}^3)$ | 0.208 | 0.258 | 0.307 | 0.352 | 0.392 | 0.429 | 0.461 | 0.489 |
| $b$ | 3.07 | 2.42 | 2.45 | 2.56 | 7.84 | 3.51 | 1.98 | 2.35 |
| $10^5\,\bar{n}_g\ (h^3\,\mathrm{Mpc}^{-3})$ | 0.463 | 13.5 | 7.02 | 3.35 | 0.0412 | 0.563 | 1.53 | 1.48 |
| $V\ (h^{-3}\mathrm{Gpc}^3)$ | 0.830 | 1.03 | 1.23 | 1.41 | 1.57 | 1.71 | 1.84 | 1.96 |

| $\bar{z}$ | 1.35 | 1.45 | 1.55 | 1.65 | 1.75 | 1.85 | 1.95 | 2.05 | 2.15 |
|---|---|---|---|---|---|---|---|---|---|
| $b$ | 3.65 | 2.40 | 2.42 | 2.08 | 2.10 | 3.33 | 3.35 | 1.72 | 1.73 |
| $10^5\,\bar{n}_g\ (h^3\,\mathrm{Mpc}^{-3})$ | 0.664 | 1.66 | 1.76 | 2.03 | 2.15 | 0.912 | 0.965 | 2.91 | 3.07 |
| $V\ (h^{-3}\mathrm{Gpc}^3)$ | 0.513 | 0.533 | 0.551 | 0.565 | 0.577 | 0.587 | 0.594 | 0.600 | 0.604 |
| $b$ | 3.65 | 2.40 | 2.42 | 2.08 | 2.10 | 3.33 | 3.35 | 1.72 | 1.73 |
| $10^5\,\bar{n}_g\ (h^3\,\mathrm{Mpc}^{-3})$ | 0.664 | 1.66 | 1.76 | 2.03 | 2.15 | 0.912 | 0.965 | 2.91 | 3.07 |
| $V\ (h^{-3}\mathrm{Gpc}^3)$ | 2.05 | 2.13 | 2.20 | 2.26 | 2.31 | 2.35 | 2.38 | 2.40 | 2.42 |

neglecting the small overlap. We effectively treat each redshift bin with mean redshift $\bar{z}$ as an independent three-dimensional survey. Our Fisher matrix is the sum of the Fisher matrices associated with each bin, $F = \sum_{\bar{z}} F_{\bar{z}}$. We translated the survey specifications used in [8] into three numbers per redshift bin: the linear galaxy bias $b$, the comoving number density of galaxies $\bar{n}_g$ and the bin volume $V$. This is sufficient to fully specify the Fisher matrix in each bin. The spherical bin volume is given by

$$V = \frac{4\pi}{3} f_{\mathrm{sky}} \left[ d_c(z_{\max})^3 - d_c(z_{\min})^3 \right], \qquad d_c(z) = \int_0^z dz\, \frac{c}{H(z)}, \qquad (\mathrm{C.5})$$

where $f_{\mathrm{sky}}$ is the sky fraction, $d_c(z)$ is the comoving distance to redshift $z$, and $z_{\min} = \bar{z} - \Delta z/2$ and $z_{\max} = \bar{z} + \Delta z/2$ are the minimum and maximum redshift of the respective bin. We use redshift bins of width $\Delta z = 0.1$ throughout.

**Table C.6** Basic specifications for DESI derived from [8], covering a sky area $\Omega = 14\,000\,\text{deg}^2$ and resulting in roughly $2.3 \times 10^7$ objects in a volume of about $61\,h^{-3}\text{Gpc}^3$

| $\bar{z}$ | 0.15 | 0.25 | 0.35 | 0.45 | 0.55 | 0.65 | 0.75 | 0.85 | 0.95 |
|---|---|---|---|---|---|---|---|---|---|
| $b$ | 1.13 | 1.39 | 1.64 | 1.81 | 1.87 | 1.89 | 1.90 | 1.82 | 1.53 |
| $10^3\,\bar{n}_g\ (h^3\,\text{Mpc}^{-3})$ | 2.38 | 1.07 | 0.684 | 0.568 | 0.600 | 0.696 | 0.810 | 0.719 | 0.558 |
| $V\ (h^{-3}\text{Gpc}^3)$ | 0.229 | 0.563 | 0.985 | 1.45 | 1.94 | 2.41 | 2.86 | 3.28 | 3.66 |

| $\bar{z}$ | 1.05 | 1.15 | 1.25 | 1.35 | 1.45 | 1.55 | 1.65 | 1.75 | 1.85 |
|---|---|---|---|---|---|---|---|---|---|
| $b$ | 1.47 | 1.49 | 1.58 | 1.62 | 1.73 | 2.01 | 1.98 | 2.56 | 4.17 |
| $10^3\,\bar{n}_g\ (h^3\,\text{Mpc}^{-3})$ | 0.522 | 0.506 | 0.454 | 0.356 | 0.242 | 0.127 | 0.0736 | 0.0289 | 0.00875 |
| $V\ (h^{-3}\text{Gpc}^3)$ | 4.00 | 4.30 | 4.56 | 4.79 | 4.98 | 5.14 | 5.28 | 5.39 | 5.48 |

**Table C.7** Basic specifications for Euclid derived from [8], covering a sky area $\Omega = 15\,000\,\text{deg}^2$ and resulting in roughly $5.0 \times 10^7$ objects in a volume of about $72\,h^{-3}\text{Gpc}^3$

| $\bar{z}$ | 0.65 | 0.75 | 0.85 | 0.95 | 1.05 | 1.15 | 1.25 | 1.35 |
|---|---|---|---|---|---|---|---|---|
| $b$ | 1.06 | 1.11 | 1.16 | 1.21 | 1.27 | 1.33 | 1.38 | 1.44 |
| $10^3\,\bar{n}_g\ (h^3\,\text{Mpc}^{-3})$ | 0.637 | 1.46 | 1.63 | 1.50 | 1.33 | 1.14 | 1.00 | 0.837 |
| $V\ (h^{-3}\text{Gpc}^3)$ | 2.58 | 3.07 | 3.52 | 3.92 | 4.29 | 4.61 | 4.89 | 5.13 |

| $\bar{z}$ | 1.45 | 1.55 | 1.65 | 1.75 | 1.85 | 1.95 | 2.05 | |
|---|---|---|---|---|---|---|---|---|
| $b$ | 1.51 | 1.54 | 1.63 | 1.70 | 1.85 | 1.90 | 1.26 | |
| $10^3\,\bar{n}_g\ (h^3\,\text{Mpc}^{-3})$ | 0.652 | 0.512 | 0.357 | 0.246 | 0.149 | 0.0904 | 0.0721 | |
| $V\ (h^{-3}\text{Gpc}^3)$ | 5.33 | 5.51 | 5.65 | 5.77 | 5.87 | 5.94 | 6.00 | |

For the photometric surveys DES and LSST, we follow [8] and define the surveys by using $(\alpha, \beta, z_*, N_{tot}, b_0) = (1.25, 2.29, 0.88, 12\,\text{arcmin}^{-2}, 0.95)$ and $(2.0, 1.0, 0.3, 50\,\text{arcmin}^{-2}, 0.95)$, respectively. These parameters are related to those employed in our forecasts as follows:

$$\bar{n}_g(\bar{z}) = \frac{N_{tot}}{V}\,\frac{\beta/z_*}{\Gamma\left[(\alpha+1)/\beta\right]}\int_{z_{min}}^{z_{max}} dz\ (z/z_*)^\alpha \exp\left\{-(z/z_*)^\beta\right\}, \qquad (C.6)$$

$$b(\bar{z}) = \frac{D_1(0)}{D_1(\bar{z}_i)}\,b_0, \qquad (C.7)$$

with gamma function $\Gamma$ and linear growth function $D_1(z)$.

For DES, we employ a survey area of $\Omega = 5000\,\text{deg}^2$ and a redshift coverage of $0.1 \leq z \leq 2.0$, while we take $20000\,\text{deg}^2$ and $0.1 \leq z \leq 3.5$ for LSST. This results in approximately $1.4 \times 10^8$ and $5.9 \times 10^8$ objects in a total survey volume of about $24\,h^{-3}\text{Gpc}^3$ and $215\,h^{-3}\text{Gpc}^3$ for the two surveys, respectively. We neglect the spectroscopic redshift error as it is expected to be comparable to (or smaller than) the longitudinal damping scale $\Sigma_\|$, but use a conservative root-mean-square photometric redshift error of $\sigma_{z0} = 0.05$ for both DES and LSST. Finally, we reiterate that, by considering galaxy clustering alone, we only take a subset of the cosmological observables into account, in particular for photometric surveys, and we therefore expect to underestimate the full power of these experiments.

## C.2.2  Future Constraints

Using these specifications, we generated forecasts for all of the cosmological parameters discussed in the main text in combination with the Fisher matrices for Planck, CMB-S3 and CMB-S4. We include both $P(k)$- and BAO-forecasts for $\Lambda$CDM (Table C.8), $\Lambda$CDM+$N_{\mathrm{eff}}$ (Table C.9), $\Lambda$CDM+$Y_p$ (Table C.10) and $\Lambda$CDM+$N_{\mathrm{eff}}$+$Y_p$ (Table C.11). As in Sect. 6.2.2, the $P(k)$-forecasts use wavenumbers up to $k_{\mathrm{max}} = 0.2\,h\,\mathrm{Mpc}^{-1}$ and marginalize over the $b_{m\leq1}$-terms of (6.17). For the BAO-forecasts, we set $k_{\mathrm{max}} = 0.5\,h\,\mathrm{Mpc}^{-1}$ and marginalize over $a_{n\leq4}$ and $b_{m\leq3}$ in each redshift bin. As we marginalize over galaxy bias, our forecasts show no improvements beyond the CMB for $\ln(10^{10}A_{\mathrm{s}})$ and $\tau$. We therefore do not include these two parameters in the following tables.

Apart from the improvements in the constraints on $N_{\mathrm{eff}}$ and $Y_p$, which we already discussed in Sect. 6.2.2, we see that mainly $\omega_b$ and $\omega_c$ benefit from combining the considered LSS surveys with CMB experiments. The sensitivities may be enhanced by factors of three (two) and more compared to Planck (CMB-S3). We note that the DESI specifications of Table C.6 are slightly more optimistic overall than what was considered in [9] resulting in roughly the same BAO-forecasts and up to about 5% better $P(k)$-forecasts.

Comparing our forecasts with the ones obtained from the BAO scale alone (combined with Planck), we see that the BOSS analysis for $\Lambda$CDM is nearly optimal, but improvements on the constraints of more than 10% can be achieved in extended cosmologies. For instance, the constraints on $\omega_b$, $n_{\mathrm{s}}$ and $N_{\mathrm{eff}}$ improve by 3% or more, and $\omega_c$ in $\Lambda$CDM+$Y_p$ even by 12%. For DESI, the obtained sensitivities can generally be increased by a larger amount, e.g. up to 15% for $\omega_b$ and $n_{\mathrm{s}}$ in $\Lambda$CDM+$N_{\mathrm{eff}}$, and for $\omega_c$ in $\Lambda$CDM+$Y_p$.

## C.3  Convergence and Stability Tests

One of the motivations for including our full list of forecasts in Appendix C.2 is to make the results reproducible. It is therefore also important that we explain how the numerical derivatives were computed in the Fisher matrix, including the employed step sizes. In this appendix, we provide this information and demonstrate that the step sizes are appropriate for the convergence and stability of our calculations.

The numerical derivatives in (6.4) and (6.21) are computed using a symmetric difference quotient or two-point stencil, $f'(\theta) = [f(\theta + h) - f(\theta - h)]/(2h)$, with

**Table C.8** Full set of forecasted $1\sigma$ constraints in a $\Lambda$CDM cosmology for current and future LSS surveys in combination with the CMB experiments Planck, CMB-S3 and CMB-S4. We do not quote the sensitivities to $\ln(10^{10}A_s)$ and $\tau$ as they are the same as in Table C.2 for all combinations

(a) Planck + $P(k)$

| Parameter | Planck | BOSS | eBOSS | DESI | Euclid | DES | LSST |
|---|---|---|---|---|---|---|---|
| $10^5\,\omega_b$ | 16 | 13 | 13 | 12 | 11 | 14 | 12 |
| $10^4\,\omega_c$ | 16 | 8.9 | 7.7 | 4.6 | 4.3 | 13 | 8.2 |
| $10^7\,\theta_s$ | 29 | 28 | 27 | 27 | 27 | 29 | 28 |
| $n_s$ | 0.0040 | 0.0033 | 0.0032 | 0.0028 | 0.0027 | 0.0037 | 0.0033 |

(b) CMB-S3 + $P(k)$

| Parameter | CMB-S3 | BOSS | eBOSS | DESI | Euclid | DES | LSST |
|---|---|---|---|---|---|---|---|
| $10^5\,\omega_b$ | 5.1 | 4.9 | 4.9 | 4.7 | 4.6 | 5.0 | 4.8 |
| $10^4\,\omega_c$ | 8.3 | 6.7 | 6.1 | 4.0 | 3.7 | 7.8 | 6.3 |
| $10^7\,\theta_s$ | 9.4 | 9.1 | 9.0 | 8.7 | 8.6 | 9.3 | 9.1 |
| $n_s$ | 0.0023 | 0.0021 | 0.0021 | 0.0019 | 0.0019 | 0.0022 | 0.0021 |

(c) S4 + $P(k)$

| Parameter | CMB-S4 | BOSS | eBOSS | DESI | Euclid | DES | LSST |
|---|---|---|---|---|---|---|---|
| $10^5\,\omega_b$ | 2.7 | 2.7 | 2.7 | 2.6 | 2.6 | 2.7 | 2.6 |
| $10^4\,\omega_c$ | 7.1 | 6.0 | 5.6 | 3.9 | 3.6 | 6.8 | 5.8 |
| $10^7\,\theta_s$ | 5.9 | 5.7 | 5.6 | 5.3 | 5.2 | 5.9 | 5.7 |
| $n_s$ | 0.0020 | 0.0018 | 0.0018 | 0.0016 | 0.0016 | 0.0019 | 0.0018 |

(d) Planck + BAO

| Parameter | Planck | BOSS | eBOSS | DESI | Euclid | DES | LSST |
|---|---|---|---|---|---|---|---|
| $10^5\,\omega_b$ | 16 | 13 | 13 | 13 | 13 | 15 | 14 |
| $10^4\,\omega_c$ | 16 | 8.7 | 8.0 | 5.1 | 5.5 | 13 | 9.4 |
| $10^7\,\theta_s$ | 29 | 27 | 27 | 27 | 26 | 29 | 27 |
| $n_s$ | 0.0040 | 0.0031 | 0.0031 | 0.0028 | 0.0028 | 0.0037 | 0.0032 |

(e) CMB-S3 + BAO

| Parameter | CMB-S3 | BOSS | eBOSS | DESI | Euclid | DES | LSST |
|---|---|---|---|---|---|---|---|
| $10^5\,\omega_b$ | 5.1 | 5.0 | 5.0 | 4.9 | 4.9 | 5.1 | 5.0 |
| $10^4\,\omega_c$ | 8.3 | 6.5 | 6.2 | 4.4 | 4.6 | 7.9 | 6.8 |
| $10^7\,\theta_s$ | 9.4 | 9.0 | 8.9 | 8.6 | 8.6 | 9.3 | 9.0 |
| $n_s$ | 0.0023 | 0.0021 | 0.0020 | 0.0019 | 0.0019 | 0.0022 | 0.0021 |

(f) S4 + BAO

| Parameter | CMB-S4 | BOSS | eBOSS | DESI | Euclid | DES | LSST |
|---|---|---|---|---|---|---|---|
| $10^5\,\omega_b$ | 2.7 | 2.7 | 2.7 | 2.7 | 2.7 | 2.7 | 2.7 |
| $10^4\,\omega_c$ | 7.1 | 5.9 | 5.7 | 4.2 | 4.3 | 6.8 | 6.1 |
| $10^7\,\theta_s$ | 5.9 | 5.6 | 5.6 | 5.2 | 5.2 | 5.9 | 5.7 |
| $n_s$ | 0.0020 | 0.0018 | 0.0018 | 0.0016 | 0.0016 | 0.0019 | 0.0018 |

**Table C.9** As in Table C.8, but for an extended $\Lambda$CDM+$N_{\text{eff}}$ cosmology

(a) Planck + $P(k)$

| Parameter | Planck | BOSS | eBOSS | DESI | Euclid | DES | LSST |
|---|---|---|---|---|---|---|---|
| $10^5\,\omega_b$ | 26 | 19 | 18 | 15 | 15 | 24 | 20 |
| $10^4\,\omega_c$ | 26 | 23 | 21 | 15 | 13 | 25 | 19 |
| $10^7\,\theta_s$ | 44 | 41 | 40 | 35 | 34 | 43 | 39 |
| $n_s$ | 0.0093 | 0.0068 | 0.0061 | 0.0039 | 0.0035 | 0.0085 | 0.0069 |
| $N_{\text{eff}}$ | 0.18 | 0.14 | 0.13 | 0.087 | 0.079 | 0.17 | 0.14 |

(b) CMB-S3 + $P(k)$

| Parameter | CMB-S3 | BOSS | eBOSS | DESI | Euclid | DES | LSST |
|---|---|---|---|---|---|---|---|
| $10^5\,\omega_b$ | 8.3 | 7.9 | 7.8 | 7.3 | 7.1 | 8.2 | 8.0 |
| $10^4\,\omega_c$ | 10 | 9.6 | 9.2 | 7.8 | 7.5 | 10 | 8.8 |
| $10^7\,\theta_s$ | 13 | 12 | 12 | 12 | 12 | 12 | 12 |
| $n_s$ | 0.0040 | 0.0037 | 0.0036 | 0.0029 | 0.0028 | 0.0039 | 0.0037 |
| $N_{\text{eff}}$ | 0.054 | 0.052 | 0.051 | 0.045 | 0.043 | 0.054 | 0.052 |

(c) S4 + $P(k)$

| Parameter | CMB-S4 | BOSS | eBOSS | DESI | Euclid | DES | LSST |
|---|---|---|---|---|---|---|---|
| $10^5\,\omega_b$ | 3.8 | 3.7 | 3.7 | 3.6 | 3.6 | 3.8 | 3.7 |
| $10^4\,\omega_c$ | 7.9 | 7.1 | 6.8 | 5.5 | 5.3 | 7.6 | 6.7 |
| $10^7\,\theta_s$ | 6.7 | 6.6 | 6.5 | 6.2 | 6.2 | 6.7 | 6.5 |
| $n_s$ | 0.0030 | 0.0029 | 0.0028 | 0.0025 | 0.0024 | 0.0030 | 0.0029 |
| $N_{\text{eff}}$ | 0.030 | 0.030 | 0.030 | 0.028 | 0.027 | 0.030 | 0.030 |

(d) Planck + BAO

| Parameter | Planck | BOSS | eBOSS | DESI | Euclid | DES | LSST |
|---|---|---|---|---|---|---|---|
| $10^5\,\omega_b$ | 26 | 18 | 18 | 17 | 17 | 22 | 19 |
| $10^4\,\omega_c$ | 26 | 26 | 26 | 26 | 26 | 26 | 26 |
| $10^7\,\theta_s$ | 44 | 43 | 43 | 43 | 43 | 44 | 44 |
| $n_s$ | 0.0093 | 0.0065 | 0.0063 | 0.0059 | 0.0059 | 0.0081 | 0.0067 |
| $N_{\text{eff}}$ | 0.18 | 0.15 | 0.15 | 0.14 | 0.14 | 0.16 | 0.15 |

(e) CMB-S3 + BAO

| Parameter | CMB-S3 | BOSS | eBOSS | DESI | Euclid | DES | LSST |
|---|---|---|---|---|---|---|---|
| $10^5\,\omega_b$ | 8.3 | 7.8 | 7.7 | 7.4 | 7.4 | 8.2 | 7.8 |
| $10^4\,\omega_c$ | 10 | 10 | 9.9 | 9.6 | 9.6 | 10 | 10 |
| $10^7\,\theta_s$ | 13 | 13 | 13 | 13 | 13 | 13 | 13 |
| $n_s$ | 0.0040 | 0.0035 | 0.0035 | 0.0031 | 0.0032 | 0.0039 | 0.0036 |
| $N_{\text{eff}}$ | 0.054 | 0.052 | 0.052 | 0.050 | 0.050 | 0.054 | 0.052 |

(f) S4 + BAO

| Parameter | CMB-S4 | BOSS | eBOSS | DESI | Euclid | DES | LSST |
|---|---|---|---|---|---|---|---|
| $10^5\,\omega_b$ | 3.8 | 3.7 | 3.7 | 3.7 | 3.7 | 3.8 | 3.7 |
| $10^4\,\omega_c$ | 7.9 | 7.2 | 7.1 | 6.5 | 6.5 | 7.8 | 7.3 |
| $10^7\,\theta_s$ | 6.7 | 6.6 | 6.6 | 6.5 | 6.5 | 6.7 | 6.6 |
| $n_s$ | 0.0030 | 0.0028 | 0.0028 | 0.0025 | 0.0025 | 0.0030 | 0.0028 |
| $N_{\text{eff}}$ | 0.030 | 0.030 | 0.030 | 0.029 | 0.029 | 0.030 | 0.030 |

**Table C.10** As in Table C.8, but for an extended $\Lambda$CDM+$Y_p$ cosmology. The constraints on $\ln(10^{10}A_s)$ and $\tau$ are the same as in Table C.3 for all combinations

(a) Planck + $P(k)$

| Parameter | Planck | BOSS | eBOSS | DESI | Euclid | DES | LSST |
|---|---|---|---|---|---|---|---|
| $10^5\,\omega_b$ | 24 | 19 | 19 | 17 | 16 | 22 | 20 |
| $10^4\,\omega_c$ | 17 | 8.9 | 7.8 | 4.7 | 4.3 | 13 | 8.6 |
| $10^7\,\theta_s$ | 33 | 30 | 29 | 27 | 27 | 32 | 30 |
| $n_s$ | 0.0082 | 0.0066 | 0.0063 | 0.0048 | 0.0044 | 0.0077 | 0.0068 |
| $Y_p$ | 0.012 | 0.011 | 0.0100 | 0.0087 | 0.0082 | 0.011 | 0.011 |

(b) CMB-S3 + $P(k)$

| Parameter | CMB-S3 | BOSS | eBOSS | DESI | Euclid | DES | LSST |
|---|---|---|---|---|---|---|---|
| $10^5\,\omega_b$ | 8.2 | 7.9 | 7.8 | 7.5 | 7.4 | 8.1 | 7.9 |
| $10^4\,\omega_c$ | 8.6 | 6.8 | 6.3 | 4.0 | 3.7 | 8.1 | 6.6 |
| $10^7\,\theta_s$ | 9.9 | 9.5 | 9.4 | 8.9 | 8.8 | 9.8 | 9.5 |
| $n_s$ | 0.0038 | 0.0035 | 0.0034 | 0.0030 | 0.0029 | 0.0037 | 0.0036 |
| $Y_p$ | 0.0037 | 0.0036 | 0.0036 | 0.0034 | 0.0033 | 0.0037 | 0.0036 |

(c) S4 + $P(k)$

| Parameter | CMB-S4 | BOSS | eBOSS | DESI | Euclid | DES | LSST |
|---|---|---|---|---|---|---|---|
| $10^5\,\omega_b$ | 3.8 | 3.8 | 3.8 | 3.7 | 3.7 | 3.8 | 3.8 |
| $10^4\,\omega_c$ | 7.2 | 6.1 | 5.7 | 3.9 | 3.6 | 6.9 | 5.9 |
| $10^7\,\theta_s$ | 6.3 | 6.0 | 5.9 | 5.5 | 5.4 | 6.2 | 6.0 |
| $n_s$ | 0.0029 | 0.0028 | 0.0028 | 0.0025 | 0.0024 | 0.0029 | 0.0028 |
| $Y_p$ | 0.0021 | 0.0021 | 0.0021 | 0.0020 | 0.0020 | 0.0021 | 0.0021 |

(d) Planck + BAO

| Parameter | Planck | BOSS | eBOSS | DESI | Euclid | DES | LSST |
|---|---|---|---|---|---|---|---|
| $10^5\,\omega_b$ | 24 | 19 | 19 | 18 | 18 | 22 | 19 |
| $10^4\,\omega_c$ | 17 | 8.7 | 8.0 | 5.5 | 5.7 | 14 | 9.4 |
| $10^7\,\theta_s$ | 33 | 29 | 29 | 28 | 28 | 31 | 29 |
| $n_s$ | 0.0082 | 0.0063 | 0.0062 | 0.0059 | 0.0059 | 0.0075 | 0.0065 |
| $Y_p$ | 0.012 | 0.011 | 0.011 | 0.0100 | 0.011 | 0.011 | 0.011 |

(e) CMB-S3 + BAO

| Parameter | CMB-S3 | BOSS | eBOSS | DESI | Euclid | DES | LSST |
|---|---|---|---|---|---|---|---|
| $10^5\,\omega_b$ | 8.2 | 7.8 | 7.8 | 7.6 | 7.6 | 8.1 | 7.9 |
| $10^4\,\omega_c$ | 8.6 | 6.6 | 6.3 | 4.4 | 4.6 | 8.2 | 6.9 |
| $10^7\,\theta_s$ | 9.9 | 9.3 | 9.3 | 8.8 | 8.9 | 9.8 | 9.4 |
| $n_s$ | 0.0038 | 0.0034 | 0.0034 | 0.0031 | 0.0031 | 0.0037 | 0.0035 |
| $Y_p$ | 0.0037 | 0.0036 | 0.0036 | 0.0035 | 0.0035 | 0.0037 | 0.0036 |

(f) S4 + BAO

| Parameter | CMB-S4 | BOSS | eBOSS | DESI | Euclid | DES | LSST |
|---|---|---|---|---|---|---|---|
| $10^5\,\omega_b$ | 3.8 | 3.8 | 3.8 | 3.8 | 3.8 | 3.8 | 3.8 |
| $10^4\,\omega_c$ | 7.2 | 5.9 | 5.7 | 4.2 | 4.3 | 6.9 | 6.1 |
| $10^7\,\theta_s$ | 6.3 | 5.9 | 5.8 | 5.4 | 5.5 | 6.2 | 5.9 |
| $n_s$ | 0.0029 | 0.0027 | 0.0027 | 0.0025 | 0.0025 | 0.0029 | 0.0028 |
| $Y_p$ | 0.0021 | 0.0021 | 0.0021 | 0.0021 | 0.0021 | 0.0021 | 0.0021 |

**Table C.11** As in Table C.10, but for an extended $\Lambda$CDM+$N_{\mathrm{eff}}$+$Y_p$ cosmology

(a) Planck + $P(k)$

| Parameter | Planck | BOSS | eBOSS | DESI | Euclid | DES | LSST |
|---|---|---|---|---|---|---|---|
| $10^5\,\omega_b$ | 26 | 20 | 19 | 17 | 16 | 24 | 21 |
| $10^4\,\omega_c$ | 49 | 40 | 35 | 23 | 21 | 45 | 34 |
| $10^7\,\theta_s$ | 89 | 76 | 70 | 53 | 50 | 84 | 69 |
| $n_s$ | 0.0093 | 0.0069 | 0.0065 | 0.0048 | 0.0045 | 0.0086 | 0.0071 |
| $N_{\mathrm{eff}}$ | 0.32 | 0.25 | 0.22 | 0.14 | 0.13 | 0.29 | 0.23 |
| $Y_p$ | 0.018 | 0.016 | 0.016 | 0.013 | 0.012 | 0.017 | 0.015 |

(b) CMB-S3 + $P(k)$

| Parameter | CMB-S3 | BOSS | eBOSS | DESI | Euclid | DES | LSST |
|---|---|---|---|---|---|---|---|
| $10^5\,\omega_b$ | 8.4 | 8.0 | 7.9 | 7.5 | 7.5 | 8.3 | 8.1 |
| $10^4\,\omega_c$ | 21 | 20 | 19 | 15 | 14 | 20 | 18 |
| $10^7\,\theta_s$ | 27 | 26 | 26 | 22 | 21 | 27 | 25 |
| $n_s$ | 0.0040 | 0.0037 | 0.0036 | 0.0030 | 0.0029 | 0.0039 | 0.0037 |
| $N_{\mathrm{eff}}$ | 0.12 | 0.12 | 0.11 | 0.094 | 0.088 | 0.12 | 0.11 |
| $Y_p$ | 0.0069 | 0.0068 | 0.0067 | 0.0060 | 0.0058 | 0.0069 | 0.0066 |

(c) S4 + $P(k)$

| Parameter | CMB-S4 | BOSS | eBOSS | DESI | Euclid | DES | LSST |
|---|---|---|---|---|---|---|---|
| $10^5\,\omega_b$ | 3.8 | 3.8 | 3.8 | 3.7 | 3.7 | 3.8 | 3.8 |
| $10^4\,\omega_c$ | 14 | 14 | 13 | 12 | 11 | 14 | 13 |
| $10^7\,\theta_s$ | 15 | 15 | 14 | 13 | 13 | 15 | 14 |
| $n_s$ | 0.0030 | 0.0029 | 0.0028 | 0.0025 | 0.0024 | 0.0030 | 0.0029 |
| $N_{\mathrm{eff}}$ | 0.081 | 0.079 | 0.078 | 0.070 | 0.067 | 0.081 | 0.078 |
| $Y_p$ | 0.0047 | 0.0046 | 0.0046 | 0.0043 | 0.0042 | 0.0047 | 0.0046 |

(d) Planck + BAO

| Parameter | Planck | BOSS | eBOSS | DESI | Euclid | DES | LSST |
|---|---|---|---|---|---|---|---|
| $10^5\,\omega_b$ | 26 | 19 | 19 | 18 | 18 | 23 | 20 |
| $10^4\,\omega_c$ | 49 | 49 | 49 | 48 | 48 | 49 | 49 |
| $10^7\,\theta_s$ | 89 | 87 | 87 | 87 | 87 | 88 | 88 |
| $n_s$ | 0.0093 | 0.0066 | 0.0065 | 0.0060 | 0.0061 | 0.0081 | 0.0068 |
| $N_{\mathrm{eff}}$ | 0.32 | 0.29 | 0.29 | 0.28 | 0.28 | 0.30 | 0.29 |
| $Y_p$ | 0.018 | 0.018 | 0.018 | 0.018 | 0.018 | 0.018 | 0.018 |

(e) CMB-S3 + BAO

| Parameter | CMB-S3 | BOSS | eBOSS | DESI | Euclid | DES | LSST |
|---|---|---|---|---|---|---|---|
| $10^5\,\omega_b$ | 8.4 | 7.9 | 7.9 | 7.6 | 7.7 | 8.3 | 8.0 |
| $10^4\,\omega_c$ | 21 | 21 | 21 | 21 | 21 | 21 | 21 |
| $10^7\,\theta_s$ | 27 | 27 | 27 | 27 | 27 | 27 | 27 |
| $n_s$ | 0.0040 | 0.0035 | 0.0035 | 0.0032 | 0.0032 | 0.0039 | 0.0036 |
| $N_{\mathrm{eff}}$ | 0.12 | 0.12 | 0.12 | 0.12 | 0.12 | 0.12 | 0.12 |
| $Y_p$ | 0.0069 | 0.0069 | 0.0069 | 0.0069 | 0.0069 | 0.0069 | 0.0069 |

(f) S4 + BAO

| Parameter | CMB-S4 | BOSS | eBOSS | DESI | Euclid | DES | LSST |
|---|---|---|---|---|---|---|---|
| $10^5\,\omega_b$ | 3.8 | 3.8 | 3.8 | 3.8 | 3.8 | 3.8 | 3.8 |
| $10^4\,\omega_c$ | 14 | 14 | 14 | 14 | 14 | 14 | 14 |
| $10^7\,\theta_s$ | 15 | 15 | 15 | 15 | 15 | 15 | 15 |
| $n_s$ | 0.0030 | 0.0028 | 0.0028 | 0.0025 | 0.0026 | 0.0030 | 0.0028 |
| $N_{\mathrm{eff}}$ | 0.081 | 0.080 | 0.080 | 0.079 | 0.079 | 0.081 | 0.080 |
| $Y_p$ | 0.0047 | 0.0047 | 0.0047 | 0.0046 | 0.0046 | 0.0047 | 0.0047 |

**Table C.12** Absolute and relative step sizes, $h$ and $h_{rel}$, used when computing the derivatives in the Fisher matrices

| Parameter | $h$ | $h_{rel}$ |
|---|---|---|
| $\omega_b$ | 0.0008 | $3.6 \times 10^{-2}$ |
| $\omega_c$ | 0.002 | $1.7 \times 10^{-2}$ |
| $100\,\theta_s$ | 0.002 | $1.9 \times 10^{-3}$ |
| $\ln(10^{10} A_s)$ | 0.05 | $1.6 \times 10^{-2}$ |
| $n_s$ | 0.01 | $1.0 \times 10^{-2}$ |
| $\tau$ | 0.02 | $3.0 \times 10^{-1}$ |
| $N_{eff}$ | 0.08 | $2.6 \times 10^{-2}$ |
| $Y_p$ | 0.005 | $2.0 \times 10^{-2}$ |

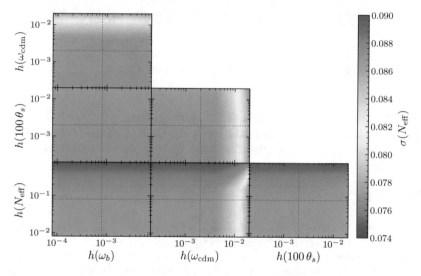

**Fig. C.3** Results of the convergence test for the $P(k)$-forecasts of DESI in the fiducial $\Lambda$CDM+$N_{eff}$ cosmology (from [1]). The spectra for the numerical derivatives were computed using a high-accuracy setting of CLASS. The dashed lines indicate the step sizes employed in our forecasts

fiducial parameter value $\theta$ and absolute step size $h$. For each parameter, we choose the step sizes given in Table C.12 resulting in relative step sizes, $h_{rel} = h/\theta$, that are generally of order $\mathcal{O}(10^{-2})$.

In Figs. C.3 and C.4, we show that our results are converged for both the $P(k)$- and BAO-forecasts. The results in these figures (as in the rest of our forecasts) use CLASS with a high-accuracy setting. We have also checked that the forecasted constraints are converged when employing the standard accuracy setting, but note that the results are slightly less stable to changes away from these values. For the $P(k)$-forecasts, we see that a sufficiently small step size is needed, but a further decrease in the step size still leads to converged results. The BAO-forecasts, by contrast, show islands of convergence where performance decreases both when the step size is increased and

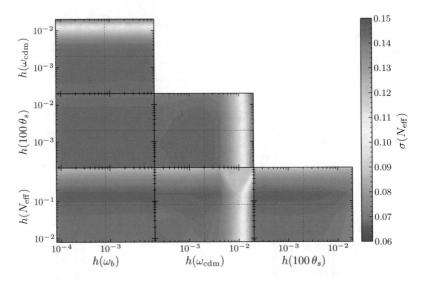

**Fig. C.4** As in Fig. C.3, but for the BAO-forecasts of DESI (from [1])

when it is decreased. This behaviour is more noticeable using the standard accuracy setting of CLASS, but likely reflects the fact that the BAO feature is itself a small effect and small step sizes are therefore more likely to produce effects comparable to numeric or modelling errors.

# References

1. D. Baumann, D. Green, B. Wallisch, Searching for light relics with large-scale structure. JCAP **08**, 029 (2018). arXiv:1712.08067 [astro-ph.CO]
2. W. Wu, J. Errard, C. Dvorkin, C. Kuo, A. Lee, P. McDonald, A. Slosar, O. Zahn, A guide to designing future ground-based cosmic microwave background experiments. Astrophys. J. **788**, 138 (2014). arXiv:1402.4108 [astro-ph.CO]
3. S. Galli et al., CMB polarization can constrain cosmology better than CMB temperature. Phys. Rev. D **90**, 063504 (2014). arXiv:1403.5271 [astro-ph.CO]
4. R. Allison, P. Caucal, E. Calabrese, J. Dunkley, T. Louis, Towards a cosmological neutrino mass detection. Phys. Rev. D **92**, 123535 (2015). arXiv:1509.07471 [astro-ph.CO]
5. K. Abazajian et al. (CMB-S4 Collaboration), CMB-S4 Science Book, 1st edn. arXiv:1610.02743 [astro-ph.CO]
6. D. Green, J. Meyers, A. van Engelen, CMB delensing beyond the B-modes. JCAP **12**, 005 (2017). arXiv:1609.08143 [astro-ph.CO]
7. P.A.R. Ade et al., (Planck Collaboration), Planck 2015 results. XIII. Cosmological parameters. Astron. Astrophys. **594**, A13 (2016). arXiv:1502.01589 [astro-ph.CO]
8. A. Font-Ribera, P. McDonald, N. Mostek, B. Reid, H.-J. Seo, A. Slosar, DESI and other dark energy experiments in the era of neutrino mass measurements. JCAP **05**, 023 (2014). arXiv:1308.4164 [astro-ph.CO]
9. A. Aghamousa et al., (DESI Collaboration), The DESI experiment part I: science, targeting and survey design. arXiv:1611.00036 [astro-ph.IM]

# Appendix D
# Broadband and Phase Extraction

In this appendix, we provide supplemental material to Chaps. 6 and 7, which is based on [1]. We describe our implementation of a robust method to extract the matter broadband power spectrum (Sect. D.1) and the computation of the phase shift template (Sect. D.2).

## D.1 Broadband Extraction

The split of the matter power spectrum into a broadband ('no-wiggle') part and an oscillatory ('wiggle') part, $P(k) = P^{\mathrm{nw}}(k) + P^{\mathrm{w}}(k)$, is not uniquely defined, but depends on the method that is being used. In the following, we describe our procedure for extracting the broadband spectrum which is robust and stable over a very large parameter space.

Computationally, it is easier to identify a bump over a smooth background than to find the zeros of oscillations on top of a smooth background. This suggests that it is convenient to sine transform the matter power spectrum to discrete real space where the oscillations map to a localized bump. We then remove this bump and inverse transform back to Fourier space.

An algorithm for the discrete spectral method was outlined in Sect. A.1 of [2]. Concretely, the relevant steps of our implementation are:

1. Provide $P(k)$: Compute the theoretical matter power spectrum $P(k)$ using CLASS for discrete wavenumbers $k$ up to a chosen $k_{\mathrm{max}}$ and log-log interpolate using cubic splines.
2. Sample $\log[kP(k)]$: Sample $\log[kP(k)]$ in $2^n$ points for an integer number $n$. These points are chosen equidistant in $k$.
3. Fast sine transform: Perform a fast sine transform of the $\log[kP(k)]$-array using the orthonormalized type-II sine transform. Denoting the index of the resulting array by $i$, split the even and odd entries, i.e. those entries with even $i$ and odd $i$, into separate arrays.

© Springer Nature Switzerland AG 2019
B. Wallisch, *Cosmological Probes of Light Relics*, Springer Theses,
https://doi.org/10.1007/978-3-030-31098-1

4. Interpolate arrays: Linearly interpolate the two arrays separately using cubic splines.
5. Identify baryonic bumps: Compute the second derivative separately for the interpolated even and odd arrays, and average over next-neighbouring array entries to minimize noise. Choose $i_{\min} = i_* - 3$, where $i_*$ is the array index of the first minimum of the second derivative. Set $i_{\max} = i^* + \Delta i$, where $i^*$ is the array index of the second maximum of the second derivative, and $\Delta i = 10$ and $20$ for the even and odd array, respectively. These shifts were obtained empirically, but are found to give reliable and stable results for a large range of $n$ and $k_{\max}$.
6. Cut baryonic bumps: Having found the location of the bumps within $[i_{\min}, i_{\max}]$ for the even and odd arrays, respectively, remove the elements within this range from the arrays. Then, fill the gap by interpolating the arrays rescaled by a factor of $(i + 1)^2$ using cubic splines. This is analogous to interpolating $r^2 \xi(r)$ instead of the correlation function $\xi(r)$ at separation $r$.
7. Inverse fast sine transform: Merge the two arrays containing the respective elements without the bumps, and without the rescaling factor of $(i + 1)^2$, and inversely fast sine transform. This leads to a discretized version of $\log[k P^{\mathrm{nw}}(k)]$.
8. Provide $P^{\mathrm{nw}}(k)$ and $P^{\mathrm{w}}(k)$: In order to cut off numerical noise at low and high wavenumbers, perform two cuts at $k_1$ and $k_2$, where $k_1 = 3 \cdot 2^{-n}$ and the value of $k_2$ is found as the trough of $|P(k) - P^{\mathrm{nw}}(k)|/P^{\mathrm{nw}}(k)$ following the smallest maximum (before the oscillation amplitude increases again due to the numerical artefacts intrinsic to the procedure). The reliably extracted no-wiggle spectrum $P^{\mathrm{nw}}(k)$ is then valid for $k \in [k_1, k_2]$. In practice, choose $n$ and $k_{\max}$ large enough initially so that $k_{1,2}$ are outside the range of wavenumbers of interest, e.g. those covered by a survey. The wiggle spectrum in this range is then given by $P^{\mathrm{w}}(k) = P(k) - P^{\mathrm{nw}}(k)$.

Examples of the broadband extraction using this procedure are shown in Fig. D.1. We see that the extraction method is unbiased, i.e. the resulting wiggle spectrum both oscillates around zero and asymptotes to zero for large wavenumbers. In addition, it is robust and stable over a large parameter space at small computation time (depending on $n$). As the position of the first BAO peak is close to the peak of the matter power spectrum, it is sensitive to how exactly the baryonic bump is removed. However, we have checked that the computed constraints on cosmological parameters are insensitive to this uncertainty. The same holds for varying the parameters $n$ and $k_{\max}$ with fixed shifts in step 5 as long as $k_{1,2}$ are outside the range of wavenumbers of interest.

## D.2    Phase Shift Measurement

In the following, we describe our method for computing the phase shift template used in the likelihood forecasts of Sect. 6.3 and in the BOSS DR12 data analysis of Chap. 7.

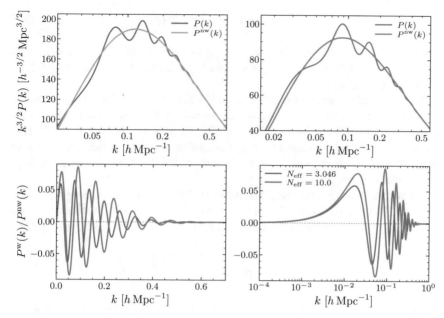

**Fig. D.1** *Top*: Extracted broadband spectrum $P^{\mathrm{nw}}(k)$ compared to the full power spectrum $P(k)$ for $N_{\mathrm{eff}} = 3.046$ (*left*) and 10 (*right*; from [1]). The spectra are rescaled by $k^{3/2}$ to exaggerate any oscillations. *Bottom*: Extracted BAO spectrum $P^{\mathrm{w}}(k)/P^{\mathrm{nw}}(k)$ for $N_{\mathrm{eff}} = 3.046$ and 10 with linear (*left*) and logarithmic (*right*) $k$-axis (from [1])

First, we compute the BAO spectrum using CLASS and the broadband extraction method detailed above for a given value of $N_{\mathrm{eff}}$. In practice, we set the primordial helium fraction $Y_p$ to the fiducial value, but the final template is independent of this choice. As discussed in Sect. 6.3.1, we keep the time of matter-radiation equality fixed at its fiducial value by changing the dark matter density $\omega_c$ according to

$$\omega_c = \frac{a_\nu + N_{\mathrm{eff}}}{a_\nu + N_{\mathrm{eff}}^{\mathrm{fid}}} \left(\omega_c^{\mathrm{fid}} + \omega_b^{\mathrm{fid}}\right) - \omega_b^{\mathrm{fid}}, \tag{D.1}$$

where $a_\nu$ is defined in (3.16). We then fit the following envelope function to the maxima of the absolute value of the BAO spectrum:

$$a(k) \equiv e(k)d(k), \quad \text{where} \quad \begin{aligned} e(k) &\equiv 1 - A_e \exp\{-a_e \, (k/k_e)^{\kappa_e}\}, \\ d(k) &\equiv A_d \exp\{-a_d \, (k/k_d)^{\kappa_d}\}. \end{aligned} \tag{D.2}$$

The parameters $A_i, a_i, \kappa_i$, with $i = d, e$, are fitting parameters, while $k_e$ is the location of the peak of $P^{\mathrm{nw}}(k)$ and $k_d$ is the wavenumber associated with the mean squared diffusion distance. These fitting functions are motivated by the modelling in [3, 4]. We define the 'undamped spectrum' as

$$\mathcal{O}(k) \equiv a(k)^{-1} P^{\mathrm{w}}(k)/P^{\mathrm{nw}}(k) \,. \tag{D.3}$$

For the fiducial cosmology, for instance, we find the following parameters: $A_e \approx 0.141$, $a_e \approx 0.0035$, $\kappa_e \approx 5.5$, $k_e \approx 0.016\, h\,\mathrm{Mpc}^{-1}$, and $A_d \approx 0.072$, $a_d \approx 0.32$, $\kappa_d \approx 1.9$, $k_d \approx 0.12\, h\,\mathrm{Mpc}^{-1}$.

Before we can measure the phase shift, we have to match the sound horizon at the drag epoch, $r_s$, to that in the fiducial cosmology to remove the change to the BAO frequency induced by $N_{\mathrm{eff}}$. By rescaling the wavenumbers as $k \to r_s^{\mathrm{fid}}/r_s\, k$, we fix $r_s k$ to the fiducial model for all wavenumbers $k$. For convenience, we also normalize the spectrum such that the amplitude of the fourth peak is the same as in the fiducial cosmology.

Finally, we can extract the phase shift as the shift of the peaks/troughs and zeros of $\mathcal{O}(k)$ relative to the fiducial cosmology, $\delta k_* = k_* - k_*^{\mathrm{fid}}$. To obtain the template $f(k)$, we sample 100 cosmologies with varying $N_{\mathrm{eff}} \in [0, 3.33]$,[1] and define

$$f(k) \equiv \left\langle \frac{1}{1 - \beta(N_{\mathrm{eff}})} \frac{\delta k_*(k; N_{\mathrm{eff}})}{r_s^{\mathrm{fid}}} \right\rangle_{N_{\mathrm{eff}}}, \tag{D.4}$$

where $\beta(N_{\mathrm{eff}})$ is the normalization introduced in (6.27). The bars in Fig. 6.6 indicate the locations of the peaks/troughs/zeros of the fiducial spectrum $\mathcal{O}(k)$ and their length shows the standard deviation in these measurements which is generally small.

# References

1. D. Baumann, D. Green, B. Wallisch, Searching for light relics with large-scale structure. JCAP **08**, 029 (2018). arXiv:1712.08067 [astro-ph.CO]
2. J. Hamann, S. Hannestad, J. Lesgourgues, C. Rampf, Y.Y.Y. Wong. Cosmological parameters from large-scale structure–geometric versus shape information. JCAP **07**, 022 (2010). arXiv:1003.3999 [astro-ph.CO]
3. B. Follin, L. Knox, M. Millea, Z. Pan, First detection of the acoustic oscillation phase shift expected from the cosmic neutrino background. Phys. Rev. Lett. **115**, 091301 (2015). arXiv:1503.07863 [astro-ph.CO]
4. T. Baldauf, M. Mirbabayi, M. Simonović, M. Zaldarriaga, Equivalence principle and the baryon acoustic peak. Phys. Rev. D **92**, 043514 (2015). arXiv:1504.04366 [astro-ph.CO]

---

[1]We restrict to this range of values of $N_{\mathrm{eff}}$ as we observed a small, but unexpected jump in the peak locations around $N_{\mathrm{eff}} \sim 3.33$. Below and above, the peak locations change coherently with $N_{\mathrm{eff}}$. This range was then chosen as we are mostly interested in smaller $N_{\mathrm{eff}}$. However, we expect the template to also be valid for larger $N_{\mathrm{eff}}$ outside the sampling range.